Utilization of Mammalian
Specific Locus Studies in
Hazard Evaluation and
Estimation of Genetic Risk

ENVIRONMENTAL SCIENCE RESEARCH

Recent Volumes in this Series

A Continuation Order Plan is available for this series. A continuation order will bring
delivery of each new volume immediately upon publication. Volumes are billed only upon
actual shipment. For further information please contact the publisher.

Utilization of Mammalian Specific Locus Studies in Hazard Evaluation and Estimation of Genetic Risk

Edited by
Frederick J. de Serres and
William Sheridan

National Institute of Environmental Health Sciences
Research Triangle Park, North Carolina

PLENUM PRESS • NEW YORK AND LONDON

Library of Congress Cataloging in Publication Data

Main entry under title:

Utilization of mammalian specific locus studies in hazard evaluation and estimation of
genetic risk.

(Environmental science research; v. 28)
"Proceedings of a workshop held March 29–31, 1982...Research Triangle Park, North
Carolina"—T.p. verso.
Includes bibliographical references and index.
1. Mutagenicity testing—Congresses. 2. Mammals—Genetics—Congresses. I. de Serres,
Frederick, J. II. Sheridan, William, date–   . III. Series.
QH465.A1U84 1983   616′.042   83-11076
ISBN-13:978-1-4613-3741-6   e-ISBN-13:978-1-4613-3739-3
DOI: 10.1007/978-1-4613-3739-3

Proceedings of a workshop held March 29–31, 1982, at the National Institute of
Environmental Health Sciences, Research Triangle Park, North Carolina

©1983 Plenum Press, New York
Softcover reprint of the hardcover 1st edition 1983

A Division of Plenum Publishing Corporation
233 Spring Street, New York, N.Y. 10013

PREFACE

   The magnitude of the threat to the human genetic material posed
by environmental agents has not as yet been fully determined. Nevertheless, the potential hazards of many chemicals have been identified by studies on lower organisms. However, too little is known
regarding the comparability or lack of it between the metabolic
pathways available in such organisms and those in man. Although at
present there is great public concern for what is considered by some
as the excessive use of laboratory animals in toxicological testing,
it seems clear that the usage of mammalian systems may be deemed
necessary. It has been proposed that cell culture systems might
suffice to meet this need, however, such approaches cannot match the
complexity of physiological occurances that are present in the intact
animal.

   For studies of genetic effects, some non-invasive human test
systems are presently available. These do not, however, meet the requirements for extensive laboratory studies. In order to assess the
risks to humans of environmental factors such laboratory investigations are essential. Therefore, for the forseeable future reliance
on experiments using laboratory animals will be necessary.

   This Volume, which contains the proceedings of a workshop which
was held at the National Institute of Environmental Health Sciences,
March 29-31, 1982, explores the existing methodologies and their
utility for risk estimations. It covers the most well developed
human systems, as well as the most widely used animal tests. It
also devotes attention to some of the more subtle effects of gene
mutations. In the final session, the question of future needs and
directions is addressed.

   In addition to those contributors represented here, the conference was widely attended by representatives of Academia, Industry
and Government agencies whose contributions in the form of lively
and probing discussions are gratefully acknowledged. We would also

like to thank Drs. F. M. Johnson and M. D. Shelby for helpful sug-
gestions, and Mrs. C. Ellington for her able assistance in the management of the meeting and the preparation of this volume.

Frederick J. de Serres
William Sheridan

Research Triangle Park
North Carolina, 1982

CONTENTS

HOW MUCH PROGRESS THIS TIME?

James V. Neel

Department of Human Genetics
University of Michigan Medical School
Ann Arbor, Michigan 48109

1. INTRODUCTION

I must confess that although I was well aware that I was responsible for a technical presentation at this meeting, it had somehow escaped my attention until a few days ago that I was also scheduled for 45 minutes of introductory remarks as well. One felicitous
result of this will be that the coffee break will come earlier than
scheduled. For one panicky moment I thought of extracting from the
sermon barrel some one of the several similar "Introductions" for
which I have been responsible in recent years, hoping that with the
information overload from which we all suffer no one would notice
this was a replay. However, looking at the roster of distinguished
and knowledgeable participants, it was clear I couldn't get away with
that, even with a bit of judicious hybridization of previous manu-
scripts. Besides, for the hybridization I couldn't decide whether
to use alternate lines or alternate paragraphs.

It is indeed difficult to say something new at the opening of
a meeting such as this. There has in recent years been a virtual
spate of workshops devoted to the general topic of "Spontaneous and
Induced Mutation: Basic and Practical Aspects." I need only mention the Bar Harbor Conference of 1976 [1], the conference in Norway
organized by Berg in 1977 [2]; Banbury Report No. 1, based on a conference held in 1978 [3], the conference organized by Hook and Porter
in Albany, N. Y. in 1980 [4], a similar conference held in Ottawa in
1980 [5], and a workshop sponsored by the March of Dimes in 1981 in
response to the mishandling of the Love Canal episode [6]. There
will be another Banbury Conference next month. All of these already
have or will result in books. In addition, there have been in recent
years at national and international scientific meetings at least a

half dozen symposia on this topic. As I look around this room, I
see a great many veterans of these gatherings. Apart from introductory material, there are 2049 pages in the six references just
cited. Friends, we've surely done a lot of talking. In addition to
meeting the general interest in mutagenesis which we all share, a
powerful motivation behind these various programs has been the need
to understand better the nature of the risks - if any - created by
the current exposures of the human species to mutagens. I wish I
could believe that our progress with respect to the human problem
was proportional to the number of the meetings we have been hold-
ing!

While we've been talking, public and political concerns have
been finding new expressions. I would like in particular to direct
your attention to a bill introduced into the Senate last year by
Senator Hatch, entitled "The Radiation Exposure Compensation Act of
1981" (S1483), which would establish a mechanism for providing compensation for (1) individuals exposed as a result of atomic bomb
tests, and (2) uranium miners. As originally introduced, the bill
was so broadly worked as to provide compensation for all residents/
workers in certain regions who had sustained any exposure above background and subsequently developed certain types of cancer. It is my
understanding that this bill is now, as a consequence of the usual
hearings, undergoing modification, by virtue of which compensation
will likely be set on the principle of 'causative odds'. Approximate estimates of the doubling dose of radiation for a variety of
malignancies are available. As I understand the modifications under
discussion, compensation will begin at a certain value for the causative odds, and increase as the causative odds increase, with maximum
compensation for some value of the causative odds yet to be determined.

Should this bill be enacted into law, I believe it is a virtual
certainty that within the next decade, there will be very substantial
pressure to enact similar legislation for persons perceived to have
been subjected to genetic risks. The genetic issues are clearly more
complex than the somatic. Is the genetic community in a position to
respond in a constructive way? If not, what are some of the issues
that need to be addressed to improve our position? In what follows,
I would like to speak in very general terms, knowing that many of
you will be dealing in the specifics in the next several days. Perhaps these generalities will assist in organizing some of our discussion. I speak to this audience from the standpoint of one deeply
committed to understanding human risks. The fact that I do not in
this presentation refer to fundamental mechanisms of mutagenesis
does not indicate a lack of interest, but rather the sentiment that
this is not the primary forum for such considerations.

The chief needs with which we are faced in dealing with the
human problem can be subsumed under two headings, namely,

1) what is the frequency and phenotypic impact of spontaneous mutation in higher eukaryotes, and

2) how do we improve the basis for extrapolating from somatic to germinal effects, and from mouse to man.

Let us consider each of these topics briefly.

2. FREQUENCY AND IMPACT OF MUTATION IN HIGHER EUKARYOTES

Despite the basic nature of the phenomenon of mutation, and the longstanding interest in its frequency, we are still a long way from understanding its frequency and impact for any species. Some of you who work with Drosophila and mice may feel that's true for man, but that for your own material, the estimates are pretty accurate. I remind you that in the case of Drosophila, as well as a variety of other eukaryotes, we're just beginning to understand the importance in spontaneous and presumably also induced mutation rates of mutator genes, transposons, and ill-defined strain differences, (the latter undoubtedly often related to defective repair enzymes). If in nature most mutations occur under the influence of what is thought of as an abnormal phenomenon - transposons or mutator genes - then that's the way the natural world is, and if such phenomena exist for man - and they presumably do - then that's the world with which the human geneticist must work and the basis for the comparisons he'd like to draw with experimental material. Yamaguchi and Mukai [7] have expressed concern that the Drosophila strain from which they and their colleagues adduced the rate of mutation resulting in electrophoretic variants [8, 9] was 'contaminated' by a mutator gene, and Mukai is now busy attempting to repeat these observations on a strain where the presence of a mutator gene is not suspected. I would respectfully suggest an alternative approach, namely, that such a study be based on 100 different lines stemming from wild-caught flies. Incidentally, I do recognize the technical problem here, that the necessary crosses might activate latent transposons (or otherwise be complicated by the phenomenon of 'hybrid dysgenesis'), but believe you get the thrust of my argument.

Even without the complication of mutator genes and transposons, there are difficulties measuring the totality of the mutations at any one locus. We do it piece-meal. The "official" mouse rates derive from the specific locus test systems. I would be reasonably certain this system picks up total or near total absences of gene product, but what proportion of simple amino-acid substitutions? Conversely, in man, we still do not know the frequency ratio of three types of mutation, namely, 1) mutation resulting in charge-change amino-acid substitutions in polypeptides, 2) mutation resulting in silent amino acid substitutions in polypeptides (both with normal or near-normal function), and 3) mutation resulting in loss of function and/or gene product. Thus, we have difficulty ex-

trapolating from the frequency of electrophoretic mutations to total
rates.

Even were we able to measure the frequency and impact of mutation at a group of loci selected for study on the basis of apparently reasonable criteria, how do we move to the total phenotypic
impact of an increase in mutation rates. One approach is to multiply the average assumed phenotypic impact of mutation at a series of
selected loci by the assumed number of mutable loci. Here, too, in
recent years, as our knowledge increased, so have the complexities
of the problem. One of the revelations of the 1960's in human genetics was the frequency of chromosomal abnormalities in newborn
infants, something like 6 per 1000 [rev. in 11]. Even more surprising was the further demonstration in the late 60's and the early
70's, that among spontaneous abortuses, 50% were characterized by
chromosomal abnormalities [12], so that it could be estimated that
approximately 6% of all recognized pregnancies were characterized
by serious chromosomal abnormalities. The proportion at the time
of fertilization can only be higher. The majority of selection
against zygotes characterized by chromosomal mutations is intra-
uterine. Recently, it has been strongly suggested that more than
half of all fertilized eggs are aborted, and, in attempting to deal
with how selection can accommodate total point mutation rates of the
magnitude that seem to be emerging for man, I have been led to the
further suggestion that of the spontaneous foetal losses not due to
anatomical abnormality of the reproductive tract or cytogenetic abnormality, the majority are genetic [13]. If this is so, a significant fraction of the impact of an increased mutation rate might
be imperceptible, in the terms by which society measures the impact
of genetic disease.

A second approach, to which we return in the next section,
makes no use of specific locus data, but measures the total impact
of mutation on a phenomenon with a genetic basis, such as prere-
productive deaths among liveborn infants. This is the approach we
have, for instance, employed in the studies in Japan on the genetic
effects of the atomic bombs [10]. It avoids the extrapolations and
assumptions of the specific locus approach, but if a doubling dose
is desired, involves a different set of assumptions, pertaining to
the proportion of the phenomenon under investigation due each generation to spontaneous mutation.

There are, then, important gaps to be filled in, as regards our
understanding of the phenotypic impact of mutation. In the past, we
investigators have tended to do that which was convenient/possible
with the species on which we were working. However, more consistent
approaches, applicable to all species, are beginning to emerge, with
exciting opportunities for parallel programs in humans and experimental organisms; we will briefly consider some of the implications
of this in the next section.

3. IMPROVING THE BASIS FOR EXTRAPOLATION

    Although I will insist that there are certain situations in-
volving questions of germinal mutation rates in which we must have
human data, I will readily agree that with respect to another group
of questions, we will surely (I hope) be forced to rely on extrapolation from animal models. I will also readily agree that with
respect to questions regarding the significance of certain human exposures, it would be most convenient to extrapolate from somatic to
germ line mutation rates. Unfortunately, for neither of these types
of extrapolation are we at present able to proceed rigorously. How
do we improve this situation?

3.1. Extrapolation from Germinal Rates in the Mouse
      to Germinal Rates in Man

    Our indications of altered mutation rates fall into two groups,
namely, what I will term "specific locus" tests, and what I will
term "epidemiological" tests. The results of the former have until
recently dominated the thinking of those working with mice, whereas
the results of the latter have, of necessity, until recently, dominated the thinking of those of us working with humans.

    Let us consider the specific locus test first. It has been the
backbone of mouse radiation genetics, and rightly so. The results
of this test are the principal basis for the extrapolation that the
genetic doubling dose for acute radiation in man is approximately
40 rem, with the true value very likely to be found between 20 and
80 rem. Now, in research on exposed human populations, there is no
way we can pursue the classical specific locus test. However, we
do begin to have its biochemical equivalent, the survey of a defined
set of proteins for mutations resulting in altered electrophoretic
mobility or enzyme activity or thermostability. We will consider
the status of these efforts later in the program. This is as close
as we will ever come to a specific locus test for the progeny of exposed humans. The techniques for such studies did not exist when
the classical specific locus test was devised. While I will scarcely
go so far as to urge that all those previous experimental studies on
radiated mice using the specific locus test be repeated now using
biochemical indicators, I do feel we need sufficient work on mice
with biochemical yardsticks to determine the comparability of the
biochemical findings with the classical finding. More importantly,
the biochemical approach permits us to work with completely homologous indicators in mice and humans, and so provides a much more solid
basis for comparing both spontaneous and induced rates in the two
species than existed previously.

    Let us turn now to the "epidemiological" tests. In the most
extensive study to date on the progeny of humans exposed to a known
mutagenic experience, the follow-up studies in Hiroshima and Nagasaki

[10, 14, 15], these are the types of tests on which, together with
cytogenetic studies, it has been thus far necessary to rely for estimates of the genetic doubling dose. In particular, we have relied
on two indicators, namely, "untoward pregnancy outcomes" and "death
rates among liveborn infants." The former is based on infants who
were stillborn and/or exhibited major congenital defect and/or died
during the first week of life, in a carefully identified cohort; the
latter is based on a continuing study of children in this cohort.
Thus, our study of the first indicator is finished, while that of
the second is ongoing. The cytogenetic studies, initiated in 1968,
also ongoing, are based on the frequency of sex chromosome aneuploids
and reciprocal translocations, both thought to be relatively neutral
with respect to implications for survival, but of course with very
real or potential phenotypic impact. There is no double of a muta-
tational component to all these indicators, but its precise magnitude is still uncertain, this introducing potential errors into the
assumptions we must make to reach a doubling dose. On the other
hand, and I want to emphasize this, these indicators are the kind
of genetic effects about which people worry, and there is no cir-
cuitous extrapolation from a specific locus test result to a human
end point. Our latest estimate of the genetic doubling dose (and
this is an evolving figure) is either 139 or 250 rem, depending on
whether one weights the various indicators in proportion to the numbers involved [16, 17]. The error to that estimate is large, not
only for statistical reasons but because of questions recently
raised concerning the radiation spectrum of the bombs. We anticipate in the next 5 years a substantial tightening up in that estimate.

When, however, we turn to the mouse data for a touchstone, we
encounter an interesting paradox. Green in 1968 [18] summarized the
results of some 22 experiments involving for the most part somewhat
inbred mice, rats, and swine exposed to relatively large doses of
radiation, in various schedules, for from one to 12 generations.
The indicators of radiation damage include percentage of sterile and
semi-sterile matings, changes in litter size as an indication of the
induction of in utero lethal and deleterious mutations, preweaning
mortality, malformations, body weight, and even activity and learn-
ing ability. In theory an increased mutation rate should impact on
all these indicators. In fact, the results of the experiments were
variable and conflicting but in general revealed little influence of
radiation on the indicators just mentioned, leading Green to write:
"Attempts to measure the effects of presumptive new mutations on the
fitness and other traits of heterozygotes ... have, in general, been
unsuccessful." Elsewhere in the same review he writes: "The limited success of these studies to date, although disheartening, should
lead investigators to design better and more successful attempts in
the future." The implication is that a study which did not yield
the results predicted by specific locus tests was a failure. I

would rather believe that these are both versions of truth, whose results must be carefully harmonized in any comprehensive consideration of the impact of radiation on populations.

3.2. Extrapolation from Somatic Cell Mutation Rates to Germinal Cell Rates in Humans or Mice

If there were a single point on which all the participants in this Workshop would probably agree, it is that studies on germ line mutations are laborious and expensive, and if there were a somatic cell "litmus paper," it would be eagerly seized upon by all interested in evaluating the consequences of possible mutagenic exposures, be they of mice or people. The recent progress in this direction will be summarized for us in due course. There is in this connection a point which must be put very strongly. It is absolutely imperative that just as soon as a reliable and convenient method to measure mutation rates in somatic cells is available, a major effort must be devoted to determining the relevance of these rates to transmitted genetic damage. It is a dubious service to inform a population exposed to a mutagen that there is an increase in sister-chromatid exchanges or mutant red or white blood cells but then have immediately to add - 'of course, we don't know what this means for your children'. As soon as the methodology is available, every study of somatic cell genetic effects should be paired with a study of germ line effects, and vice versa.

This is perhaps the best place to mention the very interesting work on the effect of mutagens on sperm cell morphology and on apparent biochemical mutations in sperm cells, the latter also to be discussed later in this meeting. Since only one among the 300,000,000 sperm in a human ejaculate is successful in effecting fertilization, there are certainly potential opportunities for sperm cell selection. While a priori there is no particular reason to suspect selection on the basis of the type of a particular enzyme in a sperm cell, such as lactate dehydrogenase, there have been so many genetic surprises lately that I'm sure we'll all be much more sold on this approach when its predictive value has been determined by associated studies of transmitted genetic changes.

4. AFTER THOUGHTS

It would be unfortunate if my remarks conveyed the impression that not much had been learned about mutation in recent years. Much has indeed been learned, but we are still in the early stages of understanding a wonderously complex phenomenon. Fortunately, as this meeting will undoubtedly document, a number of important new technologies for the study of mutation are becoming available, and the prospects for significant new insights are good. These prospects extend to studies at the DNA level, although at the risk of dis-

playing a prejudice, I would suggest that such studies, basic though
they be, will not realize their full value unless somehow linked to
phenotypic equivalents.

Those of you who have been regular attendees at these various
workshops will have recognized that none of the various questions I
have raised are especially new. Indeed, if you will go back to the
very first of the workshops cited in my opening remarks, held at
Bar Harbor with NIEHS support, you will find that in their "Introduction" the editors, Roderick and Sheridan, raised many of the same
questions.

I see in our agenda many opportunities for new and tighter linkages between experimental and epidemiological studies. By this I do
not suggest direct collaborattion so much as the kind of exchange
that will result in our making similar observations on the subjects
of our respective studies. On the other hand, periodically I must
remind my friends in experimental genetics that human populations
have a certain independence of spirit and action, not to mention
sensibilities, which render the conduct of long-range studies deli-
cate and subject to many constraints. Human populations are also
very mobile. Some kinds of observation and follow-up that at first
thought should present no problem, are really quite difficult.

The new and tighter linkages to be developed should not only
involve better bridges between the experimentalist and the genetic
epidemiologist studying human populations - we need better coordina-
tion between our various studies on human populations. The identifiable groups of children whose parents were subjected to really
noteworthy mutagenic exposure will fortunately usually be small, and
scattered all over the world. We must somehow arrange to obtain
comparable data on these groups. In this connection, I call your
attention to the preliminary steps by a World Health Organization
Consultation to suggest the desirable components of a monitoring
effort [19]; you will also be interested to know that the International Association of Human Biologists has recently formed a small
ad hoc committee to develop a program for an International Mutation
Survey.

The problem of developing these linkages is much more acute for
studies of the effects of chemical mutagens than for studies of radiation. Ionizing radiation is ionizing radiation, no matter how
delivered, but there are so very many of the chemical mutagens, with
potentially different transport and detoxification mechanisms, as
well as different spectra of genetic effects. Has the time come to
try to identify those to which humans are most conspicuously exposed,
and then plan animal experiments accordingly? Such experiments
could serve to guide the use of limited resources for genetic studies
of human populations thought to have sustained unusual chemical exposures.

All these various undertakings will not come cheaply. I have in the past chided my fellow geneticists for talking big - about the potential genetic effects of various exposures accompanying civilization and high technology - but then thinking small - about the scope of the necessary efforts. If the current societal trend to attempt to assign responsibility for all the untoward events to which human life is subject continues, the cost of implementing studies which cannot fail to lessen current uncertainties will, as I suggested earlier, likely be considerably less than the various costs attendant upon continuation of these uncertainties. On the other hand, I believe we who are active in this field really do need to give more thought to how our efforts interdigitate, with our objective being substantial financial economies. We don't need three separate studies on the potential genetic effects of a chemical mutagen, one recording pregnancy outcomes, one investigating sister chromatid exchanges, one looking for electrophoretic variants, each with high front-end costs. Rather, we need one really good, coordinated study.

Mr. Chairman, I fear I may have overstepped my charge. Many of the issues I have raised were scheduled to surface in the closing session, and I have been poaching on the assigned territory of others. On the other hand, perhaps directing early attention to some of these issues will help us focus a bit better, as the meeting unfolds, on our priorities, and thus facilitate the assignments of our speakers for the latest session.

REFERENCES

1. T. H. Roderick and W. Sheridan, eds., Methods in Mammalian Mutagenesis, Genetics 92, Pt. 1, Supplement (1979).
2. K. Berg, ed., Genetic Damage in Man Caused by Environmental Agents, Academic Press, New York (1979).
3. V. K. McElheny and S. Abrahamson, eds., Banbury Report 1, Assessing Chemical Mutagens: The Risk to Humans, Cold Spring Harbor Laboratory, New York (1979).
4. E. B. Hook and I. H. Porter, eds., Population and Biological Aspects of Human Mutation, Academic Press, New York (1981).
5. K. C. Bora, G. R. Douglas, and E. R. Nestmann, eds., Chemical Mutagenesis, Human Population Monitoring and Genetic Risk Assessment, Elsevier Biomedical Press, Amsterdam (1982).
6. A. D. Bloom, ed., Guidelines for Studies of Human Populations Exposed to Mutagenic and Reproductive Hazards, March of Dimes Birth Defects Foundation, New York (1981).
7. O. Yamaguchi and T. Mukai, Variation of spontaneous occurrence rates of chromosomal aberrations in the second chromosomes of Drosophila melanogaster, Genetics 78, 1209-1221 (1974).
8. T. Mukai and C. C. Cockerham, Spontaneous mutation rates at enzyme loci in Drosophila melanogaster, Proc. Nat. Acad. Sci., USA 74, 2514-2517 (1977).

9.   R. A. Voelker, H. E. Schaffer, and T. Mukai, Spontaneous al-
     lozyme mutations in Drosophila: Rate of occurrence and nature
     of the mutants, Genetics 94, 961-968 (1980).

10. W. J. Schull, M. Otake, and J. V. Neel, A reappraisal of the
     genetic effects of the atomic bombs: Summary of a thirty-four
     year study, Science 213, 1220-1227 (1981).

11. E. B. Hook and J. L. Hamerton, in: "Population Cytogenetics,"
     E. B. Hook and I. H. Porter, eds., Academic Press, New York
     (1977), pp. 63-79.

12. D. H. Carr and M. Gedeon, in: "Population Cytogenetics," E. B.
     Hook and I. H. Porter, eds., Academic Press, New York (1977),
     pp. 1-9.

13. J. V. Neel, The wonder of our presence here: A commentary on
     the evolution and maintenance of human diversity, Persp. Biol.
     Med. 25, 518-558 (1982).

14. W. J. Schull, M. Otake, and J. V. Neel, in: "Population and
     Biological Aspects of Human Mutation," E. B. Hook and I. H.
     Porter, eds., Academic Press, New York (1981), pp. 277-303.

15. J. V. Neel, W. J. Schull, and M. Otake, in: "Progress in Mu-
     tation Research, Vol. 3," K. C. Bora, G. R. Douglas, and E. R.
     Nestmann, eds., Elsevier Biomedical Press, Amsterdam (1982),
     pp. 39-51.

16. W. J. Schull, J. V. Neel, M. Otake, A. Awa, C. Satoh, and H. B.
     Hamilton, Hiroshima and Nagasaki: Three and a half decades of
     genetic screening, Proc., III Int. Cong. Env. Mutagens, Alan
     Liss, New York, pp. 687-700 (1982).

17. C. Satoh, A. A. Awa, J. V. Neel, W. J. Schull, H. Kato, H. B.
     Hamilton, M. Otake, and K. Goriki, Genetic effects of atomic
     bombs, Proc., VI Int. Cong. Hum. Genet., Alan Liss, New York,
     pp. 267-276 (1982).

18. E. L. Green, Genetic effects of radiation on mammalian popula-
     tions, Ann. Rev. Genet., 2, 87-120 (1968).

19. World Health Organization Consultation on Genetic Monitoring
     for Envirnomental Effects, in: "Chemical Mutagenesis, Human
     Population Monitoring and Genetic Risk Assessment," K. C. Bora,
     G. R. Douglas, and E. R. Nestmann, eds., Elsevier Biomedical
     Press, Amsterdam (1982), pp. 337-343.

HUMAN MUTAGENICITY MONITORING:

STUDIES WITH 6-THIOGUANINE RESISTANT LYMPHOCYTES

Richard J. Albertini

University of Vermont College of Medicine
Given Medical Building E305
Burlington, Vermont 05405

I. INTRODUCTION

A. Mutagenicity Testing: Screening vs. Monitoring

The field of human mutagenicity testing is maturing, with a resultant refinement in concepts and objectives. The screening of new
chemical agents in order to identify those which may cause genetic
damage is now only one form of mutagenicity testing. Screening tests
attempt to identify the mutagens among new chemicals in hopes of
achieving the ideal - i.e., the identification and elimination of
dangerous chemicals before humans become exposed to them. By contrast, there are now also monitoring tests. As the name implies,
these tests permit the monitoring of human populations for evidence
of genetic damage resulting from mutagens present in the environment.

There are literally thousands of chemical agents that are candi-
dates for screening. Because of the enormity of the screening task,
the several available tests, in practice, are usually performed in
some sort of a sequence. In order to avoid employing all tests for
all agents, a three tier approach to screening has been proposed
[1, 2]. In this proposal, tier I tests are inexpensive, have a low
potential for giving false negative results and usually use prokary-
otes as indicator organisms. By contrast, tier II tests, performed
to confirm positive tier I tests results, are expensive and usually
use cultured mammalian cells as their indicator organisms to provide
"target realism." Finally, tier III screening tests are performed
for those agents definitely deemed hazardous on the basis of lower
level tests, or for which there will be great human exposure. This
level of testing is difficult, very expensive, and currently re-

quires long-term animal studies. Tier III screening tests have two
objectives: (i) to confirm the genotoxicity of the agents under
study and (ii) to attempt to quantitate the risks in using the agents
in terms of human health.

Screening tests provide, at best, qualitative information with
respect to man as to the genotoxicity of environmental agents. Even
tier III screening tests do not provide data which allow quantitative human health risk assessments. The extrapolation of animal
data to man is a complex and uncertain science. Such extrapolation
must assume human population homogeneity as regards susceptibility
to mutagenic agents. There is now sufficient evidence to indicate
that this is not the case [3, 4].

Recently, there has been a surge of interest in developing human monitoring tests in order to provide data that might be used for
making human health risk assessments. Clearly, the objective of
monitoring is not to establish primarily the genotoxicity of an isolated agent. Rather, it is becoming generally recognized that a zero
environmental mutagen exposure for man will never be realized. Monitoring tests attempt to define the effects of a specific environment - possibly containing a suspect agent - on humans. Monitoring
tests use cells or other materials obtained directly from the body
as indicators of genetic damage.

The traditional monitoring test to detect genetic damage occurring in vivo in human somatic cells is a standard cytogenetic
test using peripheral blood lymphocytes (PBL's) [5, 6]. However, a
cytogenetic test of increasing importance for human monitoring is
that which demonstates sister chromatid exchanges (SCE's) in somatic
cells - again usually in PBL's [7-10]. Both of these methods monitor
for actual or presumed genetic damage occurring at the chromosomal
level.

There is great current interest in developing as monitoring
tests those that detect specific locus somatic cell mutants arising
in vivo in humans. Such mutagenicity tests - termed direct mutagenicity tests [11] - are now available.

B. Utility of Mutagenicity Monitoring Tests

Mutagenicity monitoring tests provide information that is not
obtained from other forms of testing. At least four advantages can
be identified:

(i) Because monitoring tests detect genetically damaged somatic
     cells that arise in vivo, these tests account for metabolic
     and/or pharmacokinetic factors that may be important in either
     enhancing or reducing the in vivo genetic effects of environ-
     mental agents.

(ii)   Mutagenicity monitoring tests allow an assessment of the
        genotoxicity of mixtures of agents to which humans are ex-
        posed. Single agents may have synergistic interactions with
        each other that are not detected by screening them singly.

(iii)   Mutagenicity monitoring tests may allow the identification of
        individuals or groups with unusual sensitivities to the muta-
        genicity of specific environmental agents. These are in-
        dividuals who may require special measures for their pro-
        tection.

(iv)   Mutagenicity monitoring test results from specific indivi-
        duals can be correlated with the health outcomes of these in-
        dividuals in order to validate tests in terms of their rele-
        vance as predictors of human disease. It is only through
        validations of this kind that laboratory test data can ever
        be used with confidence to make quantitative human health
        risk assessments.

C. Human Direct Mutagenicity Tests

     Because of their potential usefulness, a number of human direct
mutagenicity tests have been proposed. Only those purporting to detect specific locus mutations in intact somatic cells are considered
here.

     An early candidate for a human direct mutagenicity test used
mature red blood cells (RBC's) as indicators of mutation. Rare
antigen (ABO) loss variants were found in human RBC populations at
frequencies of approximately 10^{-3} [12, 13]. These were thought initially to represent mutant somatic cells. Somewhat later, rare human polymorphonuclear white blood cells (WBC's) containing an altered
glucose-6-phosphate dehydrogenase (G6PD) enzyme activity were detected in human blood samples by cytochemical means [14, 15]. These
variants also existed at frequencies of approximately 10^{-3} and were
thought to arise in vivo by somatic cell mutation. At about this
same time, infrequent fetal hemoglobin (HbF) containing RBC's (F-
cells) were detected in adult peripheral blood [14-17]. It was
postulated that these cells were the progeny of erythroid precursor
cells that had undergone mutation in a structural gene for the Beta
chain of normal Hb. Thus, the quantitation of F-cell frequencies
was proposed also as a measure of somatic cell mutants arising in
vivo in man.

     Unfortunately, on further study, all three of these tests were
rejected as valid human direct mutagenicity tests [18-20]. In none
of them could the genetic basis of the variant phenotype be established. Thus, ABO antigen loss, apparent alterations in the catalytic activity of G6PD, and persistence of HbF in RBC's, were all
able to arise from nongenetic events in the indicator cells. Pheno-

copies (i.e., nongenetic variants that mimic genetic variants) were
demonstrated for all three traits. It was never established whether
there are, in addition to the phenocopies, genetic mutants with these
phenotypes. Because this was not demonstrated, and because no methods were available to eliminate phenocopies, these proposed methods
were discarded.

Recently, Stamatoyannapoulos and co-workers have introduced a
test designed to detect mutant Hb containing variant RBC's occurring
in vivo in humans who are homozygous for normal Hb (HbA) [21-23].
The mutant Hb's differ from HbA by specific amino acids or polypep-
tide changes, and it is difficult to see how they can arise by non-
genetic means. Thus, RBC's containing a mutant Hb are presumed to
arise from somatic cell mutation in vivo in erythroid stem cells.
By using specific antibodies, it has been possible to detect rare
RBC's containing HbS, HbC, Hb Wayne and Hb Cranston. Blood samples
from normal HbA homozygotes have shown an average of 1.1×10^{-7} HbS
containing RBC's. If the system can be easily quantitated, it should
provide a useful human direct mutagenicity test.

We have been working with different indicator cells and a different genetic locus. 6-Thioguanine resistant (TG^r) peripheral
blood lymphocytes (PBL's) exist in vivo in normal humans. We have
developed and have used a test that quantitatively detects these
cells in human blood samples and have proposed it as a direct mutagenicity test for man [24, 25].

II. 6-THIOGUANINE RESISTANT PERIPHERAL BLOOD LYMPHOCYTES

A. The Lesch-Nyhan Mutation

The Lesch-Nyhan Syndrome is a rare, naturally occurring genetic
disorder that results from specific locus mutation of the X-chromosomal gene for the enzyme hypoxanthine-guanine phosphoribosyltransferase (HPRT) [26, 27]. All cells from individuals with this inherited disorder lack significant HPRT activity. HPRT converts the
normal substrates hypoxanthine and guanine to inosine monophosphate
for subsequent purine interconversions [28, 29]. It is also necessary in order to phosphorylate some purine analogs (8-azaguanine =
AG; 6-thioguanine = TG) to render them cytotoxic [30, 31].

Several years ago, we showed that lymphocytes from boys with
the Lesch-Nyhan syndrome were resistant in vitro to the inhibitory
effects of AG [32]. Specifically, AG was shown to inhibit lectin
(phytohemagglutinin = PHA) stimulated tritiated thymidine (^{3}HTdr)
incorporation in vitro into the DNA of normal, but not Lesch-Nyhan
lymphocytes. Earlier, we had shown that normal diploid human fibroblasts could undergo mutation in vitro of the gene for HPRT and
thereby develop purine analogue resistance [33]. The fibroblast
model, using the prototype Lesch-Nyhan mutation, was then developed

into a human screening test for in vitro mutagenicity testing [34].
Similarly, we suggested then that the Lesch-Nyhan prototype mutation
in human lymphocytes would be useful also for direct mutagenicity
testing [32].

B. A Method to Detect TG^r PBL's

The TG^r PBL's that arise in vivo can be demonstrated in vitro
by their resistance to AG or TG inhibition of lectin stimulated
^{3}HTdr incorporation into DNA. Autoradiography allows the rare TG^r
cells that are present in majority TG sensitive (TG^s) populations
to be counted, and their frequencies to be calculated.

The method has been reported in detail and is only summarized
here [24, 25, 35]. The mononuclear cell (MNC) fraction of whole
heparinized blood is obtained by density gradient centrifugation in
Ficoll-Hypaque. After washing, the MNC's are rapidly suspended in
tissue culture medium RPMI 1640 (Gibco) with additives (25 mM Hepes,
2 mM glutamine and 100 U Penicillin plus 100 mcg streptomycin per
ml), supplemented with 10% human serum or autologous plasma and 7.5%
dimethylsulfoxide (DMSO) for freezing. The MNC's are suspended at
densities between 5×10^6 to 10^7/ml, aliquoted at approximately 1
ml/Nunc tube, frozen under controlled conditions, and stored until
use in the vapor phase of liquid nitrogen. MNC's are now always
frozen prior to test (see below).

For assay, MNC's are rapidly thawed, washed and suspended at
approximately 10^6/ml in tissue culture medium RPMI 1640 with additives, supplemented with 20% human serum. PHA is added either as
5 mcg/ml PHA-P or 1% PHA-M. Cell suspensions in PHA are inoculated
in 10 to 50 ml volumes into replicate tissue culture flasks (25 cm^2
to 75 cm^2 depending on volume). TG, at a final concentration of
2×10^{-4} M, is added to some flasks (test cultures); medium RPMI
1640 adjusted to the pH of the TG solution is added to others (control cultures). Flasks are incubated under standard conditions of
37°C and 5% humidified CO_2 atmosphere for 30 hours, after which most
of the supernatant medium over the loose cell layer in flasks is
aspirated with a Pasteur pipette. ^{3}HTdr, 5 µC/ml of remaining medium is added and flasks are incubated for an additional 12 hours
and terminated.

To terminate, the cultured cell suspensions of each flask are
added to individual centrifuge tubes, and mixed with 4 volumes (4°C)
0.1 M citric acid per ml cell suspension, and centrifuged at $600 \times G$
for 10 minutes to prepare free nuclei. The nuclei are washed once
with fixative (7 parts methanol: 1.5 parts glacial acetic acid), and
resuspended in small volumes (approximately 200 µl) of fixative, and
refrigerated for several hours. The suspensions in fixative are
counted with a Coulter counter and applied in measured volumes to
18 × 18 mm coverslips affixed with Permount to glass microscope

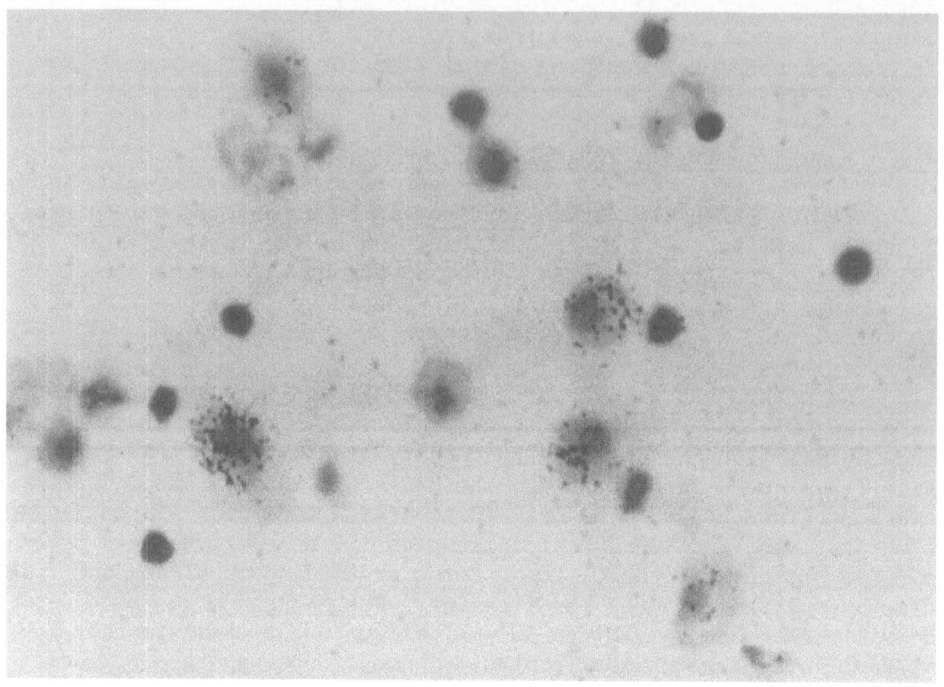

Fig. 1.   Photomicrograph of labeled nuclei from a controlled culture
          as seen under oil emersion microscopy (1000×). Nuclei are
          lightly labeled with 15 or more grains being positive.

slides. The slides are dried, stained and autoradiographed as described elsewhere [35].

     Following autoradiography, the slides are scored by light micro-
scopy. The total number of nuclei on a given coverslip is known
from the Coulter count made prior to the addition of measured volumes of nuclei in suspension. A labeling index (incidence of labeled nuclei) is determined for coverslips made with nuclei from
control (no TG) cultures by scoring the number of labeled nuclei in
5000 nuclei counted. These coverslips are scored under oil emersion
microscopy (1000×). The labeling index of the control cultures
(LI_c) is therefore:

$$LI_c = \frac{\text{No. labeled nuclei/5000 counted}}{5000}$$

     A labeling index is also determined for all coverslips made
with all recovered nuclei from all test (TG) cultures. These coverslips are viewed in their entirety at 160×, at which power the rare
labeled nuclei can be seen. Each labeled nucleus is confirmed under

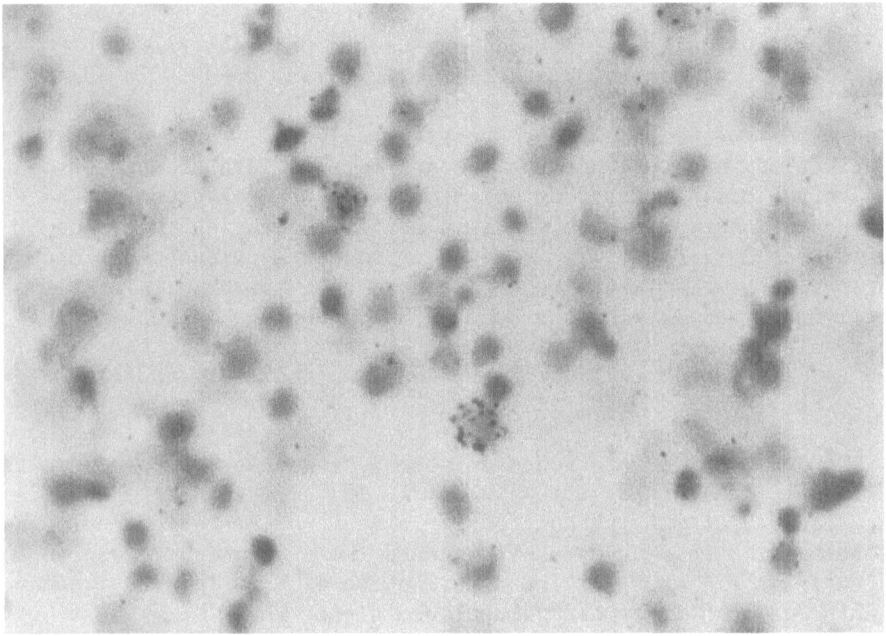

Fig. 2.   Photomicrograph of labeled nuclei from test culture as seen
          under oil emersion microscopy (1000×).

oil emersion (1000×). The total number of nuclei on coverslips is
available from the Coulter count. The labeling index of test cultures (LI_t) is therefore:

$$LI_t = \frac{\text{No. labeled nuclei on all test coverslips}}{\text{total No. nuclei on all test coverslips}}$$

The TG^r PBL variant frequency (V_f) is the ratio of these two
LI's:

$$V_f = \frac{LI_t}{LI_c}$$

Figures 1 and 2 show photomicrographs of coverslips prepared
as described. Nuclei from a lightly labeled control culture are
shown at oil emersion microscopy in Fig. 1 (1000×). Three labeled
nuclei (15 or more grains) are shown on this portion of the coverslip, and the LI_c of this control culture was 0.39. Nuclei from
the companion test culture (TG) are shown in Fig. 2 at 1000×. Two
labeled nuclei are seen for the test culture. This entire coverslip
had 38 labeled nuclei. The TG^r PBL V_f of the sample was determined
to be 7.1×10^{-6}.

C. Early Studies

After developing the method, but before we began freezing MNC's prior to test, we undertook a series of clinical studies in humans [23, 25, 26, 27]. Blood samples from boys with Lesch-Nyhan mutation showed from 100% to 23% TG^r lymphocytes [24, 37]. Thus, TG, at this concentration, was somewhat toxic even to cells with the prototype mutation. Blood samples from women who were heterozygous for the Lesch-Nyhan mutation showed TG^r lymphocyte frequencies that ranged from 10^{-3} to 5×10^{-2} [37]. Because of random X-chromosomal inactivation, such women would be expected to have approximately half of their lymphocytes showing the Lesch-Nyhan phenotype. However, apparent selection in vivo against Lesch-Nyhan blood cells had been shown earlier [38-40]. Thus, these results, too, were consistent with expectations.

TG^r lymphocytes were found in normal, nonmutagen exposed adults at a median frequency of 1.1×10^{-4} [24, 25]. By contrast, the median frequency of TG^r lymphocytes for a group of cancer patients who were being treated with cytotoxic agents (most of which are known mutagens) was 8.5×10^{-4}. The distribution of V_f's for the treated patients was such that most values were higher than the highest values seen in the normal control group. Later studies with psoriatic patients and others who were receiving 8-methyoxypsoralen and UV-A photochemotherapy (PUVA) showed that these individuals too had elevated V_f's when compared to normals [36].

However, there were unexpected results also in these early studies [25]. Many cancer and psoriatic patients had elevated V_f's as compared to normals, even before the patients were exposed to mutagens. Also, in studying some patients receiving X-irradiation therapy, TG^r PBL V_f's reached values approaching 10^{-2} – far too high to be attributed to somatic cell mutation alone. In performing longitudinal studies with normal individuals, we found a good deal of day to day variation. Finally, when we began studying MNC's cryopreserved as described in methods, we noted that the TG^r PBL V_f's were always significantly lower when determined in cryopreserved as opposed to fresh samples [41]. This all pointed to the presence of phenocopies in the system and, fortunately, also suggested a way to remove the phenocopy effect.

III. PHENOCOPIES

On reviewing these early results obtained by testing fresh cells, it appeared as though the unexpectedly high TG^r lymphocyte frequencies were found in those clinical settings where large numbers of "cycling" PBL's might be found in vivo. PBL's are usually in an arrested G_0 stage of the cell cycle in vivo, and enter into cycle only following a stimulus. In vitro, the stimulus is provided by PHA or antigen; in vivo an immunological stimulus will induce some PBL's into cycle. Cancer and the skin disorders common to the patient groups tested in the early studies might have provided this immunological stimulus in vivo.

The great majority of lymphocytes are in G_0 in the blood and when initially put into culture. They are then induced by PHA in culture to undergo a $G_0 \rightarrow G_1$ transformation with the acquisition to T-cell growth factor receptors on the activated cells [42]. This is associated with a profound metabolic alteration in the cells [43, 44]. We have postulated that it is this transformation that is inhibited in normal lymphocytes by TG in short term PHA cultures. By contrast, PBL's that are in cell cycle in the blood have already undergone this transformation in vivo. These cells too are inhibited by TG [45, 46]. However, cycling cells, not requiring the profound in vitro metabolic alteration prior to cell division, undergo one or more cell divisions in TG prior to their arrest. A fraction of these cells, although not mutant, probably become labeled in the short term cultures used to assay for TG^r lymphocytes. These cells, in TG cultures, are scored as positive thereby producing the phenocopies.

We performed experiments to determine the magnitude of the phenocopy effect by documenting and exploiting the preliminary observation that testing with cryopreserved cells results in significantly lower TG^r PBL V_f's [41]. Blood samples were obtained from 10 normal individuals. After separating the MNC's, the samples were split, with one portion of each tested fresh for its TG^r PBL V_f, and the other tested after cryopreservation as described in methods. The results of these parallel determinations, reported previously, are shown in Fig. 3 [41]. The V_f values determined with the 10 fresh samples ranged from 2.2×10^{-5} to 2.3×10^{-3}, with a mean and median value of 5.3×10^{-4} and 3.4×10^{-4}, respectively. By contrast, V_f's determined with the cryopreserved samples ranged from less than 1×10^{-6} to 2.7×10^{-5}, with a mean and median value of 8.3×10^{-6} and 9.1×10^{-6}, respectively. In every case, the V_f value determined with cryopreserved cells was significantly lower than the corresponding values determined with fresh cells. In the extreme, the decrease was more than two orders of magnitude!

In order to quantify the effects of cryopreservation on the ability of cyling, presumably nonmutant PBL's, to incorporate ^{3}HTdr in TG under the conditions of this test, we again obtained blood samples from five normal individuals [41]. After separating, the MNC's were again split, with one portion being tested fresh and the other after cryopreservation. For this experiment, however, PHA was not added to cultures. The frequency of spontaneously cycling cells in the samples was determined from the labeling index (spontaneously labeling cells) of control cultures not containing TG (LI_c). Similarly, the apparent TG^r lymphocyte frequency in the cycling cells was determined from the labeling index of test cultures containing TG (LI_t) divided by the LI_c, exactly as described for cultures con-

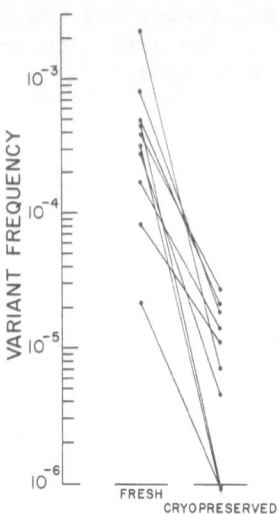

Fig. 3. A set of duplicate TG^r PBL V$_f$ values for 10 normal in-
          dividuals. Two determinations were made with cells from
          each blood sample, one using fresh and one using cryopre-
          served cells (see text). The range of V$_f$'s for fresh cells
          varied from 2.2 × 10^{-5} to 2.3 × 10^{-3}; the range of V$_f$'s for
          cryopreserved cells varied from <1 × 10^{-6} to 2.7 × 10^{-5}.
          A value of 0 was observed for three cryopreserved cell
          samples. For these, a value of 1 × 10^{-6} was used in cal-
          culating the group mean TG^r PBL V$_f$ (see text).

taining PHA. This provided an estimate of the phenocopy frequency
("apparent TG^r PBL V$_f$").

     Table 1 gives the results of this experiment, which have been
reported in different form elsewhere [41]. In four of the five samples, cryopreservation reduced to zero the frequency of labeled
lymphocytes in TG ("apparent TG PBL V$_f$") among the spontaneously
labeling cells when cultures were assayed in the usual way - i.e.,
30 hours incubation, followed by 12 hours label. By contrast,
lymphocytes labeling in TG were present in all five of the fresh
samples assayed by labeling at 30 hours with their frequencies ("apparent TG^r PBL V$_f$'s") ranging from 0.2% to 4% of the spontaneously
cycling cells. At earlier times of culture (0 hours incubation,
12 hours label), the frequencies of lymphocytes labeling in TG
among the cycling cells were greater in the cryopreserved than in
the fresh samples.

     We interpret these results to indicate that cryopreservation
moves the cycling cells that are capable of labeling in TG through
their S phase (DNA synthesis) earlier in the culture interval than

Table 1. Autoradiographically Labeled Nuclei Appearing in Non-Phytohemagglutinin (PHA) Stimulated Lymphocyte Cultures: Cultures are with and without TG; Fresh and Cryo-preserved Samples

	FRESH MONONUCLEAR CELLS					CRYOPRESERVED MONONUCLEAR CELLS				
	Time in Culture (Hours)[1]		Labeling Index		Apparent	Time in Culture (Hours)		Labeling Index		Apparent
Individual	Incubation	Label	No TG	TG[2]	TG^r PBL V_f[3] in Cycling Cells	Incubation	Label	No TG	TG	TG^r PBL V_f in Cycling Cells
Control 1	0	12	2.4×10^{-3}	2.2×10^{-4}	0.090	0	12	8×10^{-4}	2.4×10^{-4}	0.300
	20	12	1.5×10^{-3}	9.3×10^{-6}	0.006	20	12	8×10^{-4}	3.3×10^{-6}	0.004
	30	12	2.4×10^{-3}	1.0×10^{-5}	0.004	30	12	4×10^{-4}	0	0
Control 2	0	12	1.6×10^{-3}	4.9×10^{-4}	0.310	0	12	1.2×10^{-3}	7.0×10^{-4}	0.580
	20	12	2.0×10^{-3}	3.0×10^{-5}	0.150	20	12	8.0×10^{-4}	6.5×10^{-6}	0.008
	30	12	2.7×10^{-3}	2.1×10^{-5}	0.008	30	12	1.6×10^{-3}	0	0
Control 3	0	12	3.6×10^{-3}	1.0×10^{-3}	0.280	0	12	2.8×10^{-3}	1.4×10^{-3}	0.500
	20	12	2.4×10^{-3}	8.7×10^{-6}	0.004	20	12	3.2×10^{-3}	3.4×10^{-6}	0.001
	30	12	2.3×10^{-3}	4.6×10^{-6}	0.002	30	12	2.4×10^{-3}	0	0
Control 4	0	12	1.1×10^{-3}	1.6×10^{-4}	0.150	0	12	1.9×10^{-3}	--	--
	20	12	8×10^{-4}	1.8×10^{-5}	0.020	20	12	8×10^{-4}	9×10^{-6}	0.010
	30	12	4×10^{-4}	1.5×10^{-5}	0.040	30	12	2.4×10^{-3}	0	0
Control 5	0	12	2.3×10^{-3}	4×10^{-4}	0.170	0	12	4×10^{-4}	3.6×10^{-4}	0.900
	20	12	8.7×10^{-3}	4×10^{-5}	0.005	20	12	1.2×10^{-3}	7.4×10^{-6}	0.006
	30	12	6.5×10^{-3}	1.3×10^{-5}	0.002	30	12	4×10^{-4}	3.4×10^{-6}	0.009

(1) Lymphocytes were cultured in vitro without PHA; labeling was with 5 μc^3HTdr (see text)

(2) TG present in cultures at 2×10^{-4} M

(3) Apparent TG^r PBL V_f = probable frequency of phenocopies among the cycling cells (see text)

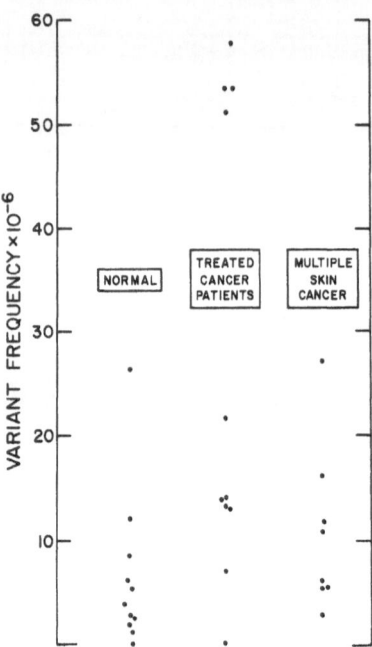

Fig. 4. TG^r PBL V$_f$ values for 11 normal individuals, 11 cancer patients who were receiving potentially mutagenic chemotherapies, and 8 adults with multiple skin cancers. Tests
were performed with cryopreserved MNC's (see text).

is the case with fresh MNC's. These cycling cells are then not in
S phase during labeling when the usual assay procedure is followed.
Thus, the frequency of labeled cells determined from cryopreserved
samples should more nearly approximate the frequency of G$_0$ cells
which are resistant to TG inhibition of their transformation to G$_1$.
We feel that this value more closely approximates the "true" TG^r PBL
V$_f$.

IV. RECENT DETERMINATIONS OF TG^r PBL V$_f$'s
     USING CRYOPRESERVED CELLS

     Studies of TG^r PBL V$_f$'s in normal individuals and in patients
exposed to known mutagenic therapies have been repeated with cryopreserved cells. Some of these results, published elsewhere, are
given in Fig. 4 [47, 48]. Blood samples from 11 normal adults
showed TG^r PBL V$_f$'s that ranged from less than 1.1×10^{-6} to $2.7 \times$
10^{-5} with group mean and median values of 6.5×10^{-6} and 3.8×10^{-6},
respectively. Nine of these 11 values were less than 10^{-5}. One of
the two values above this (2.7×10^{-5}) was for an individual who is
occupationally exposed to chemicals; the other was for an individual
with psoriasis. Another individual with psoriasis, however, had a
TG^r PBL V$_f$ of 6.7×10^{-6}. In contrast to these values, 9 of 11

blood samples from cancer patients who were receiving potentially
mutagenic chemotherapies, showed TG^r PBL V_f values greater than 10^{-5}.
The range of all values for this group was from less than 2.2×10^{-6}
to 5.8×10^{-5}, with a group mean and median value of 2.7×10^{-5} and
1.5×10^{-5}, respectively. Eight blood samples from individuals with
multiple skin cancers were tested. These individuals were not receiving chemotherapies, but may represent individuals with unusual
susceptibilities to environmental mutagens. The range of V_f values
for this group was from 2.9×10^{-6} to 2.7×10^{-5}, with 4 of the 8
values being above 10^{-5}.

Several studies are currently in progress. Normal adult TG^r
PBL V_f values determined with cryopreserved blood samples will be
much lower than we reported initially for fresh samples. Values for
individuals exposed to mutagens will still be elevated as compared
to normal values when tested with cryopreserved samples, but these
elevations will be less than reported earlier. TG^r PBL V_f elevations
of this magnitude seem more realistic and more compatible with somatic cell mutation than those previously reported.

In reviewing our data with cryopreserved cells, we do not see
an age effect for V_f's over the age range of approximately 20 to 80
years [47, 48]. However, TG^r PBL V_f's may be slightly lower for
umbilical cord blood samples. We now routinely cryopreserve all
material prior to testing.

V.   CLONING IN VITRO OF TG^r PBL

We have recently reported that TG^r PBL's can be cloned and
propagated in vitro using crude T-cell growth factors [45, 46]. This
has allowed a characterization of the variant TG^r cells and a verification by the usually accepted criteria of their mutant nature
[49]. We can now say with confidence that the TG^r lymphocytes of
human peripheral blood are mutant somatic cells.

It may be possible to automate quantitative cloning assays for
TG^r PBL's. If so, that methodology may replace labeling techniques
such as described here for detecting these mutant cells. However,
at a minimum, cloning will allow a calibration of labeling techniques such as this autoradiographic assay since, by cloning, the
"true" TG^r PBL V_f can be determined. Studies comparing V_f's determined by cloning and by autoradiography are in progress.

VI.   DISCUSSION

The variant TG^r lymphocyte assay has undergone considerable
evolution since its original description. Phenocopies, as described
here, were unrecognized initially. This resulted in our reporting
TG^r PBL V_f values that were far too high. However, as described
here, we now feel that we have identified a major source of these

interfering cells, and have been able to remove their effects by
cryopreservation. Perhaps any method that will semisynchronize
cycling cells, as does freezing, or remove them entirely, will serve
to remove the phenocopy effect.

Even though we now feel confident regarding the mutant nature
of TG^r PBL's, there remains another problem inherent in all direct
mutagenicity testing. In mutagenicity monitoring, somatic cell
mutations are the events of interest. However, in all direct tests
including the test described here, somatic cell mutants are the cells
actually scored. Mutants are related to mutations in a predictable
way only if several parameters of the cell population under study
such as growth kinetics, cell compartment sizes, mutant cell selection in vivo, etc., are known. Unfortunately, these parameters are
not known for cell populations in vivo. Furthermore, when determining the effects of mutagens in vivo, the average dose to the cells
being assayed and the level of cell killing in vivo are unknown.
Therefore, most methods developed for quantitating somatic cell mutation in vitro are not appliable in vivo. Direct mutagenicity tests
measure mutant cell frequencies only. An increase in mutation is but
one of several possible explanations for an increase in mutant cells.

Nonetheless, direct mutagenicity tests have unique advantages
for human mutagenicity monitoring. It is hoped that these tests
will eventually replace affected individuals as realistic indicators
of human health hazards. The TG^r PBL autoradiographic assay is presented as one such test.

VII.   REFERENCES

1.   B. A. Bridges, The three-tier approach to mutagenicity screen-
     ing and the concept of radiation-equivalent dose, Mutat. Res.,
     26:335-340 (1976).
2.   EPA Document: EPA 625/9-79-003. Environmental Assessment:
     short term tests for carcinogens, mutagens and other genotoxic
     agents, July 1979.
3.   C. F. Arlett and A. R. Lenman, Human disorders showing increased
     sensitivity to the induction of genetic damage, Ann. Rev.
     Genet., 12:95-115 (1978).
4.   M. Swift and C. Chase, Cancer in families with xeroderma pigmen-
     tosa, J. Natl. Cancer. Inst., 62:1415-1421 (1979).
5.   H. J. Evans and M. L. O'Riordan, in: "Handbook of Mutagen Test
     Procedures," B. J. Kilbey, M. Legator, and C. Ramel, eds., pp.
     261-274, Elsevier/North Holland, Amsterdam, New York, Oxford
     (1977).
6.   A. Brogger, in: "Genetic Damage in Man Caused by Environmental
     Agents," K. Berg, ed., pp. 87-99, Academic Press, New York
     (1979).
7.   P. Perry and H. J. Evans, Cytological detection of mutagen-
     carcinogen exposure by sister chromatid exchange, Nature, 258:

121-125 (1975).

8. D. G. Stetka and S. Wolff, Sister chromatid exchange as an assay for genetic damage induced by mutagen-carcinogens. I. In vivo test for compounds requiring metabolic activation, Mutat. Res., 41:333-341 (1976).

9. S. Wolff, Sister chromatid exchanges, Ann. Rev. Genet., 11:183-201 (1977).

10. S. Latt, J. W. Allen, W. E. Rogers, and L. A. Luerglus, in: "Handbook of Mutagen Test Procedures," B. J. Kilbey, M. Legator, W. Nichols, and C. Ramel, eds., pp. 275-291, Elsevier/North Holland, Amsterdam, New York, Oxford (1977).

11. R. DeMars, in: "Banbury Report 2: Mammalian Cell Mutagenesis: The Maturation of Test Systems," A. W. Hsie, J. P. O'Neill, and V. K. McElheny, eds., pp. 329-340, Cold Spring Harbor Laboratory (1979).

12. K. C. Atwood, The presence of A_2 erythrocytes in A_1 blood, Proc. Natl. Acad. Sci. (USA), 44:1054-1057 (1958).

13. K. C. Atwood and S. L. Scheinberg, Somatic variation in human erythrocyte antigens, J. Cell. Comp. Physiol., 52:97-123 (1958).

14. H. E. Sutton, in: "Mutagenic Effects of Environmental Contaminants," H. E. Sutton and M. I. Harris, eds., pp. 121-128, Academic Press, New York (1972).

15. H. E. Sutton, in: "Birth Defects: Proceedings of the Fourth International Conference," A. G. Motulsky and W. Lenz, eds., pp. 212-214 (1974).

16. G. Stamatoyannapoulos, W. G. Wood, T. N. Papayannopoulou, and P. E. Nute, An atypical form of hereditary persistance of fetal hemoglobin in blacks and its association with sickle cell trait, Blood, 46:683-692 (1975).

17. W. G. Wood, G. Stamatoyannapoulos, G. Lim, and P. E. Nute, F-cells in the adult: Normal values and levels in individuals with hereditary and acquired elevations of HbF, Blood, 46:671-682 (1975).

18. K. C. Atwood and F. J. Petter, Erythrocyte automosaicism in some persons of known genotype, Science, 134:2100-2102 (1961).

19. T. H. Papayannopoulou, M. Brice, G. Stamatoyannopoulos, Hemoglobin F synthesis in vitro: Evidence for control at the level of primitive erythroid stem cells, Proc. Natl. Acad. Sci. (USA), 74:2923-2927 (1977).

20. T. H. Papayannopoulou, P. E. Nute, G. Stamatoyannopoulos, and T. G. McGuire, Hemoglobin ontogenesis: Test of the gene exclusion hypothesis, Science, 197:1215-1216 (1977).

21. T. H. Papayannopoulou, T. C. McGuire, G. Lim, E. Garzel, P. E. Nute, and G. Stamatoyannopoulos, Identification of hemoglobin S in red cells and normoblasts using fluorescent anti-Hb antibodies, Br. J. Haematol., 34: 25-31 (1976).

22. T. H. Papayannopoulou, G. Lim, T. C. McGuire, V. Ahern, P. E. Nute, and G. Stamatoyannopoulos, Use of specific fluorescent antibody for the identification of a hemoglobin C in erythocytes, Am. J. Hematol., 2:95-112 (1977).

23. G. Stamatoyannopoulos, P. E. Nute, T. H. Papayannopoulou, T. G. McGuire, G. Lim, H. F. Bunn, and D. Rucknagel, Development of a somatic mutation screening system using Hb mutants. IV. Successful detection of red cells containing the human frameshift mutants Hb Wayne and Hb Cranston using monospecific fluorescent antibodies, Am. J. Hum. Genet., 32:484-496 (1980).

24. G. H. Strauss and R. J. Albertini, Enumeration of 6-thioguanine resistant peripheral blood lymphocytes in man as a potential test for somatic cell mutation arising in vivo, Mutat. Res., 61:353-379 (1979).

25. R. J. Albertini, Drug resistant lymphocytes in man as indicators of somatic cell mutation, Teratog. Carcinog. Mutagen., 1: 25-48 (1980).

26. M. Lesch and W. L. Nyhan, A familial disorder of uric acid metabolism and central nervous system function, Am. J. Med., 36:561-570 (1964).

27. J. E. Seegmiller, F. M. Rosenbloom, and W. N. Kelley, Enzyme defect associated with a sex linked human neurological disorder and excessive purine synthesis, Science, 155:1682-1684 (1967).

28. W. N. Kelley, Enzymology and Biochemistry, A. HG-PRT deficiency in the Lesch-Nyhan syndrome and gout, Fed. Proc., 27:1047-1052 (1968).

29. C. T. Caskey and G. D. Kruh, The HPRT locus, Cell, 6:1- (1979).

30. G. B. Elion, Biochemistry and pharmacology of purine analogs, Fed. Proc., 26:898-904 (1967).

31. G. B. Elion and G. H. Hitchings, in: "Advances in Chemotherapy,: A. Goldin, F. Hawkin, and R. J. Schnitzer, eds., Vol. 2, pp. 91-177, Academic Press, New York (1965).

32. R. J. Albertini and R. DeMars, Mosaicism of peripheral blood lymphocyte populations in females heterozygous for the Lesch-Nyhan mutation, Biochem. Genet., 11:397-411 (1974).

33. R. J. Albertini and R. DeMars, Detection and quantification of x-ray induced mutation in cultured human fibroblasts, Mutat. Res., 18:199-224 (1973).

34. L. Jacobs and R. DeMars, in: "Handbook of Mutagenicity Test Procedures," B. J. Kilbey, M. Legator, W. Nichols, and C. Ramel, eds., pp. 193-220, Elsevier/North Holland, Amsterdam, New York, Oxford (1977).

35. R. J. Albertini, D. L. Sylwester, and E. F. Allen, in: "Mutagenicity: New Horizons in Genetic Toxicology," J. A. Heddle, ed., pp. 305-336, Academic Press, New York (1982).

36. G. H. Strauss, R. J. Albertini, P. Krusinski, and R. D. Baughman, 6-thioguanine resistant peripheral blood lymphocytes in humans following psoralen long-wave light therapy, J. Invest. Dermatol., 73:211-216 (1979).

37. G. H. Strauss, E. F. Allen, and R. J. Albertini, An enumerative assay purine analogue resistant lymphocytes in women heterozygous for the Lesch-Nyhan mutation, Biochem. Genet., 18:529-547 (1980).

38. J. Dancis, P. H. Berman, V. Jansen, and M. E. Balis, Absence of mosaicism in the lymphocyte in x-linked congenital hyperuricemia, Life Sci., 7:587-591 (1968).
39. J. A. McDonald and W. N. Kelley, Lesch-Nyhan syndrome: Absence of the mutant enzyme in erythrocytes of a heterozygote for both normal and mutant hypoxanthine-guanine phosphoribosyltransferase, Biochem. Genet., 6:21-26 (1972).
40. W. L. Nyhan, B. Bakay, J. D. Connor, J. F. Marks, and D. K. Keele, Hemizygous expression of heterozygotes for the Lesch-Nyhan syndrome, Proc. Natl. Acad. Sci. (USA), 65:214-218 (1970).
41. R. J. Albertini, E. F. Allen, A. S. Quinn, and M. R. Albertini, in: "Population and Biological Aspects of Human Mutation: Birth Defects Institute Symposium XI," E. B. Hook and I. H. Porter, eds., pp. 235-263, Academic Press, New York (1981).
42. A. L. Maizel, S. R. Mehta, S. Hauft, D. Franzini, L. B. Lachman, and R. J. Ford, Human T-lymphocyte/monocyte interaction in response to lectin: Kinetics of entry into S-phase, J. Immunol., 127:1058-1064 (1981).
43. A. C. Allison, T. Hovi, R. W. E. Watts, and A. B. D. Webster, in: "Purine and Pyrimidine Metabolism," K. Elliot and D. W. Fitzsimmons, eds., pp. 207-224, Elsevier/North Holland/Excerpta Medica, Amsterdam (1977).
44. T. Hovi, A. C. Allison, K. O. Raivio, and A. Vaheri, in: "Purine and Pyrimidine Metabolism," K. Elliot and D. W. Fitzsimmons, eds., pp. 207-224, Elsevier/North Holland/Excerpta Medica, Amsterdam (1977).
45. R. J. Albertini, W. R. Borcherding, Cloning in vitro of human 6-thioguanine resistant peripheral blood lymphocytes arising in vivo, Environ. Mutagenis. (abstract) in press (1982).
46. R. J. Albertini, K. Castle, and W. R. Borcherding, T-Cell Cloning to Detect the Mutant 6-Thioguanine-Resistant Lymphocytes Present in Human Peripheral Blood, Proc. Natl. Acad. Sci. (USA) 79:6617-6621 (1982).
47. R. J. Albertini, D. L. Sylwester, E. F. Allen, and B. D. Dannenberg, in: "Carcinogens and Mutagens in the Environment," H. F. Stich, ed., Vol. I, CRC Press, Boca Ratan, Florida (in press) (1982).
48. R. J. Albertini, D. L. Sylwester, B. D. Dannenberg, and E. F. Allen, in: "Genetic Toxicology, An Agricultural Perspective," R. Flech, ed., Plenum Press, New York (1982) (in press).
49. E. H. Y. Chu and S. S. Powell, in: "Advances in Human Genetics," Harris and Hirshhorn, eds., Vol. 7, pp.189-258, Plenum Press, New York (1976).

DETECTION OF SOMATIC MUTANTS OF HEMOGLOBIN

George Stamatoyannopoulos[1] and Peter E. Nute[1,2]

[1]From the Center for Inherited Diseases
and [2]the Department of Anthropology
University of Washington
Seattle, Washington 98195

INTRODUCTION

The successful screening of large populations of somatic cells for the presence of cells of mutant phenotype has been the goal of several investigations [1-9]. A system for detecting somatic-cell mutants would serve many valuable ends. In addition to its utility in estimating the genetic risks to subjects exposed to mutagenic agents, the system would constitute a means for investigation of various aspects of somatic-cell biology in man, including the testing of models of proliferation and renewal of pluripotent and committed stem-cell pools [8]. We have been attempting to develop a system for the detection of rare erythrocytes that contain abnormal hemoglobins as a consequence of somatic-cell mutation. Our purpose in this paper is to summarize our efforts to date and to outline certain possibilities for further research that derive from the availability of monoclonal antibodies specific for particular globin chains.

THE SYSTEM

It is assumed that, in the globin genes of pluripotent and committed erythroid stem cells, mutations do arise. It is also assumed that such mutations are not disadvantageous to the stem cells and, hence, that rare red cells heterozygous for abnormal globin chains exist in the circulation of every individual. These erythrocytes would be appropriate targets of a system for detecting somatic mutants, provided one could screen blood samples from genetically normal individuals and count the rare cells that contain a mutant hemoglobin. In our studies we used immunochemical approaches to

achieve this goal. If antibodies specific for particular mutant
hemoglobins were raised and purified, fluorescent conjugates of
these antibodies could then be used to detect rare, mutant erythrocytes.

In the first stage of our research, we focused on raising antibodies against mutant hemoglobins. Horses were immunized with purified abnormal hemoglobins and the antibodies were isolated, by affinity chromatography, from large volumes of antisera [10-14]. Antibodies specific for Hb S [11], Hb C [12], Hb Hasharon [13], and the
frameshift mutants Hb Wayne and Hb Cranston [14] were produced. Each
of these antibodies was non-precipitating and, after conjugation with
a fluorochrome (usually fluorescein isothiocyanate, FITC), each bound
only to erythrocytes that contained the hemoglobin against which it
was raised [11-14].

In the second stage of our work, we assessed the extent to
which fluorescent anti-Hb antibodies allowed recognition of rare
mutant red cells in the blood. Mixtures of erythrocytes in which
the frequency of red cells containing a specific abnormal hemoglobin
ranged from 1×10^{-2} down to 2×10^{-6} (or, in some instances, down
to 5×10^{-6}) were prepared. Exposure of these red-cell populations
to the appropriate antibody, followed by counting of the labeled
cells under ultraviolet light, revealed labeled cells in frequencies
that were virtually identical to the frequencies of abnormal cells
in each mixture [7, 13], suggesting that the immunochemical approach
would allow detection of rare, mutant red cells.

In the third stage, erythrocytes from normal subjects were
screened in an attempt to identify red cells containing an abnormal
hemoglobin (Hb S or Hb C). Fixed preparations of erythrocytes labeled with anti-Hb S or anti-Hb C antibodies were screened under
ultraviolet light. When a cell labeled with the fluorescent antibody was detected, the intensity of fluorescence of the cell was
measured. Labeled cells whose intensity of fluorescence fell within
two standard deviations of the mean for AS (or AC) cells labeled with
the appropriate antibody were considered "putative mutant cells."
Counter-labeling of the preparation by a mutually exclusive antibody (e.g., anti-Hb C-rhodamine if the initial labeling was with
anti-Hb S-FITC) was subsequently carried out to test for the possibility of non-specific labeling. The preparations were then
stained with benzidine and positively stained cells were judged to
be erythrocytes. In this study, an average of 5×10^{7} cells from
each of 15 individuals was screened. An average of 1.1×10^{-7} cells
per subject bound anti-Hb S-FITC (or anti-Hb C-FITC), fluoresced at
appropriate intensities, failed to show non-specific labeling and
reacted positively with benzidine. In addition, a number of subjects who had been exposed to mutagens (radiation or chemotherapeutic agents) were examined. No striking differences in frequencies
of putative mutant cells were noted upon comparison of the results
with those obtained from studies of normal individuals [8].

The potential for automated screening of red cells using a
fluorescence-activated cell sorter was also explored. In collaboration with Dr. Mendelssohn and his associates, a method was devised
that permitted fluorescent labeling of hemoglobin by reacting the
antibodies with red cells in suspension. The procedure entailed the
covalent cross-linking of intracellular hemoglobin to the cell membrane, lysis of the cells to render the hemoglobin accessible to the
antibody, and passage of the suspension through a fluorescence-
activated cell sorter [15]. In spite of the rapidity with which
large numbers of cells could be screened by the cell sorter, the
results were marred by the instrument's inability to distinguish between labeled erythrocytes and artifacts. Sorted cells had to be
inspected microscopically to detect aggregates of antibody-FITC conjugates and nonspecifically labeled white cells that were included
among labeled erythrocytes. The average frequency of putative mutant cells detected in blood samples from six normal subjects (after
screening approximately 5×10^8 cells per individual) was 1×10^{-7}.

THE DESIRABILITY OF AN ANIMAL MODEL

The finding of cells with particular abnormal phenotypes at a
frequency of 1×10^{-7} raised many new questions, the most important
of which concerned the nature of the events detected. Were the
"putative mutant cells" actually products of somatic-cell mutation
or were they merely artifacts? We have dealt with this question in
previous publications [7, 8] and have concluded that one should not
consider these cells mutants in the absence of proof of mutational
origin. On the basis of previous results, we can conclude that the
frequency of cells bearing a particular variant of hemoglobin as a
consequence of somatic mutation cannot exceed 1×10^{-6} (ref. 8); the
question remains, however, whether or not the cells detected at frequencies of 1×10^{-7} are the mutant erythrocytes.

At this juncture, our work on the system for detecting somatic
mutants came to a standstill. Testing by indirect means (i.e.,
through the study of subjects exposed to mutagens) of the possibilities that the variant cells detected by the system were, in fact,
mutants was unrewarding. As noted above, there were no obvious differences between the frequencies of variant red cells in the exposed
and unexposed individuals. The failure to detect striking incre-
ments in frequencies of variant cells in individuals exposed to
mutagens is subject to several interpretations. For example, if the
pool of pluripotent stem cells is relatively small and undergoes
frequent renewal, frequencies of mutant erythrocytes would be largely
determined by mutational events in this pool. With the passage of
time mutations will accumulate so that, in an adult, the frequency
of mutant cells may surpass the mutation rate by 1000-fold or more
[7]. One would not, thus, expect acute exposure to mutagens to
produce immediate effects on the frequencies of abnormal erythrocytes in the circulation; if the frequency of a particular variant

phenotype exceeds the mutation rate by as much as 1000, the rate of
mutation subsequent to exposure to a mutagen would have to increase
by the same factor to cause a two-fold increment in the frequency of
erythrocytes with the phenotypes in question. Conversely, given a
large pool of pluripotent stem cells with a low rate of renewal, the
frequencies of mutant erythrocytes in the circulation would be, in
large measure, functions of mutational events occurring among committed stem cells [8]. While the latter set of circumstances would
favor the rapid elevation of frequencies of mutant red cells in the
mature erythron, this elevation would be short-lived, owing to the
short life span of the mature erythrocyte. Thus the effects of acute
exposure to mutagens might be missed unless the interval between exposure and screening were, fortuitously, of appropriate duration.
Discrimination between these two possibilities could, in principle,
be accomplished through systematic longitudinal studies of subjects
exposed to mutagens, but such studies could not be performed.

     Though the system of screening for hemoglobin mutants was de-
veloped for detecting somatic-cell mutants in man, the difficulties
in assessing the mutational origin of putative mutant cells underscored the need for employing an animal model in validating the system. If an animal model were available, it would be possible to
freely manipulate exposures to mutagens and test their effects on
the frequency of variant cells at prescribed intervals. From the
time of the appearance of variant cells and the frequencies they at-
tain, one could decide if the variant cells are or are not mutants.
One problem is that very few abnormal hemoglobins from animals other
than man have been characterized, and antibodies to these variant
hemoglobins are not available. This problem can, however, be overcome by using monoclonal antibodies that are specific for normal
human globin chains.

MONOCLONAL ANTIBODIES SPECIFIC FOR GLOBIN CHAINS

     In conjunction with our studies of the cellular mechanism of
the switch from Hb F to Hb A in man, we have developed monoclonal
antibodies specific for the β or γ chains of human adult and fetal
hemoglobins, as well as antibodies capable of recognizing antigenic
determinants common to α and β or β and γ chains [16, 17]. Characterization of the monoclonals includes semiquantitative assessment of the reactivity of each antibody with hemoglobins and globin
chains from various species (Figs. 1, 2). The antibody (monoclonal
3-2) recognizes human β chains as well as the β chains of a macaque,
dog and rabbit, but does not recognize the β chains of a baboon or
the β chains of mice.

     The differences in hemoglobin-antibody interactions noted above
most likely derive from structural differences in antigenic sites.
Each antibody is expected to recognize a single antigenic site that
is limited to five or six amino acid residues. During evolution,

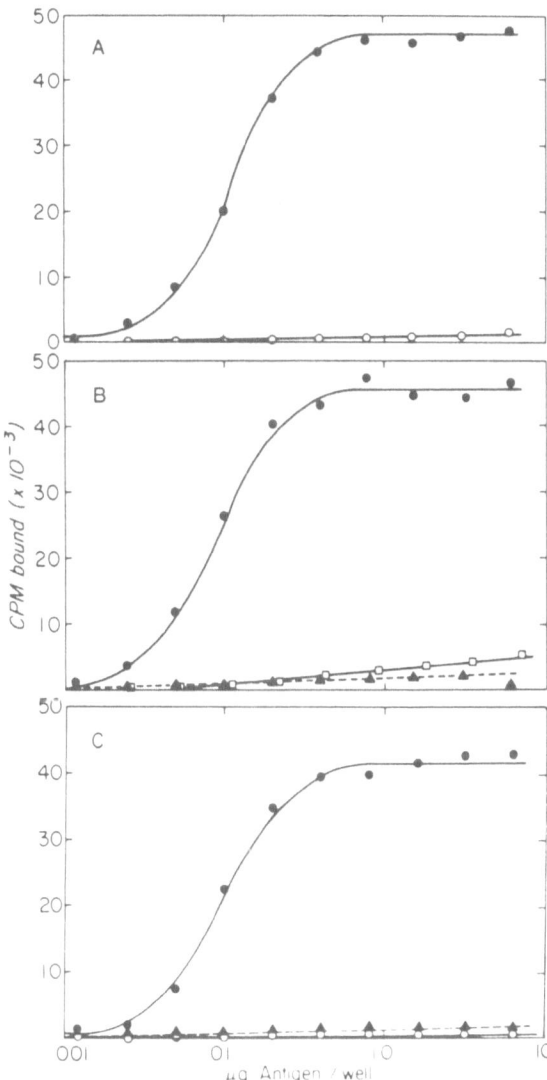

Fig. 1. Reactions of monoclonal antibody 3-2. (A) Against human
Hb A (●—●) and human Hb F (O—O); (B) against p-mercuri-
benzoate derivatives of human β (●—●) and α (▲—▲) chains,
and native Hb Bart's (γ₄) (□—□)   (C) against adult hemoglobins from Macaca nemestrina (●—●), Papio cynocephalus
(O—O), and BALB/c mouse (▲—▲). (From reference 16.)

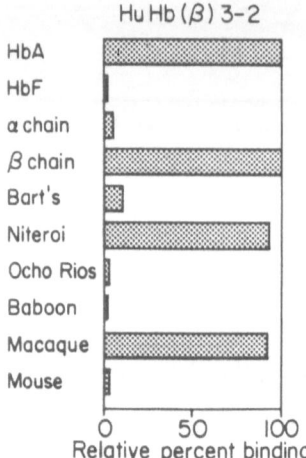

Fig. 2.   Summary of reactions of monoclonal antibody 3-2 against
          normal human macaque, baboon, and mouse hemoglobins, iso-
          lated human α and β chains, and the human variant Hbs
          Niteroi and Ocho Rios.

the globin chains of members of various lineages have diverged in
structure. Divergence of structure in antigenic sites is expected
to be reflected in the extent to which monoclonal antibodies recognize these sites. If this be the case, comparison of the structures
of the globins that are recognized by the antibody with those not
recognized might permit deduction of the position and extent of the
antigenic site. This strategy was applied to characterization of
the site with which antibody 3-2 reacts [16].

     Because antibody 3-2 reacts with human and macaque β chains
but not with baboon β chains, the antigenic site must involve a portion of the β-globin sequence in which the chains from both man and
macaque differ from those of the baboon. At only two positions, β^{43}
and β^{52}, are these conditions satisfied. Is either of these positions, identified through comparison of primary structures, included
in the antigenic site with which Hb 3-2 reacts? We reacted the
monoclonal antibody with abnormal hemoglobins that have sustained
structural alterations at these sites. Hemoglobin Niteroi is a
human variant whose β chains have sustained a deletion of a sequence
of three amino acid residues, one of which is the glutamyl residue
that normally occupies position 43. In Hb Ocho Rios, the aspartyl
residue that normally occupies position β^{52} has been replaced by an
alanyl residue. As shown in, Figs. 2 and 3, Nb Niteroi (which lacks
β^{43} Glu) reacts with the antibody to the same extent as does the
normal β chain. Conversely, Hb Ocho Rios fails to react, indicating
that the aspartyl residue normally in position β^{52} lies within the
antigenic site and is requisite to the antigen-antibody reaction.

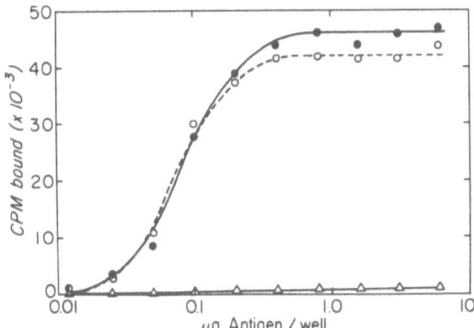

Fig. 3.   The binding of monoclonal antibody 3-2 with human Hb A
           (●—●), Hb Niteroi (○—○), and Hb Ocho Rios (Δ—Δ).   (From
           reference 16.)

Relevance to the Animal Model

The results of characterization of monoclonal antibody 3-2 are
relevant to our proposed use of an animal model in the study of
somatic-cell mutations. They suggest that monoclonal antibodies
can be used to detect the products of single nucleotide substitutions in experimental animals, provided that the mutant codons specify substitutions of single amino acids. Our reasoning is as follows:

Antibody 3-2 does not recognize the β chains of the baboon and
of Hb Ocho Rios because the amino acid sequence at positions 51
through 56 of both chains is Pro-Ala-Ala-Val-Met-Gly. The same
antibody does recognize the human β^A chain because the corresponding
sequence therein is Pro-Asp-Ala-Val-Met-Gly. If, in a baboon, a
mutation were to occur that resulted in the substitution of alanine
by aspartic acid at position β^{52}, antibody 3-2 would recognize the
abnormal baboon hemoglobin because the latter would then be identical in sequence over positions β^{51-56} to the normal human β chain.
Our data indicate that a monoclonal antibody to a site on the normal human β chain, which does not recognize the comparable portion
of a non-human β chain, will react with the non-human β chain should
the latter be appropriately altered by mutation. Antibody 3-2 could,
thus, be employed in the detection of baboon red cells that contain
the mutant hemoglobin $\alpha_2\beta_2^{52}$ Ala→Asp.

This reasoning can also be extended to, the monospecific, poly-
clonal antibodies. For example, the amino acid sequences of the
human and baboon β chains are identical from position 1 through po-

sition 8 [18].   An antibody specific for Hb S (or Hb C) must recognize a baboon β chain in which the glutamyl residue in position 6
has been substituted by a valyl (or a lysyl) residue.

PROSPECTS AND IMPLICATIONS

Use of the available monoclonal and monospecific antibodies
specific for human normal and mutant chains in screening of somatic
cells of mutagenized primates is contingent upon results of attempts
to increase the sensitivity of techniques for detecting fluorescing
erythrocytes by automated cell sorting. The technical problems inherent to this approach are described in other papers in this volume.
Work in our laboratory has led to the development of a method that
increases many-fold the fluorescent intensity of variant cells labeled with fluorescent antibodies. Further qualitative improvements
may allow conduct of the proposed experiments in animal models.

There are two additional implications of the results of the
work with monoclonal antibodies described above. First, the development of a system for detecting somatic mutants in man has required the availability of abnormal human hemoglobins. Antibodies,
specific for these variant proteins, were developed and subsequently
employed in the screening for variant red cells [11-13]. Work with
the monoclonal antibodies indicates, however, that availability of
particular abnormal hemoglobin is not a sine qua non. In theory,
monoclonal antibodies against non-human hemoglobins should bind to
cells containing specific abnormal human hemoglobins. In short, if
a non-human primate hemoglobin is employed as the immunogen, and a
monoclonal antibody that recognizes the non-human hemoglobin but not
its human counterpart is obtained, that antibody will react with a
human hemoglobin that shares, by virtue of mutation, the antigenic
site of the non-human hemoglobin.

Second, many of these procedures can be applied to non-hemo-
globin proteins -- in particular, proteins in the membranes of pro-
liferating cells. Consider a hypothetical lymphocytic protein,
protein X. Evolutionarily divergent forms of this protein in man
and a non-human primate are then designated X and X^n, respectively,
where n reflects the difference in antigenic sites. A monoclonal
antibody that recognizes X^n but not X is expected to react with one
of the sites that has contributed to the divergence of X^n from X.
Hence, a monoclonal that recognizes a surface determinant on baboon
lymphocytes but does not recognize the determinant on normal human
lymphocytes, will label those mutant human lymphocytes in which
mutation has changed the determinant so that its structure is identical to that of the baboon. Development of immunochemical methods
for detection of mutant lymphocytes would be of value; through the
isolation and culture of mutant cells, one could raise them in numbers sufficient to permit the direct chemical assessment of their
mutational origin. A question remains: will mutant surface proteins

be recognized as non-self, resulting in destruction, by the host's immune system, of the cells that bear them? They may not, provided the frequency of mutant cells is as low as 1×10^{-7}; recognition may also depend on the nature of the mutated antigenic site.

ACKNOWLEDGMENT

This work was supported in part by Grant GM-15253 from the National Institutes of Health

REFERENCES

1. K. C. Atwood, The presence of A_2 erythrocytes in A_1 blood, Proc. Natl. Acad. Sci. (USA), 44:1054-1057 (1958).
2. K. C. Atwood and S. L. Scheinberg, Somatic variation in human erythrocyte antigens, J. Cell. Comp. Physiol., 52:97-123 (1958).
3. K. C. Atwood and S. L. Scheinberg, Isotope dilution method for assay of inagglutinable erythrocytes, Science, 129:963-964 (1959).
4. K. C. Atwood and F. J. Pepper, Erythrocyte automosaicism in some persons of known genotype, Science, 134:2100-2102 (1961).
5. H. E. Sutton, in: "Mutagenic Effects of Environmental Contaminants," H. E. Sutton and M. I. Harris, eds., pp. 121-128, Academic Press, New York (1972).
6. H. E. Sutton, Somatic cell mutations, Birth Defects -- Proc. 4th Int. Conf., A. G. Motulsky and W. Lenz, eds., pp. 212-214 (1974).
7. G. Stamatoyannopoulos, in: "Genetic Damage in Man Caused by Environmental Agents," K. Berg, ed., pp. 49-62, Academic Press, New York (1979).
8. G. Stamatoyannopoulos and P. E. Nute, in: "Population and Biological Aspects of Human Mutation," E. B. Hook and I. H. Porter, eds., pp. 265-273, Academic Press, New York (1981).
9. G. H. Strauss and R. J. Albertini, Enumeration of 6-thioguanine-resistant peripheral blood lymphocytes in man as a potential test for somatic cell mutations arising in vivo, Mutat. Res., 61:353-379 (1979).
10. W. G. Wood, G. Stamatoyannopoulos, G. Lim, and P. E. Nute, F-cells in the adult: normal values and levels in individuals with hereditary and acquired elevations of Hb F, Blood, 46: 671-682 (1975).
11. Th. Papayannopoulou, T. C. McGuire, G. Lim, E. Garzel, P. E. Nute, and G. Stamatoyannopoulos, Identification of haemoglobin S in red cells and normoblasts, using fluorescent anti-Hb S antibodies, Brit. J. Haematol., 34:25-31 (1976). .
12. Th. Papayannopoulou, G. Lim, T. C. McGuire, V. Ahern, P. E. Nute, and G. Stamatoyannopoulos, Use of specific fluorescent antibodies for the identification of hemoglobin C in erythrocytes, Amer. J. Hematol., 2:105-112 (1977).

13. P. E. Nute, Th. Papayannopoulou, B. Tatsis, and G. Stamatoyan-
     nopoulos, Toward a system for detecting somatic-cell mutations.
     V. Preparation of fluorescent antibodies to hemoglobin Hasharon,
     a human α-chain variant, J. Immunol. Meth., 42:35-44 (1981).

14. G. Stamatoyannopoulos, P. E. Nute, Th. Papayannopoulou, T.
     McGuire, G. Lim, H. F. Bunn, and D. Rucknagel, Development of a
     somatic mutation screening system using Hb mutants. IV. Suc-
     cessful detection of red cells containing the human frameshift
     mutants Hb Wayne and Hb Cranston using monospecific fluorescent
     antibodies, Amer. J. Human Genet., 32:484-496 (1980).

15. W. L. Bigbee, E. W. Branscomb, H. B. Weintraub, Th. Papayan-
     nopoulou, and G. Stamatoyannopoulos, Cell sorter immunofluores-
     cence detection of human erythrocytes labeled in suspension with
     antibodies specific for hemoglobins S and C, J. Immunol. Meth.,
     45:117-127 (1981).

16. G. Stamatoyannopoulos, M. Farquhar, D. Lindsley, M. Brice,
     Th. Papayannopoulou, P. E. Nute, G. Serjeant, and H. Lehmann,
     Mapping of antigenic sites on human haemoglobin by means of
     monoclonal antibodies and haemoglobin variants, Lancet, 2:
     952-954 (1981).

17. G. Stamatoyannopoulos, M. Farquhar, D. Lindsley, M. Brice,
     Th. Papayannopoulou, and P. E. Nute, Monoclonal antibodies
     specific for globin chains, Blood (in press).

18. P. E. Nute and W. C. Mahoney, Complete primary structure of the
     β chain from the hemoglobin of a baboon, Papio cynocephalus,
     Hemoglobin, 4:109-123 (1980).

COUNTING OF RBC VARIANTS USING RAPID FLOW TECHNIQUES*

William L. Bigbee, Elbert W. Branscomb
and Ronald H. Jensen

Biomedical Sciences Division
University of California
Lawrence Livermore National Laboratory
Livermore, California 94550

INTRODUCTION

Presently there are important clinical, occupational and environmental needs for assays to estimate somatic mutation rates in human individuals. These methods are based on the enumeration of rare cells which present a variant phenotype as a result of somatic mutations having occurred in the DNA of single structural genes in precursor cells. To identify and quantitate such cells, an efficient and specific selection method must be employed; either clonogenic assay of mutant cells or immunologic detection of mutant gene products. Also, there must be an independent means for validating that the identified cells are, in fact, structural gene mutants, e.g., maintenance of the mutant phenotype under non-restrictive conditions or direct biochemical demonstration of the presence of a variant protein. In this paper, I will describe two assay systems we are developing based on high speed sorter detection of immunologically identified mutant human erythrocytes. Both take advantage of the ability of the flow sorter to rapidly screen large numbers of cells to enumerate and sort rare, presumptively mutant, cells labeled with fluorescent antibodies.

---

*Work performed under the auspices of the U.S. Department of Energy by the Lawrence Livermore National Laboratory under contract number W-7405-ENG-48 with the financial support of the National Institute of Environmental Health Sciences Interagency Agreement No. 222401-E S-00060 and the Environmental Protection Agency, Grant No. R808642-01.

DETECTION OF RARE RED CELLS IN NORMAL HEMOGLOBIN A
INDIVIDUALS LABELED WITH ANTIBODIES SPECIFIC FOR SINGLE
AMINO ACID SUBSTITUTED HEMOGLOBIN

Dr. George Stamatoyannopoulos at the University of Washington
has produced antibodies specific for the mutant human hemoglobins S
and C [1, 2]. These antibodies recognize and bind tightly to hemoglobin S or C but not to normal hemoglobin A. These two variant
hemoglobins differ from the normal hemoglobin A amino acid sequence
by single amino acid substitutions at the sixth position of the β
chain as a result of single base changes in the triplet codon corresponding to that position in the β globin gene. Hence in normal
hemoglobin A individuals, these point mutations in the β globin genes
in erythroid stem cells will give rise to circulating red cells containing hemoglobin S or C in addition to hemoglobin A. To identify
such cells, the hemoglobin S antibody was labeled with FITC and incubated with red cells fixed on slides. The preparations were then
manually examined under a fluorescence microscope for the presence
of antibody-labeled cells. Long and laborious effort revealed their
presence at a frequency of about one labeled cell in 10^7 unlabeled
cells [3]. While these positive results were encouraging, this
assay method was not practical since a single measurement was so time
consuming (about one man-month per sample) and it was not possible
to demonstrate biochemically that the labeled cells did, in fact,
contain hemoglobin S.

It was at this point we established a collaboration with Dr.
Stamatoyannopoulos to explore the potential application of flow
sorter technology to this problem. Two technical issues needed to
be addressed; (1) could a hemoglobin antibody staining method be developed for red cells in suspension and (2), could the sorter process
such samples rapidly enough to allow examination of a statistically
significant number of cells in a reasonable time. To permit antibody labeling of hemoglobin, a method needed to be devised which
would maintain the integrity of individual erythrocytes, fix intracellular hemoglobin in situ and then permit access to antibody.
Such a procedure was suggested by the work of Wang and Richards [4],
in which permeable cross-linking reagents were used to study the
topography of red cell membrane proteins. They showed that, in intact red cells, these reagents cross-linked intracellular hemoglobin
to itself and to many of the peripheral and intrinsic membrane proteins (Bands 1, 2, 3, 4.1, 4.2, 5, 6, and 7). We successfully
adapted this procedure to cross-link cells and produce ghosts which
retain about 10^6 hemoglobin molecules covalently bound to the cell
membrane, remain permeable to antibody and retain the native anti-
genicity of hemoglobin A, S, and C [5]. Such antiboby-labeled ghost
suspensions can be analyzed on the flow sorter. As shown in Fig. 1,
cross-linked ghosts prepared from cells taken from an individual
heterozygous for hemoglobin S and labeled with FITC-anti-hemoglobin
S produce a unimodal histogram with a coefficient of variation of

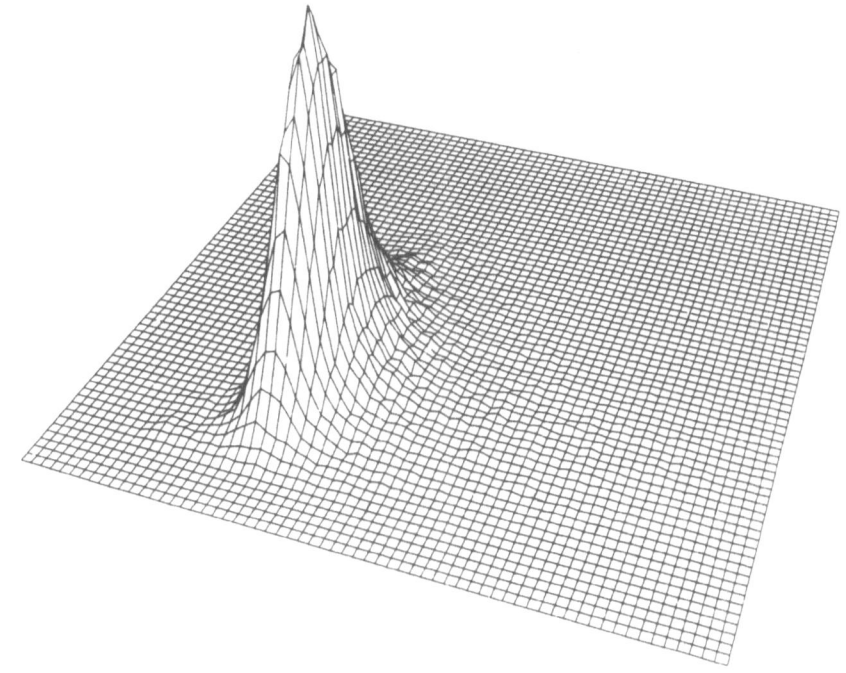

Fig. 1.   Flow histogram of cross-linked heterozygous hemoglobin AS
          red cell ghosts labeled with FITC-anti-hemoglobin S. The
          red cells were prepared as previously described [5], la-
          beled with hemoglobin S-specific antibody supplied by Dr.
          George Stamatoyannopoulos and analyzed on the LLNL-FACS
          sorter. The resulting population histogram is displayed
          with increasing fluorescence intensity along the abscissa
          and scatter intensity along the ordinate.

about 15%. These ghosts are readily detectable even with the limited
number of remaining hemoglobin molecules and direct fluorophor labeling of the primary antibody. This ease of detection is afforded
by the very efficient 488 nm laser excitation of the fluorescein and
by the performance of the flow sorter which together provide a sensitivity limit of $\sim 10^4$ fluorescein molecules per cell.

Next, the issue of sorter processing speed was examined. Assuming a background frequency of one anti-hemoglobin S positive cell
per 10^7 negative cells to be correct, one would like to analyze at
least 10^9 total cells per measurement. To do this in a timely fashion requires a sorter throughput rate of 10^6 cells per sec. From an
analysis of the limiting sample flow velocity and the sample stream/
laser intersection geometry, a sample density of about 5×10^8 cells/
ml is required resulting in about 20 cells being in the laser beam

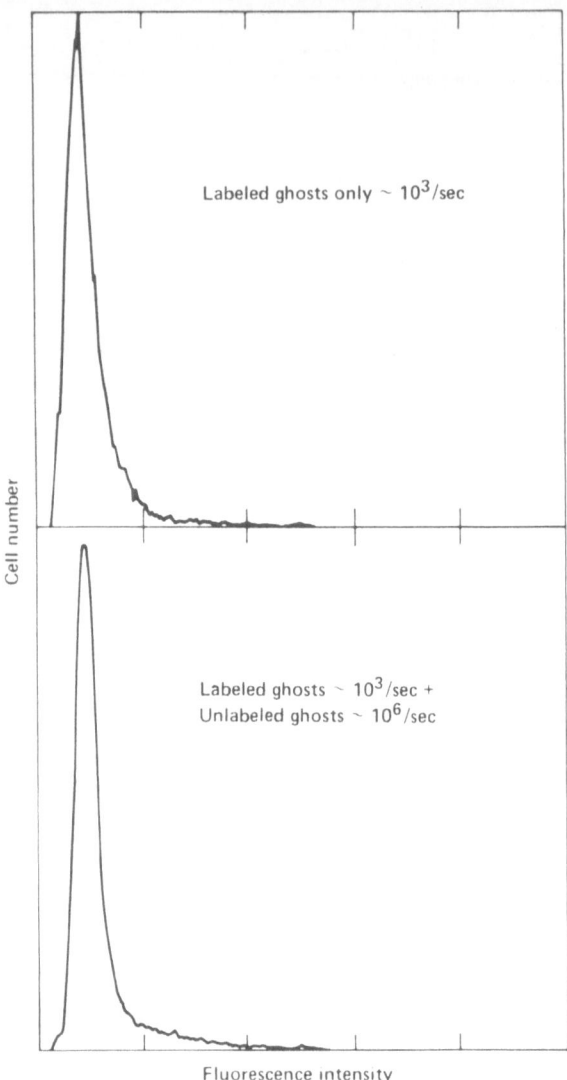

Fig. 2. High-speed detection of fluorescent red cell ghosts in the presence of a high density of non-fluorescent red cell ghosts. Fluorescent red cell ghosts were first analyzed alone at a normal rate of approximately 10^3 per sec (top panel). The fluorescent ghosts were then mixed with a thousand-fold excess of unlabeled ghosts and the mixtures analyzed at the same rate (lower panel). Under these conditions, the overall throughput rate of unlabled ghosts exceeded 10^6 per sec with no degradation in the histogram generated by the labeled ghosts.

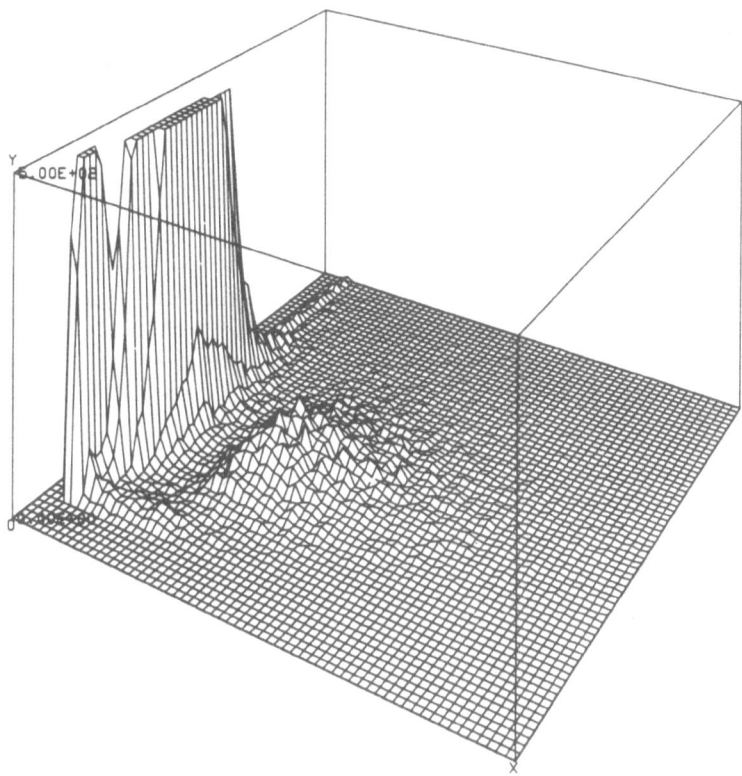

Fig. 3.   Flow histogram of a 1:10 artificial mixture of cross-linked
           heterozygous AS and normal hemoglobin AA red cell ghosts
           labeled with FITC-anti-hemoglobin S. Hemoglobin AS and AA
           cells were mixed and processed together as described in
           Fig. 1. The signals from the sub-population of antibody-
           labeled ghosts are visible as a distinct peak separated
           from signals from the normal unlabeled ghosts and fluores-
           cent debris.

at any given time. Using cross-linked red cell ghosts, we wished to
determine if rare fluorescent ghosts could be reliably detected under
such conditions. Suspensions of fluorescent ghosts in the absence
and presence of dense suspensions of unlabeled ghosts were analyzed.
As illustrated in Fig. 2, such rare ghosts could be quantitatively
detected under flow conditions resulting in an overall ghost processing rate of greater than 10^6 per sec.

To further test the specificity of antibody labeling, artificial
mixtures of variant and normal cells were prepared, processed together and examined. Microscopic observation revealed the approxi-

mate number of expected fluorescent ghosts with the remainder nearly
invisible. These ghost suspensions were then analyzed quantitatively
on the sorter. Figure 3 is a histogram obtained from a 1:10 mixture
of hemoglobin AS and AA cells incubated with FITC-anti-hemoglobin S.
By first running a calibration sample of hemoglobin AS ghosts, a
fluorescence-scatter window for antibody labeled ghosts could be determined. This window was usually defined as an area centered on
the fluorescence-scatter peak and extending ± two standard deviations
along both axes. By integrating the signals falling within this window and determining the total number of ghosts processed by the sum
of all the scatter signals, the frequency of antibody-labeled ghosts
in a mixture could be calculated and compared to the known dilution.
Artificial mixtures as dilute as one in $\sim 10^5$ could be accurately reconstructed before fluorescent background noise obscured the rare
labeled ghosts [5].

   Since the presence of this fluorescence background prevented us
from accurately counting more dilute mixtures, a detailed investigation into the potential sources of these signals was undertaken.
Noise from sorter electronics was completely absent; no signals were
detected when the flow stream was turned off. However, signals did
appear with the flow stream on even when buffer alone was used as a
sample. These signals undoubtedly arise from fluorescent objects
(cells, debris, etc.) which contaminate the sorter sample tubing.
Careful washing with detergent buffers greatly reduced this source
of signals although they could not be entirely eliminated. Using
new or thoroughly cleaned tubing reduced the frequency of such events,
under actual high speed flow conditons, to about one signal every 10-
100 sec or a total of approximately 10 to 20 false positive signals
for an entire sorter run. This contribution to the background noise
was very small however compared to that caused by the fluorescein-
ated antibody itself. To insure permeability of the cross-linked
ghosts, ghost suspensions must be maintained in hypotonic buffer at
0°C and, for reproducible staining, several hours of incubation are
required. Under these conditions, micro-precipitates of the antibody form. The measured fluorescence intensity of many of these
particles falls within the defined window for antibody-labeled
ghosts. In a typical analysis of 10^9 ghosts, about 10^4-10^5 signals
will appear in the sorting window with the vast majority being antibody precipitates. This result is obtained even in the absence of
ghosts; antibody at the same concentration incubated for the same
time under the same conditons produces the same spectrum of fluorescence artifacts. Thus, non-specific binding of the antibody to normal ghosts does not contribute significantly to the frequency of
false positive signals. The antibody-generated artifacts could be
dramatically reduced by filtration and ultracentrifugation of the
antibody solution just before sorter analysis. However, we have
been unable to devise a procedure to then re-suspend the ghosts while
discarding the pelleted precipitates. One precipitate removal
method was partially successful. Following antibody incubation,

Table 1. Frequencies of Anti-
Hemoglobin S and C
Labeled Red Cells in
the Bloods of Normal
Individuals*

	Hb S	
(LLNL)	$1.1 \times 10^{-8} - 1.1 \times 10^{-7}$	(5)
(U. of Wash.)	$4 \times 10^{-8} - 3 \times 10^{-7}$	(15)
	Hb C	
(LLNL)	$6.7 \times 10^{-8} - 2.6 \times 10^{-7}$	(3)

*The LLNL results were obtained
using the two-step sorting-automated
microscope scanning technique described in the text. The University
of Washington results, using the
manual slide-based technique, have
been reported previously [3]. The
numbers in parentheses indicate the
number of individuals assayed.

washed ghost suspensions were incubated with Sepharose 6MB beads
coupled with rabbit-anti-equine IgG. These macrobeads bound antibody micro-aggregates and could be separated from the ghosts by low-
speed centrifugation. The removal was not quantitative, however,
and reliable direct counts of artificial mixtues more dilute than
one in 10^5 could still not be obtained. A variety of other approaches, e.g., gradient centrifugation or ghost electrophoresis
prior to sorting, double labeling of ghosts to discriminate against
aggregates, addition of quenchers, etc., have been suggested but due
to additional uncertainties with the ghost method have not been pursued.

Since the unavoidable fluorescent background noise prevented a
direct machine count of the rare antibody labeled ghosts, we designed
a two-step strategy in which the sorter served as a powerful enrich-
ment device. Starting with approximately 10^9 ghosts, all objects
producing fluorescence signals falling in the defined window were
first sorted into a small conical centrifuge tube. The sample was
concentrated and the sorted objects, consisting mostly of fluorescent debris and unlabeled ghosts, immobilized on microscope slides.
The labeled ghosts were then counted manually using a fluorescence
microscope equipped with a computer controlled scanning stage.
Typically, a sorted sample was enriched approximately 3000-fold over
the initial ghost suspension and could be quantitatively scanned
under the microscope in about six hours. Positive ghosts were defined as objects which were clearly red cell ghosts under phase il-

lumination and which displayed membrane specific fluorescence comparable in intensity to the model AS ghosts. Table 1 lists our initial results using this two-step method together with the previously
reported slide-based results of Stamatoyannopoulos [4]. The table
also includes our results with the anti-hemoglobin C antibody. The
frequencies of anti-hemoglobin S labeled cells using the two techniques are consistent and the S and C frequencies comparable. In
spite of this initial success, this approach is compromised in several respects. The signals from the rare fluorescent ghosts cannot
be separated from fluorescent background artifacts, thus the slide
preparations are contaminated and difficult to analyze. The ghosts
themselves are very fragile structures and cannot be immobilized
without significant and variable losses. Thus our results are only
semi-quantitative; a large part of the order of magnitude variation
in the variant cell frequencies may be attributable to these technical problems. We believe a more quantitative procedure is required in order to detect an increase in the number of variant cells
with age or with subtle environmental mutagen exposures. Lastly, the
cells contain only about 1% residual hemoglobin which makes biochemical verification difficult.

"HARD-CELL" LABELING APPROACH

Because of the technical problems discussed above, we have explored alternatives to the ghost procedure. A potentially useful
strategy was suggested by the work of Aràgon et al. [6], in which
red cells could be made permeable to substrates of intracellular
enzymes without concomitant loss of the proteins themselves. This
was accomplished by heavy cross-linking with the same membrane permeable reagent, dimethyl suberimidate, used in the ghost procedure.
The resulting "hard cells," resistant to hypotonic lysis, are then
permeabilized by organic solvent plus detergent treatment. We have
added a third step of protease digestion in order to remove the
outer surface of residual membrane proteins to expose more of the
immobilized intracellular hemoglobin. We are presently refining
this technique for immunologic labeling; an initial result demonstrating specificity of FITC-anti-hemoglobin S binding to AS "hard
cells" is presented in Fig. 4. If accurate reconstructions of artificial mixtures can be obtained, we plan to adopt this procedure as
the one of choice for future work.

These "hard cells" possess a number of useful properties. They
are mechanically very stable and do not leach significant amounts of
hemoglobin. This physical stability will permit serial sorting of
samples. By re-sorting the initial sorted sample, we expect to obtain significantly increased enrichments of labeled cells. In model
experiments, using artificial mixtures of fluorescent and non-
fluorescent "hard cells," approximately 90% recovery efficiencies
were obtained in a two-step serial sort. Like the ghost suspensions,
these cells can be processed at throughput rates exceeding 10^6 per

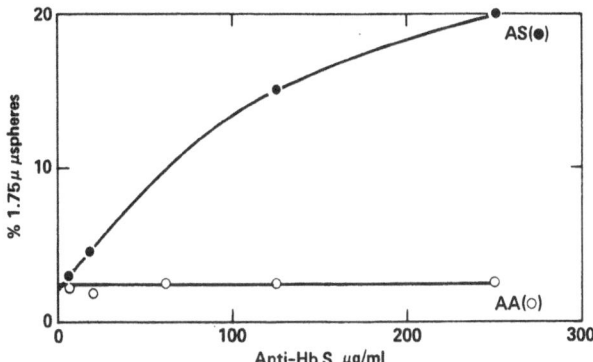

Fig. 4.   Specific binding of FITC-anti-hemoglobin S to heterozygous
          AS "hard cells." Hemoglobin AS and AA red cells were pre-
          pared as described in the text; 10^8 "hard cells" were then
          incubated for 30 min at room temperature with the indicated
          concentration of FITC-anti-hemoglobin S, washed 3 times in
          detergent buffer and analyzed on the flow sorter.   The
          plotted intensities correspond to the peak modal channels
          of the resulting histograms.

sec. In addition, these cells can be quantitatively immobilized on
microscope slides.   Secondly, since hemoglobin is now presented on
the exterior of the cell, incubation time with the antibody can be
shortened and buffers optimized for antibody stability.   This property should significantly reduce fluorescent background contamina-
tion due to precipitated antibody.   Also exterior presentation of
the antigen will permit various fluorescence amplification schemes
to be tried (sandwich antibodies, biotin-avidin, fluorescent micro-
spheres, etc.).   If a cellular fluorescence intensification method
can be found which does not also proportionately increase the number and/or the intensity of fluorescence artifacts, then cleaner
sorts of labeled cells will be obtained.   This possibility, coupled
with serial sorting, could greatly simplify, but probably not eliminate, manual microscopic examination.   Lastly, since essentially
the full cellular hemoglobin content is retained in these "hard
cells," almost 100-fold more protein can be obtained for biochemical
analysis.   By using a reversible cross-linking analog of dimethyl
suberimidate, 3,3'-dithiobispropionimidate [4], and ultra-thin gel
[7], or single cell electrophoretic techniques [8], it appears
feasible to directly characterize the hemoglobin content of the antibody labeled cells.

FUTURE OF THE HEMOGLOBIN-BASED ASSAY

     Continued progress on the development of this assay approach
is dependent on three factors: (1) success of the "hard cell"

preparative technique, (2) continued development of automated cytome-
tric analysis methods, and (3) availability of highly specific antibodies to hemoglobin variants. At LLNL there are two ongoing machine
development projects relevant to this work. The first is the construction of a high-speed sorter. This device operates at greatly
increased sample stream velocities and droplet formation rates which
should allow an order of magnitude increase in processing speed.
Secondly, a slit-laser illuminated microscope system has been constructed. This device has been used in the manual mode for examining
the sorted ghost samples but may also be capable of computer-con-
trolled slide scanning and automated detection of labeled cells.
We are also presently addressing the third point with the development of mouse monoclonal antibodies to a variety of mutant hemoglobins. This approach has already been successfully applied as
mouse monoclonal antibodies to myoglobin and to hemoglobin have been
reported [9, 10]. In no other context can the inherent advantages
of high purity, exquisite specificity and reliable production of
hybridoma-derived antibodies be better exploited than in this system. We have made a comprehensive survey of the reported human
hemoglobin single amino acid substitution, frameshift and terminator
variants and evaluated each for its applicability to this work.
From this list we are now in the process of obtaining five hemoglobin variants (S, C, Detroit, Inkster, and N-Baltimore) to begin
this effort. Ultimately, if such an approach is successful, a library of monoclonal antibodies, each specific for a different and
defined hemoglobin variant can be generated. A battery of such antibodies could then be used together to detect a variety of variant
hemoglobin-containing cells thus improving the technical ease of
the assay and its generalizability since damage at many sites, involving several mutational mechanisms, in the hemoglobin A gene
could now be detected.

THE GLYCOPHORIN A SYSTEM

     In parallel with the variant hemoglobin-based assay we are de-
veloping a second independent system based on the biochemically
well-studied protein glycophorin A [11]. Glycophorin A is a gly-
cosylated red cell membrane protein present at about $(5-10) \times 10^5$
copies per cell [12]. Its 131 amino acid residues span the membrane
with the amino-terminal portion presented on the red cell surface.
The utility of this protein as a basis for a somatic cell mutation
marker was suggested by the work of Furthmayer [13] which showed
that this protein was responsible for the M and N blood group determinants and that these determinants were defined by a polymorphism in the amino acid sequence of the protein coded for by a pair of
co-dominantly expressed alleles. The polymorphic sequence at the
amino-terminus is shown below:

```
Glycophorin A(M)    Ser-Ser(*)-Thr(*)-Thr(*)-Gly-Val-...
Glycophorin A(N)    Leu-Ser(*)-Thr(*)-Thr(*)-Glu-Val-...
                    (*) indicates a glycosylated amino acid
```

Except for the amino acid substitutions at positions one and five
of the sequence, the two proteins are identical, both in amino acid
sequence and sites and structures of glycosylation. Individuals
homozygous for the M or N allele synthesize only the A(M) or A(N)
sequence respectively, while heterozygotes present equal numbers of
the two proteins on their erythrocytes [13].

Two assays can be developed for the glycophorin A system. The
first we call our glycophorin A "gene expression loss" or "null
mutation" assay. In this approach, we wish to detect rare erythrocytes in the blood of glycophorin A heterozygotes which fail to express one or the other of the two allelic forms of the protein.
Such an approach has been described for an in vitro system using
human cells heterozygous for the multi-allelic HLA determinants [14,
15]. Using immunologic selection, these researchers have demonstrated that lymphoid cells can lose expression of one or more polymorphic HLA cell surface antigens as a result of spontaneous mutation. Exposure to radiation or chemical mutagens increases the background mutation rate by greater than two orders of magnitude. Analysis of these variants showed the majority to be single HLA gene
mutants.

The glycophorin A "gene expression loss" approach has several
inherent practical and biological advantages over the hemoglobin-
based system. First, this antigen is presented on the surface of
the red cell and is firmly anchored in the membrane; thus, cell
preparation and antibody labeling procedures are simple and straightforward. Second, since the detected mutant phenotype can result
from a variety of mutational lesions (i.e., single nucleotide
changes, insertions, deletions, frameshifts, etc., occurring either
in the glycophorin A structural gene or in its control elements),
the frequency of variant cells should be much higher (perhaps 100-
1000 times the frequency seen for a single amino acid substitution
at a single site). Hence such cells should be easier to detect and
the frequency of such cells, representing the sum of all of these
mutational mechanisms, may more accurately reflect the integrated
genetic damage in that individual.

To detect the presence of these functionally hemizygous cells
we are at present generating mouse monoclonal specific antibodies
which differentiate the M and N forms of the protein. Mouse monoclonal glycophorin A antibodies have been produced by Edwards [16]
and we have adopted a variation of his immunization protocol. First,
mice were injected with a equal mixture of homozygous MM and NN red

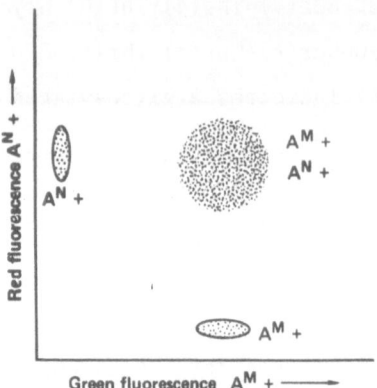

Fig. 5. Dual beam sorter, two color fluorescence detection of glycophorin A "null" variant red cells. The figure represents
a hypothetical two-color fluorescence histogram of heterozygous glycophorin A red cells incubated with monoclonal
FITC-anti-glycophorin A(M) and X-RITC-anti-glycophorin
A(N). The green and red fluorescence will be detected independently with 488 nm argon laser excitation of FITC and
568 nm excitation of X-RITC. The normal cells express both
the M and N glycophorin A sequences; they will bind both
antibodies and exhibit both green and red fluorescence.
The variant cells, lacking the expression of one allele,
will fluoresce only green or red.

cells, then boosted with purified glycophorin A(M) and A(N). The
serum was then assayed for the presence of anti-red cell antibodies.
The spleens from responding mice were then fused with SP2/0 mouse
myeloma cells and anti-red cell producing-clones were selected using
a red cell-ELISA. Positive clones were then assayed using homozygous
MM and NN cells and those showing specificity for either cell type
were selected, sub-cloned and expanded. Using this procedure, we
have isolated four clones, two of which are specific for glycophorin
A(M), one specific for A(N) and one which recognizes a shared determinant. Purified A(M)- and A(N)-specific monoclonal antibodies will
be labeled with green and red fluorophors, e.g., fluorescein isothiocyanate (FITC) and a derivative of rhodamine isothiocyanate (X-RITC)
and incubated simultaneoysly with erythrocytes from MN heterozygotes.
Variant cells, defined by binding of only one of the antibodies and
hence fluorescing green or red, will be enumerated directly as shown
in Fig. 5 using the LLNL two-color dual-beam flow sorter [17]. The
variant frequency will simply be the number of green- or red-only
cells divided by the total number of cells processed (the total
number of signals in all three peaks). Because we expect the frequency of these variant cells to be as much as 10^3 higher than the

frequency of single amino acid substitution variant cells, direct
sorter quantitation should be possible. Also adequate numbers of
variant cells should be obtainable by sorting for biochemical analysis.

   Since this assay is based on the detection of cells which fail
to express a gene product it is critical to insure that the counted
variant cells are true glycophorin A structural gene mutants. This
is important since there are both genetic and non-genetic mechanisms
which could cause the protein to fail to appear on the red cell membrane. For example, mutations outside the glycophorin A locus leading to loss of function of proteins necessary for processing, trans-
porting, glycosylating or inserting glycophorin A into the membrane
could produce apparent glycophorin A "null" cells. Non-genetic
events include loss of membrane integrity or insufficient levels of
substrate sugars for glycosylating enzymes due to metabolic anomalies. This assay is strongly protected against such false positive "phenocopies" since it requires antibody binding to one of the
glycophorin A types. The proper cell surface presentation of the
glycophorin A product of the unaffected allele insures that the rest
of the cell apparatus necessary for expression of the protein is intact. Finally, we can be reasonably assured that the variant cells
will not be selected against in vivo since erythrocytes from genetically homozygous glycophorin A "null" individuals, completely
lacking expression of the protein, appear to exhibit normal viability [18].

   Our second glycophorin A-based assay corresponds exactly to the
hemoglobin-based single amino acid substitution approach by taking
advantage of the mutational basis underlying the A(M), A(N) polymorphism. Not unexpectedly, the sequence differences at positions one
and five can arise as a result of single nucleotide changes in the
glycophorin A gene:

	Amino-acid difference		Corresponding triplet codons	
	A(M)	A(N)	A(M)	A(N)
Position 1	Ser	Leu	TC(A or G)	TT(A or G)
Position 5	Gly	Glu	GGG	GAG

Thus in homozygous MM individuals there should be rare cells containing "N-like" glycophorin A with Leu at position one or, independently, Glu at positive five. Likewise in the blood of homozygous NN people should be rare cells with "M-like" glycophorin A with
amino acids Ser or Gly at positions one and five respectively. To
detect such single substitutions our A(M)- and A(N)-specific mono-
cloncal antibodies must recognize the amino acid differences at positions one and five independently, i.e., the antibody must not depend on the presence of both differences for its binding specificity.
It is most likely that our antibodies recognize the amino acid dif-

ference at position one. The amino-terminus of the molecule appears
to be the immunodominant determinant recognized by animal anti-M and
anti-N sera [19], although some sera may be directed independently
at positive five [20]. We are presently testing the specificity of
our clones with chemically modified glycophorin A to precisely define their target antigenic sites. Given the expected specificity,
we will use these antibodies to measure the frequency of these singly
amino-acid substituted glycophorins in the blood of homozygous MM and
NN individuals. These two assays of glycophorin A variant-cells will
be particularly useful since it will be possible to compare the frequency of a single nucleotide change with the frequency of the loss
of expression of the entire gene in the same structural locus.

SUMMARY

     This paper has outlined our approaches for measuring somatic
cell mutations in human erythrocytes using high speed sorter technology. These devices are capable of processing large numbers of
cells with statistical precision and of quantitatively sorting antibody labeled, presumptively variant, cells for subsequent analysis.
All of our methods are based on immunologic detection of variant
cells and hence depend on the availability of highly specific antibody reagents. Monoclonal antibodies are ideally suited for this
purpose and are now in hand for the glycophorin A-based assays and
under development in our laboratory and elsewhere for the hemoglobin-
based assay. Armed with a battery of these hybridoma reagents it
should be possible to detect variant erythrocytes arising from an
ensemble of mutations in the α-, β-globin and glycophorin A genes.
Before practical application, these assays must be validated by measurement of reproducible background rates in "unexposed" control
populations, dose response in mutagen-exposed individuals and direct
biochemical verification of sorted variant cells.

REFERENCES

1.   Th. Papayannopoulou, T. C. McGuire, G. Lim, E. Garzel, P. E.
     Nute, and G. Stamatoyannopoulos, Identification of Haemoglobin
     S in Red Cells and Normoblasts, Using Fluorescent Anti-Hb S
     Antibodies, Brit. J. Haemat., 34:25-31 (1976).
2.   Th. Papanannopoulou, G. Lim, T. C. McGuire, V. Ahern, P. E.
     Nute, and G. Stamatoyannopoulos, Use of Specific Fluorescent
     Antibodies for the Identification of Hemoglobin C in Erythro-
     cytes, Amer. J. Hemat., 2:105-112 (1977).
3.   G. Stamatoyannopoulos, Possibilities for Demonstrating Point
     Mutations in Somatic Cells, as Illustrated by Studies of Mutant
     Hemoglobins, in: Genetic Damage in Man Caused by Environmental
     Agents, K. Berg, ed., pp. 49-62, Academic Press, New York
     (1979).

4. K. Wang and F. M. Richards, Reaction of Dimethyl-3,3-dithiobis-propiomimidate with Intact Human Erythrocytes, J. Biol. Chem., 250:6622-6626 (1975).
5. W. L. Bigbee, E. W. Branscomb, H. B. Weintraub, Th. Papayan-nopoulou, and G. Stamatoyannopoulos, Cell Sorter Immunofluores-cence Detection of Human Erythrocytes Labeled in Suspension with Antibodies Specific for Hemoglobins S and C, J. Immunol. Meth., 45:117-127 (1981).
6. J. J. Aràgon, J. E. Feliu, R. A. Frenkel, and A. Sols, Perme-abilization of Animal Cells for Kinetic Studies of Intracellu-lar Enzymes: In situ Behavior of the Glycolytic Enzymes of Erythrocytes, Proc. Natl. Acad. Sci. (USA), 77:6324-6328 (1980).
7. H. W. Goedde, H.-G. Benkmann, and L. Hirth, Ultrathin-Layer Isoelectric-focusing for Rapid Diagnosis of Protein Variants, Hum. Genet., 57, 434-436 (1981).
8. S. I. O. Anyaibe and V. E. Headings, Identification of Inherited Protein Variants in Individual Erythrocytes, Biochem. Genet., 18:455-463 (1980).
9. J. A. Berzofsky, G. Hicks, J. Fedorko, and J. Minna, Properties of Monoclonal Antibodies Specific for Determinants of a Protein Antigen, Myoglobin, J. Biol. Chem., 255:11188-11191 (1980).
10. G. Stamatoyannopoulos, D. Lindsley, Th. Papayannopoulos, M. Farquhar, M. Brice, P. E. Nute, G. R. Serjeant, and H. Lehmann, Mapping of Antigenic Sites on Human Haemoglobin by Means of Monoclonal Antibodies and Haemoglobin Variants, Lancet, ii: 952-954 (1981).
11. H. Furthmayer, Structural Analysis of a Membrane Glycoprotein: Glycophorin A, J. Supramol. Struct., 7:121-134 (1977).
12. C. G. Gahmberg, M. Jokinen, and L. C. Andersson, Expression of the Major Red Cell Sialoglycoprotein, Glycophorin A, in the Human Leukemic Cell Line K562, J. Biol. Chem., 254, 7442-7448 (1979).
13. H. Furthmayer, Structural Comparison of Glycophorins and Immuno-chemical Analysis of Genetic Variants, Nature, 271:519-524 (1978).
14. D. Pious and C. Soderland, HLA Variants of Cultured Human Lymphoid Cells: Evidence for Mutatinal Origin and Estimation of Mutation Rate, Science, 197:769-771 (1977).
15. P. Kavathas, F. H. Bach, and R. DeMars, Gamma Ray-Induced Loss of Expression of HLA and Glyoxalase I Alleles in Lymphoblastoid Cells, Proc. Natl. Acad. Sci. (USA), 77:4251-4255 (1980).
16. P. A. W. Edwards, Monoclonal Antibodies that Bind to the Human Erythrocyte-Membrane Glycoproteins Glycophorin A and Band 3, Biochem. Soc. Trans., 8:334-335 (1980).
17. P. N. Dean and D. Pinkel, High Resolution Dual Laser Flow Cytometry, J. Histochem. Cytochem., 26:622-627 (1978).
18. M. J. A. Tanner and D. J. Anstee, The Membrane Change in En(a-) Human Erythrocytes, Biochem. J., 153, 271-277 (1976).
19. H. Furthmayr, M. N. Metaxas, and M. Metaxas-Bühler, M(g) and M(c): Mutations within the Amino-Terminal Region of Glyco-phorin A, Proc. Natl. Acad. Sci. (USA), 78:631-635 (1981).

20.   W. Dahr, M. Kordowicz, K. Beyreuther, and J. Krüger, The Amino-
      Acid Sequence of the M(c)-Specific Major Red Cell Membrane
      Sialoglyoprotein - an Intermediate of the Blood Group M- and
      N-Active Molecules, Hoppe-Seyler's Z. Physiol. Chem., 362:363-
      366 (1981).

DISCLAIMER

BIOCHEMICAL APPROACHES TO MONITORING HUMAN POPULATIONS FOR

GERMINAL MUTATION RATES: II. ENZYME DEFICIENCY VARIANTS

AS A COMPONENT OF THE ESTIMATED GENETIC RISK

Harvey W. Mohrenweiser

Department of Human Genetics
University of Michigan Medical School
Ann Arbor, Michigan 48109

INTRODUCTION

The ultimate function of toxicology screening systems is to provide an accurate estimate of the human health risk associated with exposure to potentially hazardous agents. A number of test systems, with varying capabilities, have been developed and it has been proposed that combinations of several of these systems be utilized in a tier approach to estimating risk. The shortcomings of the various components of the tier system, including differences in metabolic pathways, target tissue specificity, cell replication, etc., have been discussed [1]. An additional problem is the lack of relevant data from human populations which may be used as a reference for extrapolation. That is, it is difficult to estimate human health risk utilizing data obtained in test systems in the absence of at least some data from a limited number of studies relating the nature and extent of exposure, the frequency of events induced and the associated increase in health costs in a human population. The absence of a data base is most apparent for germinal mutations, where except for genetic damage involving structural or numerical chromosomal abberations [2-4], the data on the frequency of possible mutagenic events, either spontaneous or induced, are very limited. Most previous estimates of the frequency of mutational events in human populations have utilized either the population characteristic or sentinel phenotype approach [5-7]. Recently, electrophoretic techniques have been developed which can be used to obtain data relevant to the estimation of both the background and induced mutation rate in human populations [8-10]. The current status of these electrophoretic methodologies is described by Neel et al., in a companion paper in this symposium [11].

In this presentation, I will describe our efforts to develop methods for the identification of variants characterized by a significant loss of function, which, in conjunction with efforts to develop electrophoretic methodolgies described in the previous report, are an attempt to obtain a direct estimate of the germinal mutation rate in human populations. I will focus on four areas. First, I will describe our efforts to develop a protocol for detection of a class of gene (point) mutations yielding a specific endpoint, that being enzyme deficiency variants or as they are generally called "null" variants. Second, I will present data on the relative frequency of enzyme deficiency variants in human populations. Next, I will review the current data from various test systems on the frequency of null variants, and lastly, I will discuss the possible implications of mutations to deficiency variants as a component of the human health cost.

The endpoints of gene mutations can generally be grouped into three categories: 1) proteins with altered structure but normal function, 2) proteins with altered structure and altered function, including variants devoid of function, and 3) null variants associated with the absence of recognizable protein. Different analytical techniques are necessary to detect the different types of genetic endpoints. Electrophoresis will detect category 1 events if the amino acid substitution results in a net charge change, a condition expected for 1 of 3 substitutions. Thermostability and other assays which probe tertiary structure or sequencing to determine the primary structure will be necessary to detect the "silent" amino acid substitutions. Assays which provide a quantitative measure of the amount of functional protein will provide an estimate of the frequency of category 2 and 3 events. Thus, electrophoretic and functional assays are complementary in providing an estimate of the frequency of a broad range of genetic events.

DEVELOPMENT OF METHODOLOGY FOR DETECTION OF ENZYME
DEFICIENCY MUTANTS IN HUMAN POPULATIONS

Much of the background for our approach to identicication of enzyme variants that are devoid of normal function, which we refer to as enzyme deficiency variants, is derived from the study of metabolic diseases. The extensive lists of inborn errors of metabolism caused by the loss of enzyme activity indicate, not only the prevalence of genetic events associated with loss of functional gene product [12-15] but also the utility of quantitative assays of enzymes as a tool for studying this class of genetic events [16-18]. More importantly, studies of first degree relatives of the probands confirm that quantitative assays are capable of identifying heterozygous deficiency individuals [18-21]. This latter observation is important because a heterozygous individual with a level of enzyme activity which is 50% of normal is the expected phenotype for an individual with a new mutational event leading to the loss of functional gene

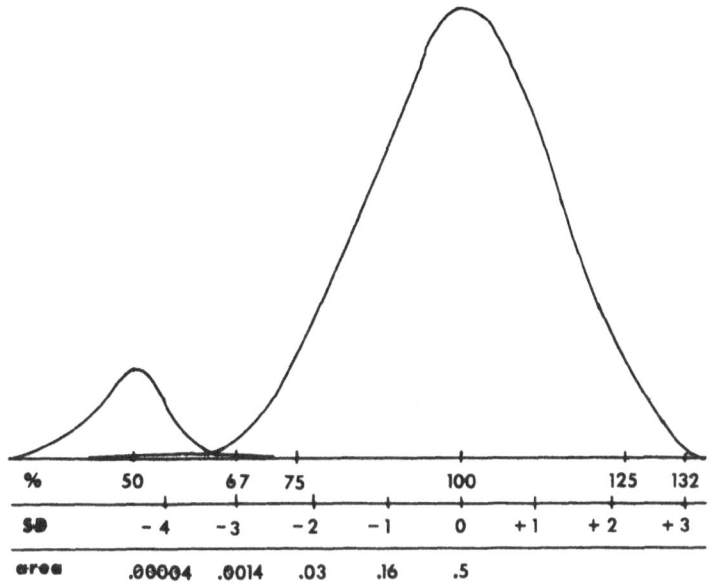

Fig. 1.   Expected distribution of enzyme activity levels in a popu-
           lation of normal and heterozygous deficient individuals.
           The coefficient of variation among individuals in each
           group is 11%.

product.   These studies also confirm the expectation that only in-
frequently are enzyme deficiency variants detectable via electrophoretic studies [20-22].

    We have classified an individual as a heterozygote (carrier)
for an enzyme deficiency allele if the level of enzyme activity is
less than 65% of the appropriate population mean and greater than 3
standard deviations below the population mean (age and sex adjusted)
[23, 24].   The coefficient of variation for the level of activity
among individuals in our sample ranges from 8-11% for the group of
enzymes currently being studied.   Thus, any individual who has a
level of activity which is less than 65% of the mean is also at
least 3 standard deviations below the mean.   As seen in Fig. 1, this
definition minimizes the overlap between the normal and heterozygous
deficient (carrier) population and minimizes the extent of misclassification, assuming that the coefficient of variation in the heterozygous (normal) group.   With these assumptions only one normal individual per thousand determinations completed would be expected to
have less than 66% of normal activity and the expectation decreases
to 4/100,000 at 55% of normal.

Table 1.   Erythrocyte Enzymes Assayed
           in Search for Deficiency
           Variants

adenylate kinase	AK
diphosphoglyceromutase	DPGM
enolase	ENOL
glucose 6-phosphate dehydrogenase	G6PD
glucosephosphate isomerase	GPI
glutamic-oxaloacetic transaminase	GOT
glyceraldehyde 3-phosphate dehydrogeanse	GAPD
lactate dehydrogenase	LDH
malate dehydrogenase	MDH
phosphoglycerate kinase	PGK
pyruvate kinase	PK
triosephosphate isomerase	TPI

The enzymes listed in Table 1 were chosen for our initial studies
because (a) in erythrocytes most of the enzyme molecules are the gene
product of a single structural locus, (b) polymorphisms associated
with significant differences in the level of enzyme activity had not
been previously reported in human populations, and (c) the level of
enzyme activity is not significantly influenced by environmental factors, e.g., the relationship between riboflavin intake and glutathione
reductase activity. Given these restrictions, it was not surprising
that the individuals in the population clustered close to the mean
and that the coefficient of variation among individuals was approximately 10%.   Thus, for these enzymes an individual with a level of
enzyme activity that is 50% of normal has a high probability of being
a carrier for an enzyme deficiency allele and as will be discussed
later, the cause of the deficiency should involve a defect in the
DNA, at ar near the structural locus for the enzyme in question.
Conversely, an individual with a null allele at the structural locus
for any of these enzymes would be expected to realize only 50% of
normal catalytic function. Finally, it was important that analytical techniques for quantitative assays could be developed which were
consistent with an extensive screening program.

To reiterate, enzyme deficiency variants, as we have defined
them, may be due to the absence of enzyme protein or the presence of
catalytically nonfunctional protein. Additionally, because the
erythrocyte is the tissue being sampled, abnormally unstable enzyme
molecules, which are degraded with a rapidity that results in a

significant reduction in enzyme activity, could be classified, by
our definition, as an enzyme deficiency variant. Although this latter type of variant would not fit the more usual definitions of a
null variant, that is, the absence of gene product (CRM$^-$) or synthesis of an inactive gene product (CRM$^+$), it is a class of enzyme
deficiency variants with potential health implications. The glu-
cose-6-phosphate dehydrogenase A$^-$ variant, a polymorphism among
Blacks which retains only 10% of normal activity in an erythrocyte
sample is an example of this latter type of enzyme deficiency variant [25]. It is possible, with appropriate additional experiments,
to identify those deficiency variants resulting from erythrocyte
enzyme instability. One can also determine, with additional experiments, which of the variants are associated with the presence
of nonfunctional protein (CRM$^+$) and which are caused by the absence
of identifiable gene product (CRM$^-$).

OPERATION OF A PROGRAM TO DETECT ENZYME DEFICIENCY
MUTANTS IN A HUMAN POPULATION

The overall strategy of this program, including the population,
the sample acqusition and the sample storage procedures, have been
described by Neel et al. [26]. The analytical techniques for detecting deficiency variants have also been described previously [23,
27]. The general data generation strategy is as follows. During
routine analysis, hemolysates are prepared from a series of 50 to
100 cord blood samples and distributed to multiple aliquots; hemoglobin concentrations are determined and enzyme activities are determined. Any assay yielding a level of activity which is less than
75% of the group mean is immediately repeated. A new hemolysate is
prepared from an aliquot of erythrocytes stored in liquid Nitrogen
for any sample still yielding an abberant result. Upon confirmation
of the abberant result, another new hemolysate, along with hemolysates from the samples obtained from the parents (also stored in
liquid Nitrogen) is analyzed for the level of activity of all 12
enzymes. If the level of enzyme activity is less than 65% of normal in the newborn and one parent, this is considered confirmation
of inheritance of an enzyme deficiency variant. Should heritable
transmission not be confirmed, standard tests for nonparentage would
be conducted. If a mutational event could not be excluded at this
point, new blood samples would be obtained from the family in order
to confirm previous results. We would examine the enzymatic activity in other cell types and continue other experiments to exclude a
nongenetic cause for the enzyme deficiency at this time. We would
also continue additional tests for non-parentage, including HLA
typing. In the absence of other possible explanations, the finding
of a child with less than 65% of normal enzyme activity, born to
parents with normal levels of activity, would be scored as a mutation to a null variant.

Table 2. Frequency of Enzyme Deficiency Variants
          in Newborn Infants from Two Ethnic
          Groups

	Black		White	
Enzyme	determinations	variants	determinations	variants
AK	65	0	611	1
G6PD	66	1(7)[a]	618	2
GOT	65	0	572	1
GPI	65	0	606	1
LDH	67	0	635	0
MDH	67	0	635	1
PGK	67	0	634	0
PK	67	0	630	1
TPI	65	0(5)[a]	606	2
	594	1(12)[a]	5549	9

[a] excluded from rare variant summary because exist in polymorphic
   frequency within ethnic group.

RESULTS FROM SCREENING HUMAN POPULATIONS
FOR DEFICIENCY VARIANTS

     Mohrenweiser [23] identified 22 newborns with levels of enzyme
activity consistent with the existence of an enzyme deficiency variant in 6141 determinations; the total frequency of enzyme deficiency
variants was 3.6/1000 determinations. In all instances, heritabil-
ity of the enzyme deficiency variant was confirmed. These data,
stratified by ethnic group, are presented in Table 2. Two variants
occurring at polymorphic frequencies were identified in the Black
population. Seven newborns with the G6PDA⁻ allele were identified
and five Black newborns were found to be carriers of the TPI 1°
allele [28]. These two polymorphic variants are excluded from our
estimation of the frequency of rare enzyme deficiency variants.
Thus, after exclusion of the polymorphic variants, 10 rare variants
were identified in 6141 determinations giving a frequency of rare
enzyme deficiency variants of 1.7/1000 determinations.

     The average level of enzyme activity for the 10 rare variants
was 53% of the mean among the newborn sample; the range being 48-
61%. Thus the "average" rare enzyme deficiency proband had a level
of activity which was approximately 4.5 standard deviations below
the mean. The average activity for all of the enzyme deficiency

variants, including TPI 1° and G6PDA⁻ in females, is greater than 4.0 standard deviations below the mean. For comparison, only one individual per 1000 determinations would be expected to have less than 66% of normal activity and less than 4/100,000 individuals from the "normal" population would be expected to have a level of activity of more than 4.5 or standard deviations below the mean.

Similar data, involving many of the same enzymes as studied in Ann Arbor, are available from a survey in Japan which is being conducted as a component of the studies at the Radiation Effects Research Foundation [29]. Thirty-four (34) deficiency variants have been identified in 11,852 determinations in the Japanese series (including approximately 6000 determinations in the radiation exposed group), yielding a deficiency variant frequency of 2.9/1000 determinations. Deficiencies involving pyruvate kinase are the most frequent variations in the Japanese study. As in the Ann Arbor study, all of these deficiency variants are inherited.

A third survey for enzyme deficiency variants has been conducted by Krietsch, Eber, and colleagues [30-32] and included approximately 3000 individuals in Germany. Their survey was restricted to five enzymes, only two of which have been studied extensively in Ann Arbor. They identified seven individuals with enzyme activity levels of less than 65% of normal. This frequency of 0.46 per 1000 determinations is quite dissimilar to the results from Ann Arbor and Hiroshima, and probably reflects differences in variant frequency among loci, as has been observed previously for both electrophoretic and deficiency variants [8, 9, 23, 29, 33].

In summary, data are currently available on the frequency of rare enzyme deficiency variants from three studies involving approximately 33,000 determinations. Fifty-two (52) rare deficiency variants have been identified, yielding a frequency of 1.6 rare enzyme deficiency variants per 1000 determinations. The total frequency, including the variants occurring as polymorphisms within particular ethnic groups, is 1.9 deficiency variants per 1000 determinations. This is a significant frequency of genetic variants at a series of loci which are generally characterized as exhibiting restricted electrophoretic variation [8, 9, 34]. With the exception of a single variant, excluded from being a mutation on the basis of nonpaternity, all of the variants have exhibited genetic transmission consistent with the existence of a null allele caused by an alteration in the structure of the DNA at or near the structural locus for the specific enzyme.

Although no new mutations involving enzyme deficiencies have been detected in these small pilot projects, several conclusions should be obvious at this point. First, enzyme deficiency variants are readily detectable in human populations at the 12 loci listed in Table 1 and enzyme deficiencies have characteristics consistent

with a genetic trait. Second, the genetic segregation and the restriction of the alteration to a specific enzyme without a general-
ized enzymopathy, is taken as evidence that a single structural locus,
encoding for the protein in question or possibly the flanking or
intervening regions, is involved in the defect.

COMPARISON OF THE FREQUENCY OF SEVERAL CLASSES
OF ENZYME VARIANTS IN HUMAN POPULATIONS

Seven of the nine enzymes studied for deficiency variants were
also screened by electrophoretic techniques in our laboratory, along
with 20 other erythrocyte enzyme loci [8]. Therefore, it has been
possible to make a direct comparison of the relative frequency of
rare electrophoretic and deficiency variants at seven loci [35].
The rare electrophoretic variant frequency is 0.7/1000 determinations at the loci at which direct comparisons were possible. Only
one of the enzyme deficiency variants was associated with a detectable alteration in electrophoretic mobility. The rare variant frequency at all 27 loci is 1.1 electrophoretically detectable variants
per 1000 determinations. Thus, the frequency of rare enzyme deficiency variants (1.7/1000) is 1.5-2.5 times the frequency of electrophoretic variants, even after exclusion of the two deficiency
polymorphisms observed in the Black population. The rare electrophoretic variant frequency at the 23 loci encoding for erythrocyte
enzymes which are monitored in the Japanese population is 1.4 per
1000 determinations [9] and again the deficiency variants (2.9/1000)
occur at more than twice the frequency of the rare electrophoretic
variants [29].

A third class of variants result from "silent" amino acid substitutions, and are not routinely detected with either electrophoretic or standard quantitative assays. Mohrenweiser and Neel [35]
employed thermostability as a probe for this class of variation.
They observed three thermostability variant (3.8 variants/1000 determinations) for every electrophoretic variant detected. This is
consistent with the estimate that 2 of 3 amino acid substitutions
involves the interchange of amino acids of similar net charge.

Summing the variant frequency observed with each technique,
electrophoresis, activity and thermostability, which provides a
minimum estimate of the three categories of genetic variation discussed in the introduction, one finds the frequency of detectable
variants in the newborn population studied in Ann Arbor is greater
than 8/1000 determinations. More than 1/3 of these variants are
associated with impaired function and extrapolating from previous
experience in clinical genetics, it would be anticipated that this
latter class of variants would be associated with significant health
costs in a homozygous or doubly heterozygous individual.

MUTATIONS TO DEFICIENCY VARIANTS IN OTHER SPECIES

It is also possible to detect deficiency variants in non-human
populations by employing appropriate mating schemes in conjunction
with electrophoretic analysis rather than direct quantitation of
functional enzyme. This approach has been employed to determine that
the frequency of null alleles at autosomal loci in two natural populations of Drosophila is 2.3-2.5 × 10^{-3} [36, 37]. This approach has
also been employed in several studies of the background and induced
mutation frequency in mice and Drosophila. The most extensive re-
ports on the frequency of spontaneous mutations to deficiency variants are the Drosophila studies of Mukai and colleagues [38, 39].
In an initial study involving 1.6 × 10^6 allele replications, 17 null
variants (frequency = 10.2 × 10^{-6}) and 3 electrophoretic variants
(frequency = 1.8 × 10^{-6}) were detected [38]. In a continuation of
this study [39] involving 1.45 × 10 additional allele replications,
5 additional null variants and 1 new electrophoretic mobility allele
were identified, while 10 of the null alleles observed by Mukai and
Cockerham [38], and thus present in the population at the initiation
of the second experiment, were not recovered. The overall frequency
(excluding the 10 null variants not recovered in the second study)
for spontaneous null mutations is 3.8 × 10^{-6} while the mutation rate
for electrophoretic variants is 1.28 × 10^{-6}. Thus the spontaneous
of null mutation rate is at least 3 times the rate for mutations involving the interchange of amino acids with dissimilar charges but
which do not significantly alter the catalytic capability of the enzyme. This should be considered a minimum estimate of the null frequency, based as it is on the assumption that null alleles do not
have a negative fitness affect in the heterozygous state. The loss
of ten null alleles in the study of Voelker et al. [39] would suggest that this assumption is not completely valid, although obviously the variants could have been lost during the course of the
experiment by chance processes also.

Similar direct estimates of the spontaneous mutation frequency
to null variants in mice are not available as only a single spontaneous mutation (of unspecified characteristics) has been reported
from the various studies employing biochemical/protein assays [40,
41]. It would seem probable though that many of the 39 spontaneous
mutations detected by the visible specific locus assay and occurring
with a frequency of 7.5-10 × 10^{-6} [42, 43] should be classified as
deficiency variants [44, 45].

Data on the relative frequency of induced null and electrophoretically detectable mutations are also available for both radiation and chemical agents. Racine et al. [46] reported seven independent null variants in Drosophila treated at a low dose rate with
γ-irradiation but did not report any electrophoretic variants. Three
of the five hemoglobin variants identified by Russell et al. [47] in
offspring of acutely radiated mice were observed to be associated

with the absence of functional hemoglobin chains. Malling and
Valcovic [48] detected 4 null mutations and no electrophoretic variants in offspring of x-irradiated male mice. Irradiation also induces visible specific locus mutations in mice [42, 49]. Presumably many of these mutations could involve the loss of specific proteins or alterations of the functional characteristics of these proteins.

     Null mutants have also been identified in mice following treat-
ment with chemical agents. Three of the five mutations observed by
Johnson and Lewis [41] in offspring of ethylnitrosourea treated
mice, at loci which could be scored for nulls, were characterized
as nulls. Null variants have also been reported among the offspring
of procarbazine [50], triethylenemelamine [51] and ethylmethane sul-
fonate [52] treated mice. Several chemical agents also induce
visible specific locus mutations, although only ethylnitrosourea induces specific locus mutations with a high frequency [53].

     The current data are not sufficient to accurately quantitate
the relative frequencies of the various classes of mutational events
generally referred to as point mutations. The data are conclusive,
however, in showing that mutational events resulting in the loss of
functional gene product do occur spontaneously. They are also induced by radiation and chemical agents and to a first approximation,
mutations to null alleles are more frequent than mutations resulting
in electrophoretically detectable alterations in the gene product.
The higher apparent frequency of null mutations reflects at least
partially, the fact that only one third of all the amino acid substitutions involve interchange of amino acids with nonidentical net
charge and are detected in electrophoretic studies. Counterbalanc-
ing this underdetection is the expectation that the spectrum of
events (deletions, base substitutions, etc.) resulting in mutations
causing an enzyme deficiency is broader than for electrophoretic
variants. Also, mutations in the noncoding regions of the gene,
either the intron or flanking region will usually result in either
a null mutation or a mutation which is not manifested in the processed messenger RNA rather than in an electrophoretic variant, thus,
the effective target size for events leading to "null" mutations is
larger than for damage which gives rise to "electrophoretic" mutations.

ROLE OF DEFICIENCY VARIANTS IN THE ESTIMATION
OF GENETIC RISK

     Enzyme deficiency variants are important as the basis for a
significant number of genetic diseases [14]. For example, the absence of enzymatic activity is associated with impaired metabolic
function for at least eleven of the enzymes of Table 1 [12, 17].
It has also been suggested that the absence of homozygous deficient
newborns in the frequency expected, given the frequency of null

heterozygotes in the populations reflects fetal loss resulting from absence of a critical enzyme [28, 30, 35].

Utilizing the previous assumption of Neel [54], that there are 5000 proteins in the human organism which are essential for normal development and/or function (or their absence or nonfunctionality is associated with disease) and the null mutation rate for Drosophila (3.8×10^{-6}) of Voelker et al. [39] one can calculate that the probability of a zygote being homozygous or doubly heterozygous at one of these 5000 loci is $1 - 0.9999962^{5000}$ or 0.019. If these assumptions are correct it is apparent that recessive gene mutations resulting in reproductive failure or genetic disease are potentially a more significant health cost than previously assumed [55, 56]. Therefore, it is important that test systems which are capable of identifying agents that induce null variants be continued as components of genetic toxicology screening programs. And for meaningful extrapolations from results in test systems to human health risks, it will be necessary to continue our efforts to obtain a data base for both background and induced mutation rates in human populations, although, hopefully with appropriate utilization of the various genetic toxicology testing systems, it will be difficult to identify large human populations at undue risk from exposure to mutagenic agents.

COSTS OF SCREENING FOR ENZYME DEFICIENCY VARIANTS

We have obtained sufficient experience in our search for mutations to enzyme deficiency variants in a human population, that we can begin to estimate the costs of a screening program utilizing this approach. As the screen for enzyme deficiency variants should be a component of a larger effort, I will assume the costs of sample acquisition are shared with an electrophoretic variant screening program as described by Neel et al. [11]. The additional expenses for the effort, in addition to sample costs, include the cost of reagents, personnel for sample preparation and analysis, and the costs of data management and storage. With the current battery of enzymes (Table 1) and the availability of analytical instrumentation capable of rapid sample throughput and computer interfacing for data analysis, management and storage, the direct costs are approximately $0.75 per locus determination. This cost is similar to the estimated cost per locus for an electrophoretic screening program which emphasizes obtaining the maximum information from each individual screened, as would be necessary for studying an exposed human population, where it is important to obtain the maximum amount of information from each offspring.

Decisions regarding utilization of various techniques in implementing a mutation monitoring program should include, in addition to the costs of operation, factors relevant to the power of the analytical approach. For the analytical approach described in this paper, the unique features include (a) identification of a class of variant which is induced by several mutagenic agents and may be the predominant lesion induced by some agents and (b) monitoring the frequency of a class of genetic events known to be associated with human health risk.

At this point, it should be apparent, that it is feasible to screen for enzyme deficiency variants and to do this at a cost which is generally similar to other approaches for obtaining data on the frequency of gene mutations in those human populations suspected to be at increased risk because of undue exposure. Therefore, this approach should be included in any effort to obtain a data base on the background and induced mutation rate in a human population, as it will provide for both a more accurate estimate of the total mutation rate and an estimate of the frequency of a class of gene mutations associated with significant health costs.

ACKNOWLEDGMENTS

Supported by Department of Energy Contract EY-77-C-02-2828.

REFERENCES

1. Committee 17, Environmental mutagenic hazards, Science, 187: 503-514 (1975).
2. D. Z. Warburton, J. Stein, J. Kline, and M. Susser, in: "Human Embryonic and Fetal Death," I. H. Porter and E. B. Hook, eds., pp. 261-287, Academic Press, New York (1980).
3. P. A. Jacobs, Mutation rates of structural chromsome rearrangements in man, Am. J. Hum. Genet., 33:44-54 (1981).
4. E. B. Hook, Contribution of chromosome abnormalities to human morbidity and mortality and some comments upon surveillance of chromosome mutation rates, Prog. Mutat. Res., 3:9-38 (1982).
5. F. Vogel, in: "Chemical Mutagenesis in Mammals and Man," F. Vogel and G. Rohrborn, eds., pp. 16-68, Springer-Verlag, New York (1970).
6. F. Vogel and R. Rathenberg, Spontaneous mutations in man, Advan. Hum. Genet., 5:223-317 (1975).
7. J. V. Neel, The detection of increased mutation rates in human populations, Persp. Biol. Med., 522-537 (1971).
8. J. V. Neel, H. W. Mohrenweiser, and M. M. Meisler, Rate of spontaneous mutation at human loci encoding protein structure, Proc. Natl. Acad. Sci. USA, 77:6037-6041 (1980).
9. J. V. Neel, C. Satoh, H. B. Hamilton, M. Otake, K. Goriki, T. Kagoeka, M. Fijita, S. Neriishi, and J. Asakawa, A search for mutations affecting protein structure in children of atomic bomb survivors: preliminary report, Proc. Natl. Acad. Sci. USA, 77:4221-4225 (1980).
10. F. Vogel and K. Altland, Utilization of material from PKU-screening programs for mutation screening, Prog. Mutat. Res., 3:143-157 (1982).

11. J. V. Neel, H. Mohrenweiser, S. Hanash, B. Rosenblum, S. Sternberg, K. H. Wurzinger, E. Rothman, C. Satoh, K. Goriki, T. Krasteff, M. Long, M. Skolnick, and R. Krezesicki, Biochemical approaches to monitoring human populations for germinal mutation rates: I. Electrophoresis. (current proceedings).

12. H. Friedmann and S. M. Rapoport, in: "Cellular and Molecular Biology of Erythrocytes," H. Yoshikawa, ed., pp. 181-259, University Park Press, Baltimore (1974).

13. A. J. Grimes and G. C. de Gruchy, in: "Blood and It's Disorders," R. M. Hardestry and D. J. Weatherald, eds., pp. 473-525, Blackwell, London (1974).

14. J. B. Stanbury, J. B. Wyngaarden, and D. S. Fredrickson, in: "The Metabolic Basis of Inherited Diseases," J. B. Stanbury, J. B. Wyngaarder, and D. S. Fredrickson, eds., pp. 2-31, McGraw Hill, New York (1978).

15. W. N. Valentine, Deficiencies associated with Embden-Meyerhof pathway and other metabolic pathways, Semin. in Hematol., 8: 348-366 (1971).

16. E. Beutler, Red cell enzyme defects as nondiseases and diseases, Blood, 54:1-7 (1979).

17. A. Kahn, J. C. Kaplan, and J. C. Dreyfus, Advances in hereditary red cell anomalies, Hum. Genet., 51:1-27 (1979).

18. K. O. Raivio and J. E. Seegmiller, Genetic diseases of metabolism, Ann. Rev. Biochem., 41:543-576 (1972).

19. S. Miwa, H. Fujii, S. Takegawa, T. Nakatsiui, K. Yamato, Y. Ishida, and N. Ninomiya, Seven pyruvate kinase variants characterized by the ICSH recommended methods, Brit. J. Haemat., 45, 576-583 (1980).

20. J.-L. Vives-Corrons, H. Rubinson-Skala, M. Mateo, J. Estella, E. Feliu, and J.-C. Dreyfus, Triosephosphate isomerase deficiency with hemolytic anemia and severe neuromuscular disease. Familial and biochemical studies of a case in Spain, Hum. Genet., 42:171-180 (1978).

21. A. G. L. Whitelaw, P. A. Rogers, D. A. Hopkinson, H. Gordon, P. M. Emerson, J. H. Darley, C. Reed, and M. A. Crawford, Congenital haemolytic anaemia resulting from glucosephosphate isomeriase deficiency: genetics, clinical picture and prenatal diagnosis, J. Med. Genet., 16:189-196 (1979).

22. H. W. Mohrenweiser and J. Novotny, An enzymatically inactive variant of human lactate dehydrogenase-LDH B GUA-1: Study of subunit interaction, Biochem. Biophys. Acta, 702:90-98 (1982).

23. H. W. Mohrenweiser, Frequency of enzyme deficiency variants in erythrocytes of newborn infants, Proc. Natl. Acad. Sci. USA, 78:5046-5050 (1981).

24. H. W. Mohrenweiser, Frequency of rare enyme deficiency variants: Search for mutational events with human health implications, Prog. Mutat. Res., 3:159-162 (1982).

25. A. Morelli, U. Benatti, G. F. Gaetami, and A. DeFlora, Biochemical mechanisms of glucose 6-phosphate dehydrogenase deficiency, Proc. Natl. Acad. Sci. USA, 75:1979-1983 (1978).

26. J. V. Neel, H. W. Mohrenweiser, C. Satoh, and H. B. Hamilton, in: "Genetic Damage in Man Caused by Environmental Agents," K. Borg, ed., pp. 29-47, Academic Press, New York (1979).

27. S. Fielek and H. W. Mohrenweiser, Erythrocyte enzyme deficiencies assessed with a miniature centrifugal analyzer, Clin. Chem., 205:384-388 (1979).

28. H. Mohrenweiser and S. Fielek, Elevated frequency of carriers for triosephosphate isomerase deficiency in newborn infants, Ped. Res., 16:960-963 (1982).

29. C. Satoh, A. A. Awa, J. V. Neel, W. J. Schull, H. Kato, H. B. Hamilton, M. Otake, and K. Goriki, Genetic effects of atomic bombs, Proc. Int. Cong. Human Genet., in press (1982).

30. S. W. Eber, B. H. Belohradsky, and W. K. G. Krietsch, A case for triosephosphate isomerase testing in congential non-spherocytic hemolytic anemia, J. Pediat., in press (1983).

31. S. W. Eber, M. Dunnwald, B. H. Belshradsky, F. Bidlingmaier, H. Schievelbein, H. M. Weinman, and W. K. G. Krietsch, Hereditary deficiency of triosephosphate isomerase in four unrelated families, Eur. J. Clin. Investig., 9:195-202 (1979).

32. W. K. G. Krietsch, H. Krietsch, W. Kaiser, M. Dunnwald, G. Kuntz, I. Duhm, and T. Bucher, Hereditary deficiency of phosphoglycerate kinase: a rare variant in erythrocytes and leucocytes not associated with haemolytic anaemia, Eur. J. Clin. Investig., 7:427-425 (1977).

33. H. Harris, D. A. Hopkinson, and E. B. Robson, The incidence of rare alleles determining electrophoretic variants: data on 43 enzyme loci in man, Ann. Hum. Genet. (Lond.), 37:237-253 (1974).

34. P. T. Wade-Cohen, G. S. Omenn, A. G. Motulsky, S. H. Chen, and E. R. Giblett, Restricted variation in the glycolytic enzymes of human brain and erythrocytes, Nature, 241:229-233 (1973).

35. H. W. Mohrenweiser and J. V. Neel, Frequency of thermostability variants: Estimation of total "rare" variant frequency in human populations, Proc. Natl. Acad. Sci. USA, 78:5729-5783 (1981).

36. C. H. Langley, R. A. Voelker, A. J. Leigh-Brown, S. Ohnishi, B. Dickson, and E. Montgomery, Null allele frequency at allozyme loci in natural populations of Drosophila melanogaster, Genetics, 99:151-156 (1981).

37. R. A. Voelker, C. H. Langley, A. J. Leigh-Brown, S. Ohnishi, B. Dickson, E. Montgomery, and S. C. Smith, Enzyme null alleles in natural populations of Drosophila melanogaster: Frequencies in a North Carolina population, Proc. Natl. Acad. Sci. USA, 77:1091-1095 (1980).

38. T. Mukai and C. C. Cockerham, Spontaneous mutation rates at enzyme loci in Drosophilia melanogaster, Proc. Natl. Acad. Sci. USA, 74:2514-2517 (1977).

39. R. A. Voelker, H. E. Scheffer, and T. Mukai, Spontaneous allozyme mutations in Drosophilia melanogaster: Rate of occurrence and nature of the mutants, Genetics, 94:961-968 (1980).

40. W. Prestsch and D. Charles, in: "Electrophoresis 1979: Adv. Methods, Biochemical Clinical Appl." B. J. Radola, ed., pp. 817-824, DeGruyter, Berlin (1980).

41. F. M. Johnson and S. E. Lewis, Electrophoretically detected germinal mutations induced in the mouse by ethylnitrosourea, Proc. Natl. Acad. Sci. USA, 78:3138-3141 (1981).

42. W. L. Russell and E. M. Kelly, Specific locus mutation frequencies in mouse stem-cell spermatogonia at very low radiation dose rates, Proc. Natl. Acad. Sci. USA, 79:539-542 (1982).

43. A. G. Searle, Mutation induction in mice, Adv. Radiat. Biol., 4, 131-207 (1974).

44. L. B. Russell, Definition of functional units in a small chromosomal segment of the mouse and its use in interpreting the nature of radiation-induced mutations, Mutat. Res., 11:107-123 (1971).

45. L. B. Russell, W. L. Russell, and E. M. Kelly, Analysis of the albino-locus region of the mouse, Genetics, 91:127-139 (1979).

46. R. R. Racine, C. H. Langley, and R. A. Voelker, Enzyme mutants induced by low-dose-rate γ-irradiation in Drosophila: Frequency and characterization, Environ. Mutagen., 2:167-177 (1980).

47. L. B. Russell, W. L. Russell, R. A. Popp, C. Vaughan, and K. B. Jacobson, Radiation-induced mutations at mouse hemoglobin loci, Proc. Natl. Acad. Sci. USA, 73:2843-2846 (1976).

48. H. V. Malling, and L. R. Valcovic, Biochemical specific locus mutation system in mice, Arch. Toxicol., 38:45-51 (1977).

49. W. L. Russell, Mutation frequencies in female mice and the estimation of genetic hazards of radiation in women, Proc. Natl. Acad. Sci. USA, 74:3523-3527 (1977).

50. F. M. Johnson, G. T. Roberts, R. K. Sharma, F. Chasalow, R. Zweidinger, A. Morgan, R. W. Hendren, and S. E. Lewis, The detection of mutants in mice by electrophoresis: Results of a model induction experiment with procarbazine, Genetics, 97: 113-124 (1981).

51. E. R. Soares, TEM-Induced gene mutations at enzyme loci in the mouse, Environ. Mutagen., 1:19-25 (1979).

52. J. B. Bishop and R. J. Feuers, Development of a new biochemical mutations test in mice based upon measurement of enzyme activities II. Test results with ethyl methanesulfonate (EMS), Mutat. Res., 95:273-285 (1982).

53. W. L. Russell, E. M. Kelly, P. P. Hunsicker, J. W. Bangham, S. C. Maddux, and E. L. Phipps, Specific-locus test shows ethylnitrosourea to be the most potent mutagen in the mouse, Proc. Natl. Acad. Sci. USA, 76:5818-5819 (1979).

54. J. V. Neel, Mutation and disease in man, Canad. J. Genet. and Cytol., 20:295-306 (1978).

55. C. O. Carter, Contribution of gene mutations to genetic disease in humans, Prog. Mutat. Res., 3:1-8 (1982).

56. C. O. Carter, in: "Prog. Genetic Toxicol." D. Scott, B. A. Bridges and F. H. Sobels, eds., pp. 1-14, Elsevier/North Holland Press, Amsterdam (1977).

BIOCHEMICAL APPROACHES TO MONITORING HUMAN POPULATIONS

FOR GERMINAL MUTATION RATES: I. ELECTROPHORESIS*

James V. Neel, Harvey Mohrenweiser, Samir Hanash,
Barnett Rosenblum, Stanley Sternberg,
Karl-Hans Wurzinger, Edward Rothman, Chiyoko Satoh,
Kazuo Goriki, Todor Krasteff, Michael Long,
Michael Skolnick, and Raymond Krzesicki

Departments of Human Genetics, Statistics
and Pediatrics
University of Michigan,
Ann Arbor, Michigan 48109

and Radiation Effects Research Foundation
Hiroshima 730, Japan

The advent of relatively inexpensive and convenient techniques
for identifying genetic variants of proteins on the basis of differences in molecular charge, thermostability, or, for enzymes, activity has opened a new chapter in the study of mutation. In this presentation, we shall speak to the progress being made in our understanding of human mutation rates and in monitoring programs because
of the ability to readily detect charge and size variations in proteins; in a companion presentation, Dr. Mohrenweiser will address
the progress being made with reference to the search for mutations
affecting thermostability or activity levels.

1. TECHNICAL COMMENTS

The advance that has made it convenient to detect charge and
size differences between proteins is of course the development of a
variety of simple techniques which apply the principle of electro-

---

*These studies have been supported in part by Department of Energy
Contract E(11-1)(2828), National Cancer Institute Program-Project
5-P01-CA-26803, and the Radiation Effects Research Foundation.

phoresis to complex protein mixtures. We assume you are all fam-
iliar with the various types of one-dimensional electrophoresis,
now using for the most part starch or acrylamide as the supporting
matrix, and employing either protein stains or, for enzymes, appro-
piate substrates coupled with dye reduction techniques to reveal
protein position on the gel. More recently, the technique of two-
simensional electrophoresis, as described by O'Farrell [1] and Klöse
[2], is finding increasing acceptance as a way to detect protein
variants. In this procedure, the proteins of a cell type or tissue
are solubilized and then, in first dimension, separated on the
basis of charge by isoelectric focusing, on cylindrical gels approximately 1-2 mm in diameter. The focused gel - a flaccid, trans-
parent "noodle" - is then sealed in place across the top of a slab
acrylamide gel, and the proteins are further separated in the second
dimension on the basis of molecular weight, by electrophoresis in
the presence of dodecyl sulfate. Not all the proteins present in
the first-dimension gel enter the second-dimension gel. Since the
solubilization mixture contains urea in a 4-8 M concentration, the
proteins are dissociated to their individual subunits and denatured
to a variable extent. Thus activity stains are generally of no
value in revealing the final position of enzyme proteins. For nu-
cleated cells, proteins can be labelled with isotopes and their position identified by autoradiography. Alternatively - and of necessity for non-nucleated cells or plasma proteins - stains can be used
to identify protein position. The recently developed, highly-sensi-
tive silver strains [3-5] have resulted in a major advance in the
ability to visualize unlabelled polypeptides.

2.   THE RESULTS TO DATA WITH ONE-DIMENSIONAL ELECTROPHORESIS

     One-dimensional electrophoresis has been applied to ,the study
of mutation in two different contexts, namely, 1) in the estimation
of background ("spontaneous") rates, and 2) in an effort to detect
exposure-related increased rates in select populations, with concurrent collection of control data. With respect to the estimation
of background rates in industrialized populations, there are to date
three principal efforts, each of which will be briefly characterized:

2.1.   The Studies of United Kingdom Caucasoids

     Harris and colleagues [6], summarizing the extensive studies
of electrophoretic variants performed at the Galton Laboratory over
a period of some 10 years, estimated that in view of the family
studies performed on individuals with rare electrophoretic variants,
the material included the equivalent of 133,478 locus tests for mutation. The tests involved 43 different enzyme loci. No putative
examples of mutation were encountered.

2.2. Studies on Newborn Infants in the United States

Elsewhere we have described the circumstances under which for
the past 6 years we have in Ann Arbor been collecting placental cord
blood samples from newborn infants and venous samples from their
parents. The samples are being examined with respect to rare electrophoretic variants of 36 different proteins; family studies are
carried out whenever one is encountered [7]. At last summary, a
total of 111,587 locus tests had been performed, with no apparent
instances of mutation [8, 9]. It should be noted that both this
program and the one described in the next section can be readily
adapted to the continuous monitoring of populations for changes in
mutation rates.

2.3. Studies on Newborn Infants in West Germany

Altland [10] has recently summarized theresults of a search for
mutations in newborn West German infants, a search which utilized
the blood-stained piece of filter paper routinely submitted to a
state laboratory for testing for phenylketonuria. At present only
the 5 hemoglobin polypeptides are being analyzed for variants (the
2α, 2γ, and the β), but Altland is exploring the feasibility of adding some 5 other gene products to the battery of tests. An esti-
mated 225,000 locus tests have been performed but for technical reasons family studies are still incomplete on the γ-chain variants.
Until these studies are complete, the number of locus tests must be
considered to be 135,000. One putative mutation, involving the α-
chain, has been encountered. The appropriate genetic studies, including HLA determinations, reveal no evidence for a discrepancy
between stated and biological parentage.

Let us turn now to studies of populations presumed to be at
risk of increased mutation rates. To date there are two such studies.

2.4. Studies of the Children of Atomic Bomb Survivors

The principal such effort is the study of the children being
born to atomic bomb survivors. The current status of these studies
has recently been described in considerable detail [11-13]. In the
present context we are concerned only with that portion of the study
which involves the search for mutations altering electrophoretic mobility. A battery of 28 protein indicators is being employed in the
search. At the time of last summary, a total of 419,666 locus tests
had been performed on children born to what are termed "proximally
exposed" parents, for whom, from the radiation histories collected
by the Radiation Effects Research Foundation and the body-surface-
organ attenuation tables of Kerr [14], the average conjoint parental
gonadal dose of radiation is estimated to have been 59 rem. (These
dosage estimates are now being reevaluated because of questions concerning the radiation spectrum of the atomic bombs.) Two putative

mutations have been observed. The first is a slowly migrating variant of glutamate pyruvate transaminase. The second is a slowly migrating variant of phosphoglucomutase-2. Rather extensive studies
reveal no evidence of a discrepancy between legal and biological
parentage as regards the putative mutants. The control material in
this study involves children born to parents now residing in Hiroshima and Nagasaki who were either beyond the zone of radiation at
the time of the bombings ("distally exposed") or were in fact absent
from the city when the event occurred. A total of 282,848 locus
tests have been performed on this group of children, with no evidence for mutation [13].

2.5.   Studies of Children in the Marshall Islands

In addition to the data from the study in Japan, we can report
here for the first time on the result of a second, very limited study
on children born to the Micronesian inhabitants of the Marshall
Islands who were exposed to radiation in consequence of fall-out from
the testing of nuclear weapons on Bikini in 1954 [cf. 15]. The exposed natives can be divided into four groups, 64 on Rongelap receiving an estimated air gamma dose of 175 R; 18 exposed on Alignae,
estimated to receive 69 R; 28 on Rongerik, estimated to receive 78
R; and 157 on Utirik, whose air gamma dose was estimated at 14 R.
In this situation, the children at risk of increased mutation rates
constitute those born subsequent to the exposure of one or both
parents, whereas the 'control' children were those either born before the exposure of their parents or to parents who were 'visiting'
on other atolls at the time of the critical test and so missed exposure. It was clear from the inception of these studies that -
given our requirement that samples be available from both parents -
sample size would be so very small that no decisive insights could
be anticipated but, since the necessary blood samples could be obtained in conjunction with other ongoing studies, advantage was taken
of this opportunity to add to the data on the genetic effects of radiation. Through the courtesy of Dr. Robert Conard and his associates at the Brookhaven National Laboratory, samples were returned
to the Ann Arbor laboratory on 3 different occasions. A list of
the 25 proteins examined for electrophoretic variants will be found
in [16]. A total of 1835 gene products from the children of exposed
were tested, and 1897 from controls. Samples size was even smaller
than feared, nor, with the virtual cessation of the long-time surveillance program of the Marshallese, is this sample ever likely to
be extended. No mutations were encountered. In this situation, one
or both parents might be exposed. We estimate that the average conjoint parental surface exposure was 97.6 R. A gonad dose has not
been calculated, but in the dosimetry studies conducted in connection with the Hiroshima-Nagasaki experience, roughly half of the
surface gamma dose reaches the gonad, the precise figure varying
with sex and position at the time of the bombings. Some additional

Table 1. The Results of Five Efforts to Detect by the Direct
          Method Electrophoretic Variants due to Spontaneous Muta-
          tion and/or Mutagenic Exposures

POPULATION	CONTROL SERIES		AT RISK SERIES		REFERENCE
	Equivalent locus tests	Mutations	Equivalent locus tests	Mutations	
United Kingdon (London)	113,478	0			(6)
Japan (Hiroshima/Nagasaki)	365,738	0	419,666	2	(13)
United States (Ann Arbor)	111,587	0			(8) and unpubl.
West Germany (Hessen)	135,000	1			(10)
Marshall Islands	1,897	0	1835	0	This paper
	727700	1	421501	2	

radiation was received from the absorption of isotopes; an average
gonad dose from the source has not been calculated.

The results of these five studies are summarized in Table 1.
To data there are 727,700 locus tests based on populations not known
to be at particular mutagenic risk, with one putative mutation. At
face value, this is a "rate" of 0.13×10^{-5}. By contrast, 421,501
locus tests on children whose parents sustained a mutagenic exposure
have yielded 2 putative mutations.

At this point we should mention two other efforts in progress
to use electrophoresis to detect the effects of exogenous agents on
mutation rates, both efforts in their early stages. The one is directed at the children born to Japanese war-time workers with mustard
gas, these workers already well known to those interested in environmental carcinogenesis because of their increased rates of broncho-
genic, gastric, hepatic, and other malignancies [17-19]. The other
effort stems from the suspicion that the chemical mutagen of greatest importance in our environment is tobacco smoke; we are attempting to obtain smoking histories on all the parents of the children
being examined in the Hiroshima-Nagasaki study, so that the data
being collected can be analyzed from this point of view as well as
that provided by the radiation histories of the parents.

2.6. Studies of Tribal Populations

There is another set of observations to be considered in this
context, namely, those on electrophoretic variants in non-indus-
trialized/tribal/mostly tropical dwelling populations. Recent population studies have yielded extensive data on the frequency and kinds
of rare electrophoretic variants in tribal Amerindians, Australians,
and New Guinese. On the assumptions: 1) of an equilibrium situation, 2) that the tribe constitutes a self-contained breeding unit
(i.e., a genetic isolate), and 3) that the electromorphs encountered
as heterozygotes are essentially neutral with respect to survival
and reproduction, the mutation rate necessary to maintain the observed number of variants in the population can be calculated, given
estimates of tribal size, of the average number of different alleles
per locus, and of average survival time, in generations, for a neutral mutation destined for ultimate loss. (It is customary to dis-
regard the common polymorphisms in this calculation.) Through simu-
lation based on an Amerindian tribe, the Yanomama, we have arrived
at an estimate of average survival time of 5.6 generations for a
mutation ultimately to be lost, introduced in an adult [20, 21].
(It is necessary to use an estimate based on the adult generation
because the very young children are usually not sampled in field
work.) This approach to estimating mutation rates, which can be
pursued by any of three different methods [22-24], has yielded an
average estimate of 1.6×10^{-5}/locus/generation for Amerindians [21],
0.7×10^{-5} for Australian aborigines [25] and 0.4×10^{-5} for the
natives of Papua New Guinea [26]. The unweighted average of the
three estimates on these populations is 0.9×10^{-5}/locus/generation.
The errors of these estimates, which in addition to the usual sampling error include error terms introduced by uncertainties in the
assumptions, are large, but, we have argued, are probably not sufficiently large to permit an overlap of the unweighted average with
the results of the direct approach presented in the preceding sections [27]. We consider that this possibility, that these tropical
dwelling/tribal/non-industrialized populations have higher mutation
rates than the temperate-dwelling civilized/industrialized populations on whom the direct estimates are based, is supported by the
fact that the total frequency of the kinds of rare variants on which
these estimates are based is 6 times higher in the tribal populations studied. This fact is difficult to explain other than on the
basis of higher mutation rates, since the principal alternative explanation, of higher selective values for electromorphs under tribal
conditions, seems much less plausible [27]. Incidentally, the data
on Amerindians, when last summarized, included 94,796 locus tests
for mutation by the direct approach; none was encountered [8]. The
possibility of regional differences in mutation rates, for reasons
completely unknown, makes us very reluctant to use other-than-Japan-
ese controls in our plan to incorporate in due time these biochemical data into our effort to derive an estimate for the genetic
doubling dose for radiation in the study in Japan (cf. 12).

A principal criticism of the indirect approach is that because of intertribal migration, tribes cannot be considered to be breeding isolates. While genetic exchange between tribes in the past cannot be excluded, especially in the remote past when the tribes found in the various areas under study were coming into being, we have argued that the manner in which the so-called private polymorphisms are largely restricted to single tribes suggest a high degree of tribal isolation [28]. Furthermore, if a particular electromorph is found in more than one tribe, it is scored only once, being assigned to the tribe in which it has the highest frequency. But since such arguments and precautions can never be proven to be adequate, we can also pursue an empiric approach to this criticism. Let us argue that to all intents and purposes, South America, peopled at least some 20,000 years ago, has been a genetic isolate for all this time and should be considered one large interbreeding unit. Denevan [29] estimates the number of Indians in South America at the time of first contacts as 20,000,000. From our own field work, we find 48% of an Indian population to be in the reproductive generation, aged 15-40, so that N, the number in the generation of reproduction at time of first contact, becomes 9,600,000. Let us assume that the 11 relatively undisturbed South American tribes we have studied are, to a significant approximation, genetically representative of the situation at time of contact. We can then calculate a mutation rate for South American Indians in general by the approach of Rothman and Adams [24]. From data in Neel and Rothman [21], we find we have studied 2761 adults in the reproductive age with reference to an average of 26 gene products (polypeptides), in whom we encountered an average of 0.81 variants (other than widely distributed polymorphisms) per locus. Fed into the appropriate equation, the resultant minimal estimate of electrophoretic mutations/locus/generation is 0.7×10^{-5}. One reason why this is a minimal estimate is that the calculation assumes a population at numerical equilibrium, whereas the Amerindian population must be seen as slowly expanding since the peopling of South America. It is noteworthy that this estimate is identical with the rate adduced by Neel and Thompson [30] as most consistent with the frequency and numbers in which electrophoretic variants (other than common polymorphisms) occurred in Amerindian populations.

3.   SOME GENERAL OBSERVATIONS ON THE USE
     OF ONE-DIMENSIONAL ELECTROPHORESIS TO STUDY MUTATION

The experience to data warrants five general comments concerning the use of 1-D electrophoresis to study human mutation rates.

1. The feasibility of the approach is now beyond doubt. Since in the work with humans, the study population maintains itself, we are spared a principal cost sustained by the experimental mammalian geneticist. We estimate that in the study of cord blood samples in Ann Arbor, the cost per locus determination is $0.75. This figure

includes all direct costs, such as sample acquisition, electrophoretic analysis, and data management. We believe this cost can be
lowered as we increase the use of "multiple system gels." The cost
is less in the method of Altland [10], (cf. also 31), but this latter
method of collecting samples will not usually be appropriate to the
study of high risk populations. We would welcome the opportunity to
compare these costs with those incurred in studies on mice.

2. Since in general only a single electrophoretic technique is
employed in the study of any particular protein in a screening program, any estimate of the frequency of electrophoretic variants is
minimal. This fact should bias downwards both the direct and indirect estimates of mutation rates.

3. Because of the occurrence of genetic polymorphisms in some
of the systems under study, some variants (and mutations) with electrophoretic mobilities similar to common polymorphisms will be
missed. We propose that this is a minor source of error. Assume,
as is the case for our study, that one is employing a battery of 36
protein indicators, with genetic polymorphisms absent for 29 of the
proteins but present for 7. All variants encountered in the non-
polymorphic systems are subject to family studies, but for systems
characterized by common polymorphisms, family studies are not conducted on variants conforming to common polymorphisms. We will
further assume only two common alleles in the polymorphic systems
(the usual situation) and that 'gain-of-charge' mutations (and polymorphisms) are equal to 'loss-of-charge' mutations. Finally, we
will assume any gain (loss)-of-charge mutation in a polymorphic system precisely mimics the common polymorphism. This latter is a very
conservative assumption. Then the frequency of mutations missed because of failure to do family studies on the common polymorphisms
should be $1-(0 \times 29 + .5 \times 7)/36 = 0.097$. At a time when debate
still centers around an order of magnitude for human mutations rates,
investigators such as Stevenson and Kerr [32], Cavalli-Sforza and
Bodmer [33], and Yasuda [34] suggesting total rates of 10^{-6}, a 10%
error seems acceptable.

4. Estimates of mutation rates derived from the electrophoretic approach lend themselves to cautious extrapolation to "total"
mutation rates for the loci in question. Depending on the polypep-
tide involved, for each mutation resulting in an amino acid substitution which causes a change in molecular charge, there should be 2
or 3 resulting in electrophoretically 'silent' amino acid substitutions. Some empiric support to this calculation comes from our
demonstration that thermolability variants not associated with electrophoretic findings, which we presume for the most part to result
from silent amino acid substitutions, occur with roughly 3 times
the frequency of electrophoretic variants [35]. In addition, the
studies of Mukai and Cockerham [36] and Voelker and colleagues [37]
suggest that for each spontaneous mutation resulting in a charge

change, there are some 3 resulting in marked loss of enzyme activity.
We consider this an absolute minimum estimate of the relative frequency in this species of such mutations (which may involved introns
and control sequences as well as exons). The techniques employed by
these investigators involved allowing chromosomes to accumulate
"null" mutations of a battery of isozymes for several hundred generations before the breeding tests to expose the presence (or absence)
of any nulls. Although homozygotes for nulls of many isozyme loci
of Drosophila are homozygous viable (cf. 38), the nulls investigated
by O'Brien may be a selected set, consistent with survival as homozygotes through the larval stage; further, it is difficult to believe that homozygosity is without any deleterious effect (else why
the persistence of the enzyme?).

   Now, there is a large literature on the heterozygous effect on
viability of deleterious and lethal genes. For example, thirty years
ago Stern and colleagues [39] showed that on the average the viability
of heterozygotes for lethals was 0.965, with no difference between
spontaneous and induced lethals, and Hiraizumi and Crow [40] found
an average viability of 0.974 in flies heterozygous for lethal or
semi-lethal chromosomes isolated from nature. It therefore seems
probable that at least some of the null mutations in the studies of
Mukai and colleagues are associated with heterozygote disadvantage.
Voelker et al. [37] found that at least 10 nulls present in the
strain 3 years earlier when analyzed by Mukai and Cockerham [36]
were no longer recovered. This can best be explained by chance or
selective loss. We suggest the bias in their estimates could be as
high as 100%. There are as yet no data on the ratio of electromorphs: nulls among spontaneous mutations in the mouse, but in 3
series involving chemical induction, the ratio is roughly 1:2 [41-
43]. This ratio must of course be applied with caution to spontaneous events. There is no reason to assume the same ratio of mutations resulting in electromorphs to mutations resulting in marked
loss of enzyme activity in all species, but let us from the fore-
going assume that in man that ratio is at least 1:2. Then one mutation resulting in an electrophoretic variant would in an appropriate series suggest that at least four other mutations had occurred. Although we will never be guilty of calculating a rate on
a fraction with a numerator of one, someone is bound to point out,
on the basis of the data just quoted, that the observed frequency of
spontaneous electrophoretic mutations in man, of 0.13×10^{-5}/locus/
generation, should correspond to a total (not electrophoretic) rate
of approximately 0.7×10^{-5}, with an error of equal magnitude! Incidentally, to allay some of the justifiable concern about appearing to operate with numerators of one, I should mention that we now
have at least one electrophoretic mutation in the children of non-
irradiated parents in the study in Japan, and we are aware of several more possible spontaneous electrophoretic mutations in the unpublished data of others.

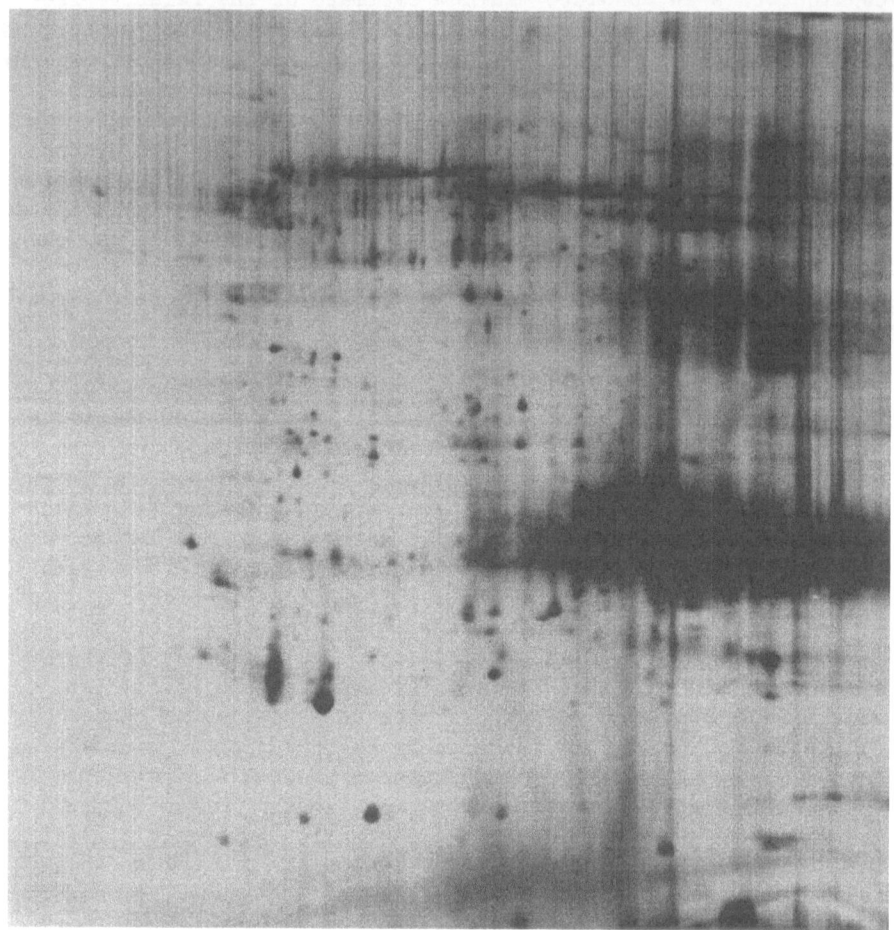

Fig. 1. Photograph of a 2-D preparation of an erythrocyte cytosol
         prepared from a sample from an adult male (2840-2), stained
         by the Sammon's procedure [4].

     5. Now that truly comparable interspecific indicators of mu-
tation are available, it is impressive, how similar the mutation
rates are appearing to be across species of higher eukaryotes. The
Drosophila rate for electromorphs was 0.5×10^{-5}/locus/generation
in the studies of Tobari and Kojima [44] and $.1 \times 10^{-5}$/locus/genera-
tion in the studies of Voelker et al. [37]. Although there are some
uncertainties about both those rates, let us for the moment simply
use the average of the two series, of 0.3×10^{-5}. Pretsch and
Narayanan [45] report one spontaneous mutation in 67,678 locus tests
in mice; Drs. Johnson and Lewis [43], who will be speaking elsewhere

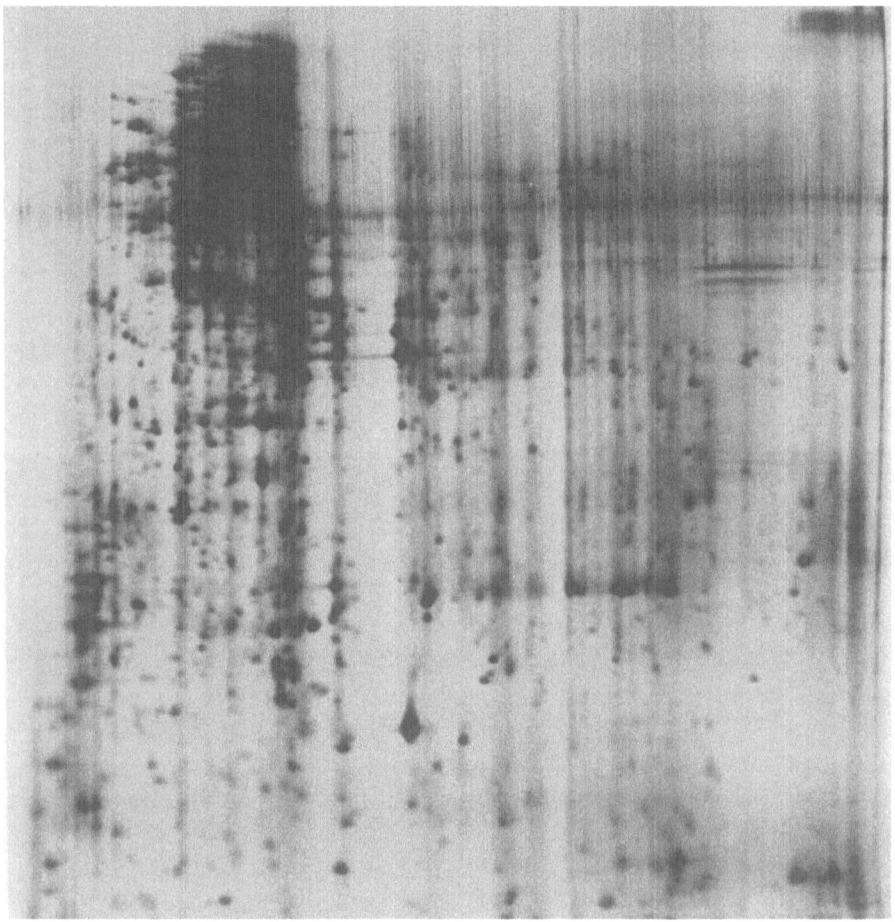

Fig. 2.   Photograph of a 2-D preparation of solubilized erythrocyte
          membranes from an adult female (2925-1), stained by the
          Sammon's procedure [4].

on this program, have yet to encounter a spontaneous mutation in
334,572 locus tests. These two series yield a rate of $0.2 \times 10^{-5}/$
locus/generation, an estimate of course just as shaky as the human
estimate, which, I remind you, was 0.1×10^{-5} for electromorphs.
Given the longer life cycle, higher mean body temperature, and
greater exposure to mutagens characteristic of humans, we must assume our species has evolved very superior repair mechanisms, a development which we believe accounts in part for the evidence we have

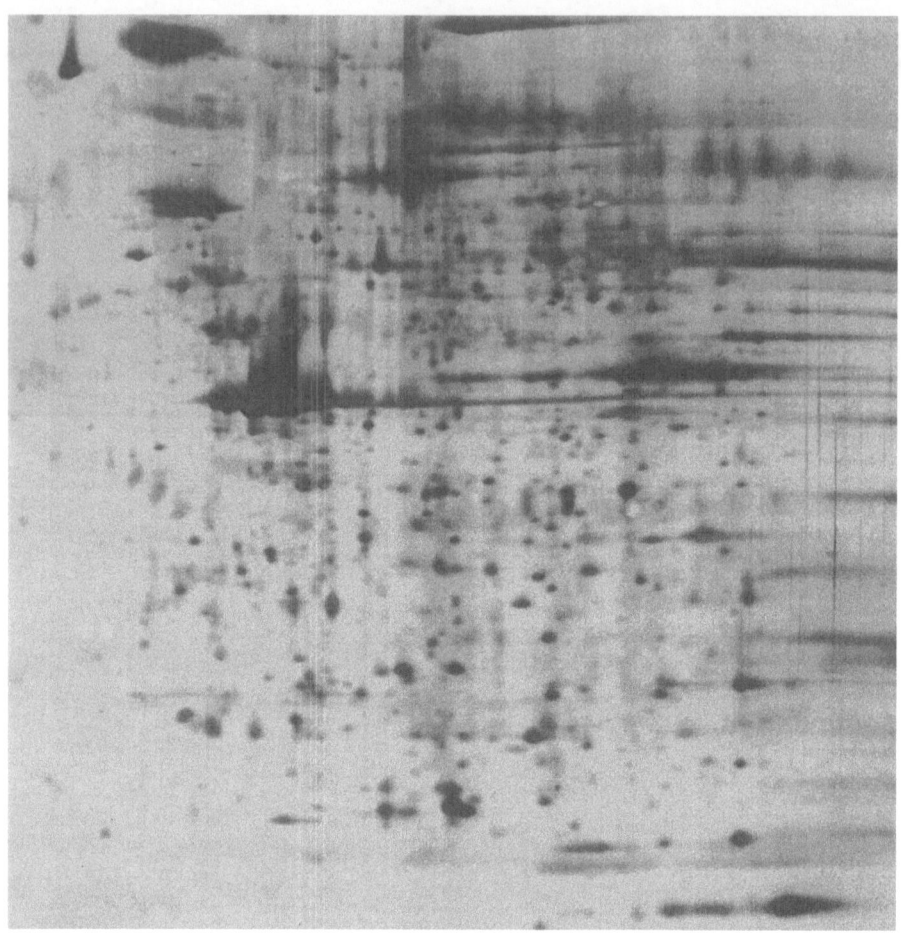

Fig. 3. Photograph of a 2-D preparation of platelets, obtained from
         an adult male (Mi. Lo.), stained by the Sammon's procedure
         [4].

presented elsewhere that the genetic doubling dose of radiation may
be substantially higher in humans than in the mouse [12].

4.   THE DEVELOPMENT OF TWO-DIMENSIONAL ELECTROPHORESIS
     FOR MONITORING PURPOSES

4.1.   The Pilot Study in Ann Arbor

     As mentioned earlier, the program supplying blood samples for

Fig. 4. Photograph of a 2-D preparation of human plasma from an
adult male (Jo. St.), stained by the Sammon's procedure
[4].

the study of mutation with one-dimensional electrophoretic techniques will also supply the necessary aliquots for the work on 2-D
gels now under way in our laboratory. The development of the 2-D
gel technology is still in its early stages; we do not anticipate
"production runs" for another year. On the other hand, the progress
to data is sufficiently encouraging that it merits discussion at
this time, under 7 headings.

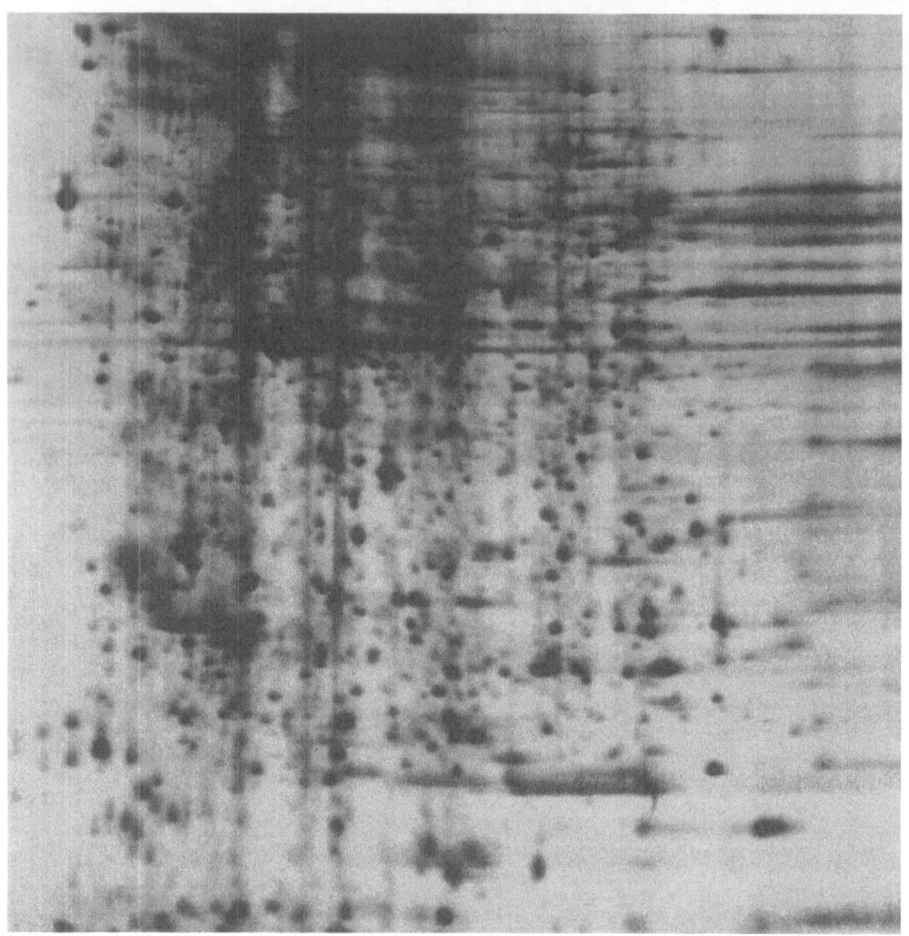

Fig. 5.   Photograph of a 2-D preparation of the non-polymorphonu-
          clear leucocytes of a peripheral blood sample from an adult
          male (2958-2), stained by the Merril procedure [3].

4.1.1. Types of Preparations

Since in human monitoring programs involving newborn infants,
we are essentially limited to the use of blood samples, we have expended considerable effort on exploring the relative merits of different types of preparations. A particular effort has been directed
towards the erythrocyte membrane, since the occurrence of so many
genetic polymorphisms associated with this membrane should provide
abundant material to validate the sensitivity of the 2-D procedure

to the demonstration of known genetic variants [46]. Figures 1-5
illustrate the current state of the art with reference to erythrocyte
cytosols, erythrocytes membranes, platelets, plasma, and the non-
polymorphonuclear leucocyte fraction resulting from a Ficoll-Hypaque
centrifugation, when corresponding 2-D patterns are visualized with
various types of silver stains. From 100 to 500 protein species can
be clearly visualized in each gel, but there is of course considerable overlap in the proteins visualized in some of the preparations.
A problem with the leucocyte preparations is that the sensitive
silver stains in fact reveal too many proteins for a convenient
analysis; we are attempting to determine whether further fractionation of these cells is appropriate within the context of a screening
operation.

4.1.2. Constraints Placed on the Analysis of the Preparations

At the outset, we propose to restrict our search for genetic
variation and mutational events to protein moieties meeting three
requirements. First, they must of course be reproducibly present
in parents and offspring, with variation following genetic logic.
Second, they should be present in sufficient quantity that in a
heterozygous individual, the two gene products will both be present
in amounts well above the arbitrary density threshold established
for our computer programs. Third, there should be a sufficient
relatively "clear" area to either side of the moiety that a shift in
its mobility in the horizontal axis within a band of some 8 mm (the
range within which variants characterized by a gain or loss of 1 or
2 charges should occur) would not superimpose the shifted spot on
another major protein. We estimate that these conventions (which
may be relaxed with further experience) will reduce the number of
moieties suitable for analysis to 50-100 per gel, again with some
overlap across gels from different preparations. This overlap is
in one sense desirable, since variation in a particular protein in
a particular preparation should be verified by variation in the same
protein on another preparation.

4.1.3. Demonstration of Ability of Technique to Detect Known Genetic Variants

We have recently demonstrated that 13 of 17 gene products visualized on one-dimensional electrophoresis could also be visualized
on standard O'Farrel-type gels [47]. These 17 products included 5
classified as normal and 12 classified as variants. The addition
of a second pH gradient made it possible to separate all but one of
the variant proteins. Given the well-known fact that no one electrophoretic technique detects all charge-change variants, we feel
the method has the requisite sensitivity.

4.1.4. Development of Image Enchancement Routines

There is often uneven staining and considerable "streaking" on
2-D gels. A principal reason for the latter is overloading with
respect to certain proteins, difficult to avoid in view of the wide
range of protein concentrations. For instance, when dealing with
erythrocyte cytosols, a gel properly loaded for most of the protein
constituents is grossly overloaded with respect to hemoglobin. A
second reason for streaking is post-translational modification of
proteins. This is of course more prevalent in aging cells, such as
erythrocytes, or in blood plasma, then it is in rapidly metabolizing
cells, such as fibroblasts, and so constitutes more of a problem in
our type of study than it should to someone utilizing cultured
cells. Computer programs have been developed for background normalization and streak removal [48].

4.1.5. Development of Standard "Maps," with Due Allowance for Genetic Variation and "Noise" Factors

We are now well launched on an effort to develop standard genetic "maps" of the position of the protein moieties meeting our requirements on the types of preparations illustrated earlier. Our
entire program is directed towards the identification of genetic
variants present in a child but not in either parent. To be successful, we must identify and eliminate from consideration, among
the moieties otherwise meeting the constraints listed above under
2), those which are foetal proteins, (not normally represented in
adults), degradation products, or "noise," the latter including
proteins whose entry into the gel is unpredictable. Satisfactory
progress is being made.

4.1.6. Development of Image Analysis Programs Appropriate to the Detection of Mutational Events in 2-D Gels

The visual search of thousands of gels for events with a frequency of the order of 10^{-6} would be intolerably tedious. Two types
of image analysis programs are being explored, in an effort to re-
lieve much of this tedium [48, 49]. In the first, unique constellations of spots well distributed throughout the gel are identified,
and algorithms are now at hand capable of superimposing corresponding constellations in father, mother, and child, and identifying
unique spots in the child. In the second approach, which does not
require operator definition of constellations, Gabriel-type graphs
are constructed for various of the gels. The criterion for Gabriel
graph construction is that two "nodes" (i.e., spots) in the gel are
connected by an edge if and only if the circle defined by the diameter (which is also the potential edge) connecting the two nodes
contains no other nodes of the graph. A Gabriel graph is a member
of a family of graphs in which, at one extreme, all nodes are inter-

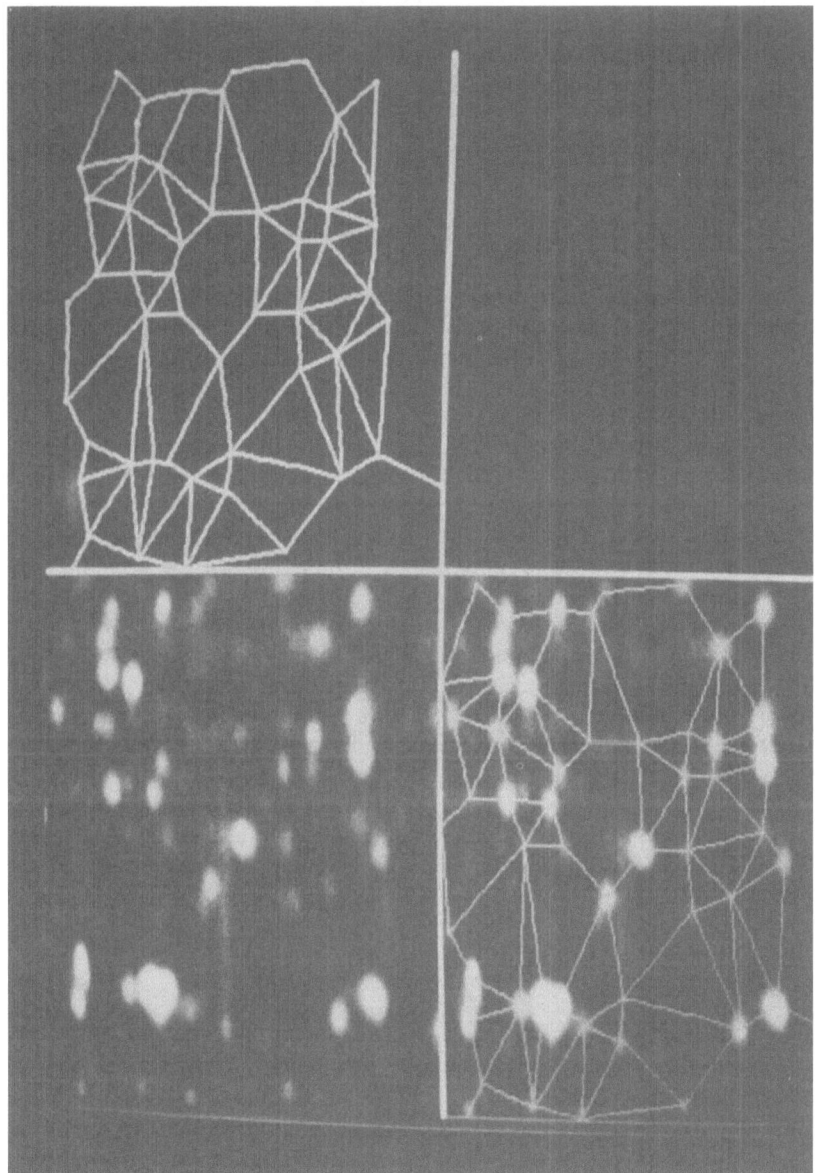

Fig. 6. An example of a Gabriel graph. Upper left, CRT display of
         the section of the 2-D autoradiograph of non-polymorphonu-
         clear leucocytes on which the graph is based. Upper right,
         the corresponding Gabriel graph. Bottom left, graft super-
         imposed on the original display.

connected, and, at the other extreme, the nodes are connected by a minimal spanning tree, such as finds widespread usage in genetic phylogenies. Such a graph is shown in Fig. 6. With the above-defined conventions, each spot is associated with 2 or more vectors plus a "position" which together should provide each spot a unique identity. In the absence of mutation, the Gabriel graphs for any given region of 2-D gels derived from a mother-father-child trio should show only differences consistent with Mendelian inheritance.

4.1.7. The Characterization of Putative Mutations

It is imperative that steps be taken to characterize any apparent mutant, to ensure the relationship of the variant protein to the moiety from which it is thought to be derived. Our preliminary studies indicate that high pressure liquid chromatography will be very useful in this respect, not only for the isolation of small amounts of the protein after elution from a gel but for further characterization studies. This same technique may enable us to fractionate a complex protein mixture on the basis of say, hydrophobicity, thus obtaining cellular fractions which when run on a gel, would not present as cluttered a landscape as we now obtain with, for example, the leucocyte preparations.

The limits of this technology are yet to be determined. If we consider a 2-D gel only as a collection of 1-D gels, it is clear that even if we had to resort to visual inspection, important data on mutation rates could be obtained. The initial constraints we have placed on the use of these gels for the study of mutation are really rather severe. We hope, as our familiarity with modern image analysis techniques in this context improves, to relax these constraints and significantly increase the amount of information to be extracted per preparation.

4.2. The Introduction of the Technology to the Follow-Up Program in Japan

Just last month, the decision was reached to initiate a pilot study of the applicability of the 2-D methodology to the follow-up studies in Hiroshima and Nagasaki. For a number of reasons, relating primarily to the resources available, the program, at least initially, will assume a somewhat different form than the Ann Arbor program. Specifically, rather than the degree of automation of gel reading we are attempting to develop in Ann Arbor, the reading of the gels will be visually. On the other hand, we hope to develop for the NEC ACOS 450 computer available to us in Japan an interactive program such that the findings for each gel can be entered directly into the computer files, without the intervention of the time-honored lab books. In addition, we will independently analyze photographs of a subset of these gels in Ann Arbor, in a double-blind study of the efficacy of the two methods of analyzing the gels.

The chief practical problem confronting us both in Ann Arbor and Hiroshima is achieving the necessary success rate for research-type gels. The quantity of lymphocytes or platelets which can be recovered from a 10 ml blood sample is usually sufficient for two gels when the silver-stain technology is employed. The preparation of really good gels is a demanding art. Most laboratories would be happy with a 50% success rate. Since in Ann Arbor neither we nor the parents of a newborn infant propose to traumatize the child with a follow-up venapuncture, this means there is a 1 in 4 chance we will not have satisfactory leucocyte or platelet gels. The problem is only slightly less acute in Japan. The same odds apply to the parents, although they represent a more renewable resource. The issue of adequate material is of course less acute with erythrocyte membranes or hemolysate or plasma.

If the technique of 2-dimensional gel electrophoresis can be brought on line for the study of mutation, it has the potentiality, overall, for an order of magnitude improvement in the efficiency with which one can search for mutations affecting protein structure. There is another compelling reasons for bringing this technology on line. The relatively few studies to date have suggested much less heterozygosity with respect to the proteins visualized by this method - say 1-2% [50-54] - than with respect to those visualized thus far in studies employing 1-D gels - say 6-7% [55]. We are not yet convinced that the difference is as great as reported, but should it be confirmed, one is forced to postulate either greater selection against variants of the types of proteins visualized by the 2-D methodology or lower mutation rates for the genes encoding for these protein types. If the latter proves to be the case, it could force a substantial revision of our current thinking concerning spontaneous and - by inference - induced mutation rates.

5. CONCLUDING COMMENTS

In conclusion, we would suggest, on the basis of this presentation, and that by Mohrenweiser to follow, that the feasibility of the biochemical approach to monitoring human populations for genetic damage, is now solidly established. This is perhaps the best place to meet a commonly raised question, the relevance of electrophoretic findings to human health problems. It is true that electrophoretic variants of human proteins do not usually have significant health effects, either as heterozygotes or homozygotes, although there are exceptions, such as hemoglobins S and C, and the unstable hemoglobins and those with abnormal oxygen affinities. On the other hand, mutations resulting in electrophoretic variants can serve as indicators of the frequency of occurrence of mutation resulting in nulls, and with respect to the medical significance of these, there is no doubt. After all, these variants, when homozygous, result in humans in the multitudinous inborn errors of metabolism.

The question of cost-effectiveness of the electrophoretic approach is quite a different matter. In my opinion, this is a societal rather than a scientific judgement. Society alone can decide whether the cose of <u>not</u> improving our information base, in uncertainties resulting in legal actions and regulations of dubious validity, will exceed the cost of implementing appropriate studies.

REFERENCES

1. P. H. O'Farrell, High resolution two-dimensional electrophoresis of proteins, J. Biol. Chem., 250:4007-4021 (1975).
2. J. Klöse, Protein mapping by combined isoelectric focusing and electrophoresis of mouse tissue. A novel approach to testing for induced point mutations in mammals, Humangenetik, 26:231-243 (1975).
3. C. R. Merril, D. Goldman, S. A. Sedman, and M. H. Ebort, Ultrasensitive stain for proteins in polyacrylamide gels shows regional variation in cerebrospinal fluid proteins, Science, 211: 1437-1438 (1981).
4. D. W. Sammons, L. D. Adams, and E. E. Nishizawa, A silver-based color development system for staining of polypeptides in polyacrylamide gels, Electrophoresis, 2:135-141 (1981).
5. W. Wray, T. Boulikas, V. P. Wray, and R. Hancock, Silver staining of proteins in polyacrylamide gels, Anal. Biochem., 118: 197-203 (1981).
6. H. Harris, D. A. Hopkinson, and E. B. Robson, The incidence of rare alleles determining electrophoretic variants: Data on 43 enzyme loci in man, Ann. Hum. Genet., Lond., 37:237-253 (1974).
7. J. V. Neel, H. W. Mohrenweiser, C. Satoh, and H. B. Hamilton, in: "Genetic Damage in Man Caused by Environmental Agents," K. Berg, ed., pp. 29-47, Academic Press, New York (1979).
8. J. V. Neel, H. W. Mohrenweiser, and M. H. Meisler, Rate of spontaneous mutation at human loci encoding for protein structure, Proc. Nat. Acad. Sci. USA, 77:6037-6041 (1980).
9. J. V. Neel, The wonder of our presence here: A commentary on the evolution and maintenance of human diversity, Persp. Biol. Med., 25, 518-558 (1982).
10. K. Altland, M. Kacmpter, M. Forssbohm, and W. Werner, Monitoring for changing mutation rates using blood samples submitted to PKU screening. Proc., VI Int. Cong. Hum. Genet., Alan Liss, New York, pp. 277-287 (1982).
11. J. V. Neel, C. Satoh, H. B. Hamilton, M. Otake, K. Goriki, T. Kageoka, M. Fujita, S. Neriishi, and J. Asakawa, Search for mutation affecting protein structure in children of atomic bomb survivors: Preliminary report. Proc. Nat. Acad. Sci. USA, 77:4221-4225 (1980).
12. W. J. Schull, M. Otake, and J. V. Neel, A reappraisal of the genetic effects of the atomic bombs: Summary of a thirty-four year study, Science, 213:1220-1227 (1981).
13. C. Satoh, A. A. Awa, J. V. Neel, W. J. Schull, H. Kato, H. B. Hamilton, M. Otake, and K. Goriki, Genetic effects of atomic

bombs, Proc., VI Int. Cong. Hum. Genet., Alan Liss, New York, pp. 267-276 (1982).

14. G. D. Kerr, Organ dose estimates for the Japanese atomic-bomb survivors, Hlth. Phys., 37:487-508 (1979).

15. R. A. Conard, et al., A twenty-year review of medical findings in a Marshallese population accidentally exposed to radioactive fallout, Brookhaven Natonal Laboratory Publication 50424, pp. ix and 154, Upton, Brookhaven National Laboratory (1975).

16. J. V. Neel, R. E. Ferrell, and R. A. Conard, The frequency of "rare" protein variants in Marshall Islanders and other Micronesians, Am. J. Hum. Genet., 28:262-269 (1976).

17. S. Wada, A. Yamada, Y. Nishimoto, S. Tokuoka, M. Miyanishi, S. Katsuta, and H. Umiza, Neoplasms of the respiratory tract among poison gas workers, Hiroshima Med. J., 16:728-745 (1963).

18. S. Wada, M. Miyanishi, Y. Nishimoto, S. Kambe, and R. W. Miller, Mustard gas as a cause of respiratory neoplasia in man, Lancet, 1:1161-1163 (1968).

19. Y. Nishimoto, Personal communication.

20. F. H. F. Li, J. V. Neel, and E. D. Rothman, A second study of the survival of a neutral mutant in a simulated Amerindian population, Am. Nat., 112:83-96 (1978).

21. J. V. Neel and E. D. Rothman, Indirect estimates of mutation rates in tribal Amerindians, Proc. Nat. Acad. Sci. USA, 75: 5585-5588 (1978).

22. M. Kimura and T. Ohta, The average number of generations until extinction of an individual mutant gene in a finite population, Genetics, 63:701-709 (1969).

23. M. Nei, Estimation of mutation rates from rare protein variants, Am. J. Hum. Genet., 29:225-232 (1977).

24. E. D. Rothman and J. Adams, Estimation of expected number of rare alleles of a locus and calculation of mutation rate, Proc. Nat. Acad. Sci. USA, 75:5094-5098 (1978).

25. K. K. Bhatia, M. M. Blake, and R. L. Kirk, The frequency of private electrophoretic variants in Australian aborigines and indirect estimates of mutation rate, Am. J. Hum. Genet., 31: 731-740 (1979).

26. K. K. Bhatia, N. M. Blake, S. W. Serjeantson, and R. L. Kirk, The frequency of private electrophoretic variants and indirect estimates of mutation rate in Papua New Guinea, Am. J. Hum. Genet., 33:1121122 (1981).

27. J. V. Neel and E. D. Rothman, Is there a difference between human populations in the rate with which mutation produces electrophoretic variants? Proc. Nat. Acad. Sci., USA, 78:3108-3112 (1981).

28. J. J. Neel, in: "Population Structure and Genetic Disorders," A. Eriksson, ed., pp. 173-193, Academic Press, London (1980).

29. W. M. Denevan, ed., The Native Population of the Americas in 1492, University of Wisconsin Press, Madison (1976).

30. J. V. Neel and E. A. Thompson, Founder effect and number of private polymorphisms observed in Amerindian tribes, Proc.

Nat. Acad. Sci. USA, 75:1904-1908 (1978).

31. F. Vogel and K. Altland, in: "Progress in Mutation Research,"
     K. C. Bora, G. R. Douglas, and E. R. Nestman, eds., Vol. 3, pp.
     143-157, Elsevier Biomedical Press, Amsterdam (1982).

32. A. C. Stevenson and C. B. Kerr, On the distribution of frequen-
     cies of mutation to genes determining harmful traits in man,
     Mut. Res., 4:339-352 (1967).

33. L. L. Cavalli-Sforza and W. Bodmer, The Genetics of Human Popu-
     lations, Freeman and Co., San Francisco (1971).

34. N. Yasuda, An average mutation rate in man, Jap. J. Hum. Genet.,
     18:279-287 (1973).

35. H. W. Mohrenweiser and J. V. Neel, Frequency of thermostability
     variants and estimation of the total "rare" variant frequency
     in human populations, Proc. Nat. Acad. Sci. USA, 78:5729-5733
     (1981).

36. T. Mukai and C. C. Cockerham, Spontaneous mutation rates at
     enzyme loci in Drosophila melanogaster, Proc. Nat. Acad. Sci.
     USA, 74:2514-2517 (1977).

37. R. A. Voelker, H. E. Schaffer, and T. Mukai, Spontaneous al-
     lozyme mutations in Drosophila: Rate of occurrence and nature
     of the mutants, Genetics, 94:961-968 (1980).

38. S. J. O'Brien, On estimating functional gene number in eukary-
     otes, Nature (New Biol.), 242:52-54 (1973).

39. C. Stern, G. Carson, M. Kinst, E. Novitski, and D. Uphoff, The
     viability of heterozygotes for lethals, Genetics, 37:413-449
     (1952).

40. Y. Hiraizumi and J. F. Crow, Heterozygous effects on viability,
     fertility, rate of development, and longevity of Drosophila
     chromosomes that are lethal when homozygous, Genetics, 45:1071-
     1083 (1960).

41. E. R. Soares, TEM-induced gene mutations at enzyme loci in the
     mouse, Env. Mut., 1:19-25 (1979).

42. F. M. Johnson, G. T. Roberts, R. K. Sharma, F. Chasalow, R.
     Zweidinger, A. Morgan, R. W. Hendren, and S. E. Lewis, The de-
     tection of mutants in mice by electrophoresis: Results of a
     model induction experiment with procarbazine, Genetics, 97:
     113-124 (1981).

43. F. M. Johnson and S. E. Lewis, Electrophoretically detected
     germinal mutations induced in the mouse by ethylnitrosourea,
     Proc. Nat. Acad. Sci. USA, 78:3138-3141 (1981).

44. Y. N. Tobari and K. Kojima, A study of spontaneous mutation
     rates at ten loci detectable by starch gel electrophoresis in
     Drosophila melanogaster, Genetics, 70:397-403 (1972).

45. W. Pretsch and K. R. Narayanan, Erfassung von Genmutationen
     bei Mäusen durch isoelectric Fokussierung, Hoppe-Seyler's Z.
     Physiol. Chem., 360, 345 (1979).

46. F. M. Johnson and S. E. Lewis, Personal communication.

47. B. B. Rosenblum, S. M. Hanash, N. Yew, and J. V. Neel, Two-
     dimensional analysis of red cell membranes, Clin. Chem., 28:
     925-931 (1982).

48. L. A. Wanner, J. V. Neel, and M. M. Meisler, Separation of alleleic variants by two-dimensional electrophoresis, Am. J. Hum. Genet., 34:209-215 (1982).

49. M. Skolnick, S. Sternberg, and J. V. Neel, Computer programs for adapting two dimensional gels to the study of mutation, Clin. Chem., 28:969-978 (1982).

50. M. Skolnick, An approach to completely automatic comparison of two-dimensional electrophoresis gels, Clin. Chem., 28:977-986 (1982).

51. E. H. McConkey, B. J. Taylor, and D. Phan, Human heterozygosity: A new estimate, Proc. Nat. Acad. Sci. USA, 76:6500-6504 (1979).

52. K. E. Walton, D. Styer, and E. I. Gruenstein, Genetic polymorphism in normal human fibroblasts as analyzed by two-dimensional polyacrylamide gel electrophoresis, J. Biol. Chem., 254: 7951-7960 (1979).

53. A. J. L. Brown and C. H. Langley, Reevaluation of level of genic heterozygosity in natural populations of Drosophila melanogaster by two-dimensional electrophoresis, Proc. Nat. Acad. Sci. USA, 76:2381-2384 (1979).

54. S. C. Smith, R. R. Racine, and C. H. Langley, Lack of genic variation in the abundant proteins of human kidney, Genetics, 96:967-974 (1980).

55. H. Hamaguchi, A. Ohta, R. Mukai, T. Yabe, and M. Yamada, Genetic analysis of human lymphocyte proteins by two-dimensional gel electrophoresis: 1. Detection of genetic variant polypeptides in PHA-stimulated peripheral blood lymphocytes, Hum. Genet., 59:215-220 (1981).

56. H. Harris, The Principles of Human Biochemical Genetics, 3rd revised ed., Elsevier/North Holland Biomedical Press, Amsterdam (1980).

THE DETECTION OF ENU-INDUCED MUTANTS IN MICE BY ELECTROPHORESIS

AND THE PROBLEM OF EVALUATING THE MUTATION RATE INCREASE

F. M. Johnson[1] and S. E. Lewis[2]

[1]Laboratory of Genetics
National Institute of Environmental Health Sciences
Research Triangle Park, North Carolina 27709

[2]Chemistry and Life Sciences Group
Research Triangle Institute
Research Triangle Park, North Carolina 27709

ABSTRACT

Electrophoretic methods of mutant detection in mice are described. Spermatogonial exposure of parental male mice to a 250
mg/kg dose of ethylnitrosourea resulted in greatly elevated frequencies of electrophoretically detectable mutants in their F_1
progeny. Two major categories of mutants were observed, deficiency
mutants and electrophoretic mobility mutants. Deficiency mutants
were induced to higher average frequencies than electrophoretic mobility mutants. Since no newly arisen mutants were found in control groups, the spontaneous background mutation rates are not known.
However, various comparisons with previous data suggest that ENU
caused a 1000- or greater-fold increase in the spontaneous mutation
rate at some loci.

A small prenatal recessive-lethal test was also conducted, but
resulting data do not substantiate the very high recessive lethal
frequency that was predicted from the frequencies of extreme deficiency (null) mutants. Recessive lethal-mutable loci may have
lower spontaneous mutation rates (or be less ENU-mutagen sensitive)
than some null-mutable loci. The degree to which such mutations as
those detected by electrophoresis contribute measurably to harmful
genetic change is difficult to assess since no obvious detrimental
effects were associated with any of the induced mutants in heterozygous condition. It is possible that the combined effects of many
deficiency mutations with small effects may be as significant as

dominant mutations that cause severe abnormalities. However, the
number of null-mutable loci of functional importance may not be
large. An expanded set of loci at which biochemical genetic variation and other mutants can be detected is recommended for future
investigations of induced mutation in mice.

INTRODUCTION

The technique of electrophoresis has been used for some time
for the analysis of genetic variation in populations [1]; however,
only recently has application of the method been extended to identify variation attributable to newly induced germinal mutations in
mice [2, 3, 4, 5]. Since electrophoretically identified mutations
are detected on the basis of alterations in specific gene products,
the method is potentially useful for providing material to investigate molecular characteristics of mutant genes and functional effects
resulting from mutation. We became interested in using ethylnitrosourea (ENU) to induce a number of electrophoretic mutations for
such studies after the compound was demonstrated to be a potent in-
ducer of visible specific-locus mutations [6]. Some of our results
have already been published [5, 7, 8, 9], but detailed genetic, developmental and physiological studies remain to be done on most of
the mutants. All of the mutants we describe in this paper behave
as heritable Mendelian factors.

We (Lewis and Johnson, these proceedings) have discussed the
known characteristic features of induced and spontaneous mutants
detected during electrophoretic screening. In this paper we address
primarily the frequency aspects of ENU-induced mutants in comparison
with control and ther data from mice and Drosophila. In addition,
preliminary data are also presented from a recessive lethal
analysis performed on a small group of animals from the ENU experiment.

METHODS

Details of the electrophoretic procedures, loci screened, and
the methods for calculating mutant gene frequencies from electrophoretic data have been published previously [3, 6, 7, 8]. Two
strains of mice are used, C57BL/6J (B6) and DBA/2J (D2). Males are
mutagen-treated and mated with females of the other strain, and parents and progeny are then analyzed for genetic variation at 21 loci.
Concurrent control matings are includeed in all experiments (see
above references for details). All results from mutagen-exposed
animals in this report are for 250 mg/kg doses of ENU administered
as single intraperitoneal injections and are limited to progeny conceived ten or more weeks after treatment. The control data presented are cumulative results from our first procarbazine experiments [4] to the present.

Table 1.   Null and Reduced-Activity Mutants Induced by ENU in
           C57BL/6J and DBA/2J Mice

Strain	Locus	No. Nulls	No. Reduced Activ. Mut.	No. Loci Tested	Mutant Frequencies		
					Nulls	Red. Activ.	Combined
B6	Pep-3	3	0	669	4.48×10^{-3}	---	4.48×10^{-3}
B6	Pep-2	0	1	669	---	1.49×10^{-3}	1.49×10^{-3}
B6	Es-1	1	0	669	1.49×10^{-3}	---	1.49×10^{-3}
B6	Total	4	1	2007	1.99×10^{-3}	4.98×10^{-4}	2.49×10^{-3}
D2	Mod-1	1	2	1577	6.34×10^{-4}	1.27×10^{-3}	1.90×10^{-3}
D2	Pgm-1	1	0	1577	6.34×10^{-4}	---	6.34×10^{-4}
D2	Pep-7	1	0	1577	6.34×10^{-4}	---	6.34×10^{-4}
D2	Pep-3	1	0	1577	6.34×10^{-4}	---	6.34×10^{-4}
D2	Total	4	2	6308	6.34×10^{-4}	3.17×10^{-4}	9.51×10^{-4}

Ten of the 21 loci have electrophoretically expressed allelic
differences between the strains and provide the capability for easily detecting both mobility-shift mutants and mutants with decreased
activity. Since the latter may not be detected as readily when the
electrophoretic patterns are the same in the parents and the offspring, it is advantageous to use two strains of mice with electrophoretic differences to detect deficiency-type mutants.

The recessive lethal analysis [11, 12] was conducted by mating
F_1 males from control and ENU-treated B6 parents to AH/eJ females
and subsequently backcrossing the F_2 females produced from this cross
to the F_1 male parents. The F_1 males utilized for recessive lethal
analysis were a sample of those animals also used for mutant detection by electrophoresis. The F_2 females were dissected shortly before the expected time of birth and the uteri examined for live and
dead embryos. Generally, the combined backcross progeny originating
from a given male F_1 serves to test that male for the presence of a
lethal factor evidenced by a frequency of 1/8 dead backcross embryos
produced in addition to normal background embryonic lethality.

RESULTS AND DISCUSSION

Induced mutations resulting in deficiency of gene product (absence or reduction of activity) were found at three loci (Pep-2,
Pep-3, and Es-1) in B6 and four loci (Mod-1, Pgm-1, Pep-3, Pep-7)

Table 2. Ranges of Induced-Mutant Frequencies Counting
          12 Loci and 3 or 4 Loci (in C57BL/6J and DBA/2J,
          Respectively) per F_1 Animal Analyzed

Strain	Type of Mutant	Mutant Frequencies	
		w/11 Loci	w/3-4 Loci
B6	Nulls	5.4×10^{-4}	19.9×10^{-4}
B6	Nulls & Red. Activ.	6.8×10^{-4}	24.9×10^{-4}
D2	Nulls	2.3×10^{-4}	6.3×10^{-4}
D2	Nulls & Red. Activ.	3.5×10^{-4}	9.5×10^{-4}

in D2. Mutant frequencies for those loci at which mutants were
found are shown in Table 1. Null mutants were somewhat more frequent than reduced-activity mutants and deficiency mutations overall
were more frequent in B6 than in D2. The differences, however, are
not statistically significant. Considering all 11 loci at which it
was technically possible to observe null and reduced-activity mutants, calculated average frequencies were to 6.8 × 10⁻⁴ for B6 and
3.5 × 10⁻⁴ for D2. The denominator in the calculation is 11 times
the number of F_1 offspring to take into account the 10 loci with
allelic differences between the strains plus one for the Pep-7 and
Pep-2 mutants found in D2 and B6, respectively, in the other set of
loci. These freuencies are thus determined on the assumption that
11 loci in each strain are mutable by ENU. If these loci are not
all ENU-mutable, the average frequencies of mutation at the mutagen-
sensitive loci, of course, are underestimated. On the other hand,
when only the loci at which mutants were found are counted, the frequencies are overestimated if all loci are ENU-mutable and only the
limited sample size was responsible for the failure to detect mutants
at other loci. The range of values produced by independently applying both considerations is shown in Table 2.

Summarizing from Tables 1 and 2, induced-mutant frequencies are
approximately 3.5 × 10⁻⁴ to 24.9 × 10⁻⁴ for deficiency mutants, using
the low value from D2 and the high from B6, i.e., approximately one
mutant in 3000 loci to one mutant in 400 loci. Assuming 30,000 loci
for the haploid mouse genome [13] and that all loci mutate to the
extent indicated by the sample of 12 loci, our results suggest ENU
is capable of producing deficiency mutants at a rate of at least 10
per gamete from exposed spermatogonia. From the fraction of loci
that proved to be deficiency-mutable, i.e., 6/12, and assuming from
this that half the genome of the mouse may respond similarly to ENU
exposure, the estimated upper limit for the genomic rate is 30,000/2÷
400 = 37 deficiency mutants/gamete.

Table 3.   Electrophoretic-Mobility Mutants Induced by ENU
          in C57BL/6J and DBA/2J Mice

Strain	Locus	No. Mutants	No. Loci Tested	Mutation Frequencies
B6	Pep-2	1	669	1.5×10^{-3}
B6	Idh-1*	1	669	1.5×10^{-3}
B6	Two Loci	2	1338	1.5×10^{-3}
B6	21 Loci	2	14049	1.4×10^{-4}
D2	Mod-1*	2	1577	1.3×10^{-3}
D2	Hba	1	1577	6.3×10^{-4}
D2	Pgm-2	1	1577	6.3×10^{-4}
D2	Idh-1	1	1577	6.3×10^{-4}
D2	4 Loci	5	6308	7.9×10^{-4}
D2	21 Loci	5	33117	1.5×10^{-4}

*One reduced-activity mutant at Idh-1 and one at Mod-1 were too indistinct to
determine if mobility was also altered. These mutants are not included in the
Table.

Electrophoretic mobility mutants were found at low loci (Pep-2
and Idh-1) in B6 and four loci (Mod-1, Hba, Pgm-2, and Idh-1 in D2
(Table 3). Mobility-mutant frequencies are presented in Table 3.
Although most of the 21 loci involved in electrophoretic analysis
are known to be genetically variable as evidenced by natural polymorphisms and, thus, are spontaneously mutable, the extent to which
these loci are mutable by ENU is not known. In Table 3 induced-
mutant frequencies are presented on the basis of the set of loci
that provided detectable mutants and also from the total set of 21
loci. As shown in Table 3, the frequency of induced-mobility mutants
ranges from 1.5×10^{-4} (21 loci in D2) to 1.5×10^{-3} (2 loci in B6).
Thus, mobility mutants appear to be induced about half as frequently
as deficiency mutants by ENU in mice, comparing mid-range frequency
estimates.

No newly arisen mutants of spontaneous origin were found by
electrophoresis. Without spontaneous mutant frequencies it is not
possible to directly calculate a rate increase attributable to ENU
exposure. Data obtained from other approaches to detect spontaneous
mutants, however, provide some limited opportunities for comparison.

Voelker, Mukai, and Schaffer [14] found the spontaneous null-
mutation rate to be 3× the mobility-mutation rate in Drosophila.

If the same relationship applies in mice, our results suggest that
ENU may increase the rate of spontaneous null mutation and mobility
mutation about equally, or perhaps cause a slightly greater increase
in the null-mutation rate than the mobility-mutation rate.

Schlager and Dickie [15] found the recessive spontaneous-muta-
tion rate at 5 specific coat-color loci (a, b, c, d, and ln) in mice
to be 8.9×10^{-6}. This values is midway between the spontaneous
mutation rate (7.5×10^{-6}) determined by Russell [10] and the rate
of 10×10^{-6} provided by Carter, Lyon, and Phillips [16] from the a,
b, c, d, se, p, and s loci. These loci may be more mutable than
most other loci in the mouse since at another group of 26 unselected
loci Schlager and Dickie [15] found that the spontaneous recessive
rate was 0.67×10^{-6}. Considering these 26 loci and the 5 specific
loci together, the average spontaneous rate becomes 0.8×10^{-6}.
Schlager and Dickie [15] claim that it should have been possible for
them to detect 160 recessive mutants based on Green's [17] tabulation of mutations in mice. If the loci at which no mutants were
found are included, the average spontaneous rate for recessive visible
mutations becomes about 2×10^{-7}, with some variation depending on
how the calculation is made. A lower rate would result if mutants
observed after this study were included.

If it is assumed that deficiency mutants detected by electrophoresis occur spontaneously at the same rate and represent metabolic
defects of the same type as those recessive mutations that cause
morphological alterations, the frequency of induced deficiency mutants may be compared with the spontaneous rate for recessive-visible
mutants. However, the large variation in spontaneous mutability
among loci for recessive-visibles suggests that similar variation
would be expressed among loci analyzed by electrophoresis. If so,
the sample of loci (from the set of loci analyzed by electrophoresis)
which showed mutants by electrophoresis may represent the most ENU-
mutable loci and it is then appropriate to compare these induced mutants frequencies with the frequencies of spontaneous mutants at the
most mutable visible markers. Thus, with 8.9×10^{-6} as the spontaneous mutation rate and 2.49×10^{-3} (from D2) and 9.51×10^{-4} (from
B6) as ENU-induced rates, a mutation rate increase of 107–208 fold
can be calculated.

Comparing all spontaneously mutable loci capable of producing a
recessive visible mutation and the 11 loci in the electrophoretic
system at which deficiency mutations should have been detectable
shows a much greater mutation rate increase of 1750–3400 fold. Obviously, there is uncertainty in all the estimates, but so far as
existing data permit comparison, a mutation-rate increase of 100 to
several thousand fold after ENU exposure would seem to be a distinct
possibility for at least some loci in the mouse genome.

Frequencies of induced null mutants and spontaneous recessive-lethal mutants might also be considered for comparison. Nulls at loci with an essential role in development will result in a recessive-lethal effect. Spontaneous recessive-lethal frequencies in Drosophila have been determined through a variety of investigations [18-27] and are believed to average about 3.0×10^{-6} per locus. A spontaneous frequency of null mutations detected by electrophoresis in Drosophila has been estimated to be 3.86×10^{-6} [14]. There is a large range in spontaneous recessive-lethal frequencies estimated for the mouse, $(2.9-10) \times 10^{-3}$ lethals per gamete, and there is considerable uncertainty as to the number of loci spontaneously mutable to a recessive-lethal condition, i.e., 500-10,000 [28-33]. A few highly mutable loci could help explain the larger estimated per gamete rates and the smaller estimates for the number of mutable loci. At mid-range values the spontaneous rate of recessive lethal mutations is 1.2×10^{-6} mutants per locus. In comparison, the range of ENU-induced null mutant frequencies (Table 2) is 192-1658 times as large; roughly equivalent to the values obtained when spontaneous recessive visibles are compared with frequencies of induced deficiency mutants.

If there are many ENU-mutable loci, i.e., 5000, in the mouse which have critical function and are inducible at the highest average rates indicated for nulls by electrophoresis (6.34×10^{-4}-1.99×10^{-3}), one would predict ENU-induced recessive lethal mutants to occur at a rate of 3-10 per gamete.

Our recessive lethal data (not shown) consisted of evaluation of 5 control F_1 males (an F_1 animal tested is equivalent to one male gamete evaluated) and 12 F_1 males from ENU-treated B6 males. There was no evidence of pre-implantation death in the treated group (considering total live plus dead average litter sizes in both groups) and only a suggestion of one possible induced recessive lethal acting to cause intrauterine death between implantation and birth. However, the lethal gene (indicated by 31% embryonic death) could have been pre-existing in one or the other of the parents before treatment, since none of the sibs of the F_1 individual involved were among the tested group. While the experiment is one of very small scale, the results are sufficient to contradict the massive recessive-lethal effects of ENU predicted from the null-mutant frequencies. We conclude as a result of the recessive-lethal analysis that there are probably fewer than 5000 loci induced to mutate to a lethal condition at the same rate nulls are induced by ENU. The number could be as low as 140 and still be consistent with the data. This, of course, does not argue against the possibility of many thousands of recessive lethal-mutable loci that are much less mutable by ENU than the loci at which nulls were induced.

Most of the mutants induced by ENU have been subjected to genetic analysis, but none of them have been found to be lethal or

even detrimental in the heterozygous or homozygous state under standard laboratory conditions. As nulls and enzyme deficiencies are
the basis of many known diseases [34], such mutations are potentially
harmful. There are explanations that may account for the absence of
observed harmful effects for the ENU-induced mutants. First, many
of the enzymes, such as the peptidases and esterases have broad substrate specificities and are represented by multiple loci which may
permit one enzyme to carry on the function for another. Second,
normal husbandry practices provide relatively luxurious environmental conditions such that mutations which might be deleterious in
the wild would not necessarily be obviously harmful under normal
laboratory conditions. It also is conceivable, perhaps as a result
of natural selection, that functionally important loci may tend to
be much less null/lethal mutable than other loci.

The spontaneous occurrence of many times more slightly to mod-
erately deleterious mutations than lethal and severely detrimental
mutations as provided from Drosophila studies (cf., e.g., Greenberg
and Crow [35], Mukai [36], Temin [37], and Mukai et al. [18] is consistent with the finding of a number of loci in mice at which ENU-
induced nulls do not cause drastic harmful effects. Nevertheless,
polygenic mutations that have relatively mild effects in homozygous
condition have been found to be subject to natural selection as
heterozygotes in the laboratory, and polygenic mutations and null
mutations may be selected against almost as strongly as recessive
lethals after they arise in Drosophila in nature [14, 18]. Such
mutants as the nulls and reduced activity mutants detected by electrophoresis, therefore, may contribute some genetic detrimental
effects in the F_1 and later generations though an obviously harmful
effect is not expressed in the laboratory.

The health risks resulting from induced mutations are clearly
difficult to specify when detrimental effects are not associated
with the mutational events detected. Indeed, in the extreme case
of mutations being induced only at loci with redundant or vestigial
functional roles, it is possible that genetic health risks could be
small or non-existent even at high induced rates. ENU-induced rates
are exceedingly high when only those loci at which mutants were found
are considered, and predictions of damaging effects could be made on
the basis of observed mutant frequencies, estimated average effects
on fitness, and other assumptions. The exercise may be instructive
for the purposes of exploring the potentials for genetic damage, but
reliable risk estimates are impossible with the uncertainty of
whether or not selective forces act upon the mutant individuals.

We believe it reasonable to suspect that some of the mutants,
particularly the nulls, detected by electrophoresis will eventually
prove to be disadvantageous in heterozygotes under some environmental circumstances and that some proportion will be found to be
harmful under virtually all conditions. However, for the present

we consider it premature to formulate a quantitative expression for the damaging effects to be expected from increased mutation rates caused by ENU.

For dominantly expressed ENU-induced mutations there may be less question of biological impact than with the recessive deficiency mutants and the electrophoretic mobility mutants. Ehling et al. [38] reported a considerable increase in the frequency of genetically determined dominant cataracts in the F_1 progeny of ENU-treated parents, and the presence of cataracts is obviously a harmful effect in man. However, Ehling et al. [38] also found that the increase in frequency of recessive mutations at certain visible markers was greater than the increase in cataract frequencies. Extrapolation to genomic effects from recessive visible mutations presents much the same problem as the electrophoretically detected mutants. Recessive deficiency and visible mutations may be representative of a large number of mutants, the impact of which might be greater than that of mutants causing dominantly expressed abnormality as indicated by genetic load theory [39, 40] and experiments with Drosophila [18, 19, 35-37]. Unfortunately, the level of effort expended on experimentation with the mouse has not been sufficient to provide data comparable with that from Drosophila [41], and much more work is required to determine the extent to which genetic load concepts are applicable to mouse mutageneis.

The estimation of human genetic risk from dominant mutations induced in mice may be in error if those loci capable of mutating to produce a dominantly expressed effect are more mutagen sensitive in the mouse than in man. Thus, mutability and the functional importance of the loci at which dominant mutations arise could be considered as we have done with respect to electrophoretically detected mutants.

For the future, we suggest increasing the number of enzyme markers in the electrophoretic system and including a number of loci of established functional significance to mammals generally. At the same time it would be useful to include the visible recessive markers of Russell's [10] system, dominant mutation indicators and others, such as those addressed by Drs. Searle, Ehling, and Selby (these proceedings). Conveniently, the T-stock employed by Russell and others is similar to D2 at electrophoretically expressed loci (unpublished results) and B6 is a commonly used genetic background on which variant alleles are placed (James Womack, personal communication). This approach might provide increased efficiency, but perhaps more importantly also information on inter-locus variation in mutagen sensitivity and the effects of mutations at different loci. Even if efficiency of detection is reduced as a result of including relatively expensive assays, it would be counterproductive to an understanding of genetic risk to allow efficiency to be a deciding criterion. However, later it may become possible to reduce

mutation detection efforts to a few indicator methods, if some can
be found that are representative of induced genetic damage to the
genome generally.

Genetic testing of compounds by existing methods is a subject
where administrative and technical decisions deserve careful thought.
As Dr. Lyon (these proceedings) has described, one can perform genetic risk estimates using various assumptions and presently existing sources of data if limitations are accepted. This being the
case, the question of what to do about compounds found to be more
or less mutagenically active then ENU needs to be asked. To what
extent can genetic test data be used for regulatory purposes when
the level of damage associated with increased mutation rates is im-
perfectly understood. We hesitate to recommend large scale adoption
of the present electrophoretic method for testing because costs will
be reduced and interpretability increased as expanded and improved
systems are developed. It would be useful for comparative purposes,
however, to have available a number of mutants represenative of the
action of mutagens other than ENU, so some continued level of routine
application of the available methods can also be justified. Radiation would be appropriate to use in obtaining a series of biochemically expressed mutants for comparison with the ENU-induced mutants,
and perhaps choosing a chemical that is capable of causing small deletions would also be informative. Dr. William Dove (personal communication) has suggested the use of diepoxyoctane, a compound demonstrated to be mutagenic in Neurospora [42].

The problem of measuring accurately the spectrum of gene-dam-
aging effects that accompany increased mutation rates has no apparent easy answer. The solution we suggest is more work dedicated to
detecting and characterizing spontaneous and induced mutants by a
variety of approaches, utilizing human material to the extent possible (cf., Drs. Neel and Mohrenweiser, these proceedings) and the
mouse as the experimental organism of choice for reasons of economy
and relevance to humans.

CONCLUSIONS

ENU exposure resulted in greatly elevated mutant frequencies
at some loci but no evidence of damaging effects was found to be
associated with the mutations. Extensive recessive lethal effects
predicted from the induced-mutant frequencies were not found.

The data suggest the possibility that loci with non-critical
functions (or non-critical sites within functionally important loci)
may be the most common targets of ENU action. If so, grossly exag-
gerated risk estimates could result from comparisons of average
mutant frequencies in mutagen-treated and control groups.

Alternatively, if mutants such as those detected by electrophoresis reflect alterations with deleterious affects too small to
be measured under usual laboratory conditions, then neglecting to
consider these effects might ignore a critically important component
of genetic risk.

Since the extent to which ENU causes damage to the genome cannot be specified from the data, it is not possible to derive a meaningful quantitative expression for the genetic risks of ENU exposure.

Possible differences in mutability varying with functional importance of the targets of mutagen action in the genome is a consideration that could be applied to all dominant and recessive mutant characteristics of unknown genetic risk estimates derived
therefrom.

More work needs to be directed toward characterizing spontaneous and induced mutations at a number of loci. Use of the mouse
for detecting mutants with different forms of phenotypic expression
is the most efficient and cost-effective approach to obtaining experimental data from the laboratory that is applicable to man.

ACKNOWLEDGMENTS

The authors thank Rosemary Batten, Lois Barnett, Deborah
Leverton, and Chris Worthy for excellent technical assistance and
Raye Powell for typing the manuscript.

REFERENCES

1.   R. C. Lewontin, The genetic basis of evolutionary change,
      Columbia University Press, New York (1974).
2.   H. V. Malling and L. R. Valcovic, A biochemical specific locus
      mutation system in mice, Arch. Toxicol., 38:45-51 (1977).
3.   E. R. Soares, TEM-induced gene mutations at enzyme loci in the
      mouse, Environ. Mutagenesis, 1:19-25 (1979).
4.   F. M. Johnson, G. T. Roberts, R. K. Sharma, F. Chasalow, R.
      Zweidinger, A. Morgan, R. W. Hendren, and S. E. Lewis, The de-
      tection of mutants in mice by electrophoresis: Results of a
      model induction experiment with procarbazine, Genetics, 97:113-
      124 (1981).
5.   F. M. Johnson and S. E. Lewis, Electrophoretically detected
      germinal mutations induced in the mouse by ethylnitrosourea,
      Proc. Natl. Acad. Sci. USA, 78:3238-3141 (1981).
6.   W. L. Russell, E. M. Kelley, P. R. Hunsicker, J. W. Bangham,
      S. C. Maddux, and E. L. Phipps, Specific locus test shows ethyl-
      nitrosourea to be the most potent mutagen in the mouse, Proc.
      Natl. Acad. Sci. USA, 76:5818-5819 (1979).

7. F. M. Johnson and S. E. Lewis, The human genetic risk of air-
    borne genotoxics: An approach based on electrophoretic tech-
    niques applied to mice, in: "Brookhaven Symposium on the
    Genotoxic Effects of Airborne Agents," R. R. Tice, D. L. Costa,
    and K. M. Schaich, eds., pp. 595-606, Plenum Press, New York
    (1981).

8. F. M. Johnson, S. E. Lewis, L. Barnett, W. C. Worthy, and R.
    Batten, Problems in genetic risk assessment: The detection of
    transmissible point mutations in mice by electrophresis. Latin-
    American Course in Genetic Toxicology (1982), in press.

9. F. M. Johnson and S. E. Lewis, Mutation rate determinations
    based on electrophoretic analysis of laboratory mice, Mutation
    Res., 82:125-135 (1981).

10. W. L. Russell, X-ray induced mutations in mice, Cold Spring
    Harbor Symp. Quant. Biol., 16:327-336 (1951).

11. K. G. Luning, Test of recessive lethals in the mouse, Mutation
    Res., 27:357-366 (1975).

12. A. G. Searle, Mutation induction in mice, Adv. Radiation Biol.,
    4:131-207 (1974).

13. M. F. W. Festing, Inbred Strains in Biomedical Research, Ox-
    ford University Press, New York (1979).

14. R. A. Voelker, H. E. Schaffer, and T. Mukai, Spontaneous allo-
    zyme mutations in Drosophila melanogaster, Genetics, 94:961-
    968 (1980).

15. G. Schlager and M. M. Dickie, Spontaneous mutations and muta-
    tion rates in the house mouse, Genetics, 57:319-330 (1967).

16. T. C. Carter, M. F. Lyon, and R. J. S. Phillips, Genetic hazard
    of ionizing radiations, Nature, 182:409 (1958).

17. M. C. Green, Mutant genes and linkage, in: "Biology of the
    Laboratory Mouse," E. L. Green, ed., 2nd edition, pp. 87-150,
    McGraw-Hill, New York (1966).

18. T. Mukai, S. I. Chigusa, L. E. Mettler, and J. F. Crow, Muta-
    tion rate and dominance of genes affecting viability in
    Drosophila melanogaster, Genetics, 72:335-355 (1972).

19. J. F. Crow and R. G. Temin, Evidence for the partial dominance
    of recessive lethal genes in natural populations of Drosphila,
    Amer. Naturalist, 98:21-33 (1964).

20. B. Wallace, Distance and the allelism of lethals in a tropical
    population of Drosophila melanogaster, Amer. Naturalist, 100:
    565-678 (1966).

21. B. Wallace, Mutation rates for autosomal lethals in Drosophila
    melanogaster, Genetics, 60:389-393 (1968).

22. B. Wallace, Spontaneous mutation rates for sex-linked lethals
    in the two sexes of Drosophila melanogaster, Genetics, 64:
    553-557 (1970).

23. B. Wallace, E. Zouros, and C. Krimbas, Frequencies of second
    and third chromosome lethals in a tropical population of
    Drosophila melanogaster, Amer. Naturalist, 100:245-251 (1966).

24. B. H. Judd, M. W. Shen, and T. C. Kaufman, The anatomy and func-
    tion of a segment of the x-chromosome of Drosophila melanogaster,
    Genetics, 71:139-156 (1972)

25. B. Hochman, The fourth chromosome of Drosophila melanogaster, in: "The Genetics and Biology of Drosophila," M. Ashburner and E. Novitski, eds., Vol. 1b, pp. 903-925, Academic Press, London (1976).

26. A. Schalet, G. Lefevre, Jr., The proximal region of the x-chromosome, in: "The Genetics and Biology of Drosophila," M. Ashburner and E. Novitski, eds., Vol. 1b, pp. 848-896, Academic Press, London (1976).

27. S. Abrahamson, F. E. Wurgler, C. De Jongh, and H. U. Meyer, How many loci on the x-chromosome of Drosophila melanogaster can mutate to recessive lethals, Environ. Mutagenesis, 2:447-453 (1980).

28. M. F. Lyon, Some evidence concerning the "mutational load" in inbred strains of mice, Heredity, 13:341-352 (1959).

29. A. G. Searle, Effects of low-level irradiation on fitness and skeletal variation in an inbred mouse strain, Genetics, 50:1159-1178 (1964).

30. K. G. Luning and A. G. Searle, Estimates of the genetic risks from ionizing radiation, Mutation Res., 12:291-304 (1971).

31. K. G. Luning, Spontaneous recessive lethal mutations in the mouse, Mutation Res., 27:367-373 (1975).

32. K. G. Luning and A. Eiche, X-ray induced recessive lethal mutations in adult and foetal female mice, Mutation Res., 92:169-180 (1982).

33. T. C. Carter, Recessive lethal mutations induced in the mouse by chronic γ-irradiation, Proc. Roy. Soc. Ser. B., 147:402-411 (1957).

34. J. V. Neel, H. Mohrenweiser, C. Satoh, and H. B. Hamilton, A consideration of two biochemical approaches to monitoring human populations for a change in germ cell mutation rates, in: "Genetic Damage in Man Caused by Environmental Agents," K. Berg, ed., pp. 29-47, Academic Press, New York (1979).

35. R. T. Greenberg and J. F. Crow, A comparison of the effect of lethal and detrimental chromosomes from Drosophila populations, Genetics, 45:1153-1168 (1960).

36. T. Mukai, The genetic structure of natural populations of Drosophila melanogaster, I. Spontaneous mutation rate of polygenes controlling viability, Genetics, 50:1-19 (1964).

37. R. T. Temin, Homozygous viability and fertility loads in Drosophila melanogaster, Genetics, 53:27-46 (1966).

38. U. H. Ehling, J. Favor, J. Kratochvilova, and A. Neuhauser-Klaus, Dominant cataract mutations and specific locus mutation in mice induced by radiation or ethyl nitrosourea, Mutation Res., 92:181-192 (1982).

39. H. J. Muller, Our load of mutations, Am. J. Human Genet., 2:111-176 (1950).

40. N. E. Morton, J. F. Crow, and H. J. Muller, An estimate of the mutational damage in man from data on consanguinous marriages, Proc. Natl. Acad. Sci. USA, 42:855-863 (1956).

41. E. L. Green, Genetic effects of radiation on mammalian popula-
     tions, Ann. Rev. Genet., 2:87-120 (1968).
42. T. Ong and F. J. de Serres, Mutation induction by difunctional
     alkylating agents in Neurospora crassa, Genetics, 80:475-482
     1975).

RELATION OF MOUSE SPECIFIC-LOCUS TESTS TO OTHER

MUTAGENICITY TESTS AND TO RISK ESTIMATION*

William L. Russell

Biology Division
Oak Ridge National Laboratory
Oak Ridge, Tennessee 37830

1. INTRODUCTION

Several years ago, when concern over the possible human genetic hazards from chemicals was first becoming widespread in the scientific community, markedly different views of possible risk were being expressed. At one extreme, there were predictions that, because detoxification systems in mammals are so effective, noxious chemicals would be prevented from reaching the gonads, or the genes in the germ cells, and would therefore present much less risk to the human population than penetrating ionizing radiation. At the other extreme, there were calculations, based on the mutagenic effects of caffeine in lower organisms, that coffee consumption alone could be causing a vastly greater genetic hazard than that from the total exposure to ionizing radiation.

Where do we stand today? Much information has been obtained from a large number of mutagenicity tests in different organisms with a wide array of chemicals. Nevertheless, we are still not in a position to make an informed guess about the total genetic hazard to the population from chemicals. We are not even able to make a precise estimate of genetic risk for a single chemical, although it now seems safe to conclude that some chemicals have only negligible or no mutagenic effect in mice.

---

*Research sponsored jointly by the National Institute for Environmental Health Sciences under Interagency Agreement 222 Y01-ES-10067 and by the Office of Health and Environmental Research, U.S. Department of Energy, under contract W-7405-eng-26 with the Union Carbide Corporation.

One of the major difficulties in reaching estimates of risk
from the wide array of chemicals to which man is exposed is that results from the simple, relatively inexpensive, mutagenicity tests,
for example, those obtained with Salmonella, have given poor correlation with the results from tests, such as the specific-locus test,
that measure mutation frequency in the offspring of treated mammals.
The nature of this problem, and possible ways to deal with it in the
near future, will be the main topic of this paper.

Most of the mouse data cited here came from the specific-locus
method that we have been using for over 30 years [1], which involves
seven visible genetic markers that can be rapidly scored. However,
there are already indications that the conclusions reached are
likely to receive support, at least in general, from other specific-
locus tests in mice [2], and perhaps from other tests for transmitted
mutations in mice, for example, those screening for dominant skeletal
defects [3] or cataracts [4].

Our first experience of comparing results from the mouse specific-locus method with those from other organisms was with ionizing
radiation. Even when the mouse results were compared with those
from Drosophila, a striking lack of parallelism, both quantitative
and qualitative became apparent [5]. The induced mutation rate per
locus averaged approximately one order of magnitude higher in the
mouse; the dose-response curve, which was linear in Drosophila,
turned out to be non-linear in both sexes in the mouse; the mutational response of both sexes in the mouse also showed marked effects
of dose rate and dose fractionation, effects that were absent in
Drosophila; and the arrested oocyte stage in the mouse proved to be
highly resistant to mutation induction even by acute doses of x-rays
and neutrons. Thus, what had long been thought to be basic principles of radiation genetics, based on the Drosophila findings,
proved not to be valid for the mouse.

Having had this experience, we could expect, before we even
started working with chemicals, that the lack of parallelism between mammals and lower organisms might be even greater for chemical mutagenesis than it had proved to be for radiation mutagenesis.
Metabolic differences involving alteration, activation, detoxification, or elimination of a chemical before it reached the gonads were
obvious possible complications. In addition, because mouse germ
cells had shown a capacity to repair even radiation-induced damage,
as well as marked differences in repair capabilities in different
germ-cell stages, it seemed likely that the responses to chemicals
that did reach the germ cells in active form might be quite complex.
Repair enzymes and other factors might be involved that were not
predictable from other organisms. It is hardly surprising, therefore, that our expectations regarding the degree of non-parallelism
between results from mice and lower organisms have proved correct.

The most striking feature of the results from the first batch

of chemicals tested with the mouse specific-locus method was the
weak or zero response in mouse spermatogonial stem cells to compounds that were potent mutagens in other test systems [6, 7].
These and later results showed, for example, that ethylmethanesul-
fonate (EMS), hycanthone, and diethylnitrosamine gave no significant
increase over control mutation frequencies even when injected at
sublethal levels [8]. Furthermore, of the compounds that did give
positive results in spermatogonia, none, until recently, approached
the effectiveness of radiation when comparisons were based on maximum tolerated doses.

     Some investigators speculated that perhaps the specific-locus
test was detecting only deficiencies, or other chromosomal events,
and not gene mutations. For various reasons, this did not seem a
plausible hypothesis. For example, among the specific-locus mutations induced in spermatogonia by radiation, there were several that
had phenotypic effects intermediate between wild-type and the null
effect produced by a deficiency.

     It was also speculated that, granting that the specific-locus
method detects intragenic events, perhaps these have to be massive
damages within the gene in order to produce the externally visible
effects scored by the method. This led to the suggestion that a
test that would measure enzyme differences or other biochemical endpoints resulting, for example, from simple base-pair changes, might
reveal a much higher mutation rate than that obtained by our specific-locus method. This also seemed implausible. At least one,
and probably more, of the coat color changes used in our method result from enzyme differences. A mutational effect that is externally
visible can have just as simple a biochemical origin as one that requires a biochemical test. That our specific-locus method is not
underestimating the mutation rate that would be detected using direct biochemical end-points has now been demonstrated by Johnson
and Lewis [2]. Their specific-locus test, employing an electrophoretic approach, has given a mutation rate per locus for ethylnitrosourea which is slightly less than that obtained by our specific-locus test.

     A third speculation was that the low mutational responses were
due to failure of the chemicals, or their active metabolites, to
reach the gonads. However, it is perfectly clear for at least some
compounds that have produced no mutagenic effect in spermatogonia
that this cannot be attributed to failure to reach the testis. For
example, EMS, which has produced no mutations in spermatogonia, does
produce mutations in postspermatogonial stages. Furthermore, measurements with radioactively labelled EMS have shown that it reaches
the testis [9] and that it ethylates DNA in spermatozoa [10].
Hycanthone, which also has given no mutations in spermatogonia, has
been shown by means of labelled compounds to reach the testis [11].
It is also known to reach the ovary.

None of the above speculations can, therefore, account for the weak or zero response of mouse spermatogonia to many chemicals known to be potent mutagens in other systems. A remaining hypothesis, one I have entertained for some time, is that mouse spermatogonia are resistant to mutation induction by many chemicals, possibly because of an efficient repair system. This seemed plausible in view of the fact that some repair even of radiation-induced mutation was apparently occurring, as evidenced by the dose-rate effect. In addition, the type of damage produced by at least some chemicals might be more easily repaired than the damage from radiation.

The alternative views, which I have described above as implausible, have, nevertheless, gained a foothold in a curious semantic way. Our specific-locus method is sometomes described as an "insensitive" test, simply because it often shows a weak or zero response to chemicals that are potent mutagens in other systems. This use of the term "insensitive" is made either without much thought as to the cause of the weak mutational response, or because of a belief in one or the other of the first two speculations, above, that I have regarded as untenable. Calling the test "insensitive" implies that it is not measuring the true genetic risk to man, and that, for this, one must use other tests, presumably ones where the chemical is more mutagenic. But the evidence to be presented in this paper indicates that the weak mutational responses are not due to any failure of the specific-locus method. For chemicals that reach the gonads, the explanation may lie in the resistance of the spermatogonia, which apparently have an efficient capacity to repair mutational damage. It is, therefore, not correct to ascribe the weak mutational responses to "insensitivity" of the specific-locus test; instead, one should refer to mouse spermatogonia as "resistant" to mutation induction. This implies that human spermatogonia may also be resistant, and, therefore, that the specific-locus test is actually estimating the human genetic risk, and not failing to do so because of any "insensitivity."

The results obtained recently with ethylnitrosourea (ENU) and related compounds have a strong bearing on the points already discussed as well as on other aspects of the relation of the specific-locus test to other tests and to risk estimation.

2. RESULTS WITH ETHYLNITROSOUREA

2.1. Mutagenic Effectiveness

The extreme mutagenic effectiveness of ENU in the mouse specific-locus test on spermatogonia was first reported in 1979 [12], and was confirmed independently [4]. In more extensive data published later [13], a dose of 250 mg/kg of ENU injected intraperitoneally yielded 72 mutations at the specific loci in 10,146 offspring derived from cells that were stem-cell spermatogonia at the

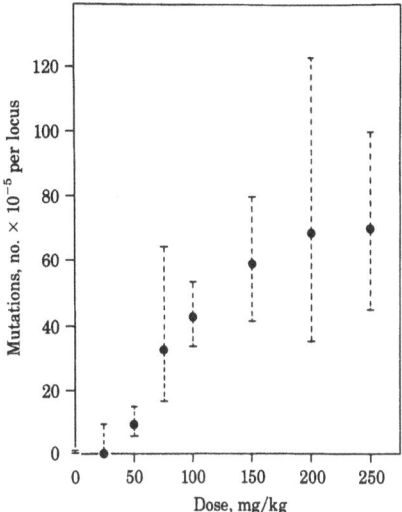

Fig. 1. Frequencies and 90% confidence limits of ENU-induced spe-
         cific-locus mutations in mouse spermatogonia. (From Russell
         et al. [14].)

time of treatment. This is equivalent to one mutation per 141 offspring examined. The induced mutation frequency is 8 times that obtained with 600 R of acute x-rays, 23 times that from 600 R of
chronic gamma rays, 24 times the highest rate obtained before with
any other chemical, and 134 times the spontaneous mutation rate.
This finding clearly sets to rest the speculation that the mouse
specific-locus test might be a poor detector of chemically-induced
mutational changes.

     For hazard evaluation, it is important to know how mutation fre-
quency is affected by such factors as dose response and germ-cell
stage in both sexes. Prior to the findings with ENU, investigation
of these factors by the specific-locus method was difficult for all
chemicals tested because of their weak mutagenic effect. Only limited progress had been made, and only with a few compounds. The
extreme effectiveness of ENU has made this type of investigation
possible with relative ease.

2.2. Dose-Response Curve

     The results from ENU have provided the most extensive dose-
response curve yet obtained for induction of specific-locus mutations in mouse stem-cell spermatogonia by any chemical [14]. Seven
doses, ranging from 25 to 250 mg per kg of body weight were tested,
and, after allowing for clustering, at least 120 independent mutations were scored. The results are shown in Fig. 1. In the lower

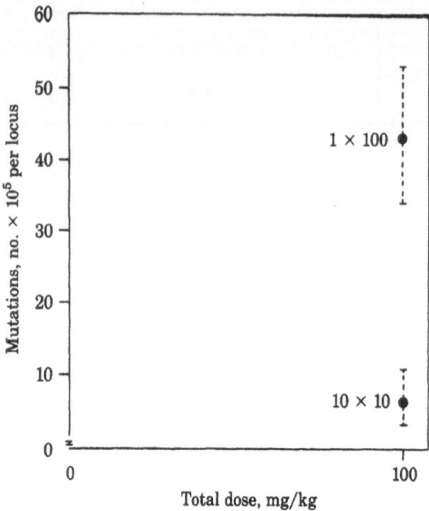

Fig. 2.   Frequencies and 90% confidence limits for specific-locus
         mutations induced by single and fractionated doses of ENU
         in mouse spermatogonia.   The interval between dose frac-
         tions was one week.   (From Russell et al. [15].)

portion of the curve, below a dose of 100 mg/kg, the data fall statistically significantly below a maximum likelihood fit to a straight
line.   Independent evidence, described in the next section, indicates
that, over this dose range, ENU reaches the testis in amounts directly proportional to the injected dose.   It is concluded that the
spermatogonia are capable of repairing at least a part of the mutational damage when the repair process is not swamped by a high dose.

2.3.   Effect of Dose Fractionation

In order to estimate what happens at still lower doses, it was
necessary to resort to a fractionation experiment.   The specific-
locus mutation frequency from a single dose of 100 mg/kg of ENU was
compared with that from a total dose of 100 mg/kg fractionated into
doses of 10 mg/kg injected at weekly intervals [15].   The results
are shown in Fig. 2.   It is obvious that fractionation of the dose
yields a mutation rate that is significantly lower than that produced by the single dose ($P < 1 \times 10^{-9}$).   The induced mutation frequency with the fractionated dose is only 13% of that with the single
exposure.

Carricarte and Sega [16] have measured the amount of unscheduled DNA synthesis occurring in mouse spermatids after intraperi-

toneal injection of ENU. Over the range from 100 to at least 10
mg/kg, and possibly lower, unscheduled DNA synthesis is directly
proportional to the amount of ENU injected. In an additional study,
Sega [17] has shown that DNA ethylation in the testis after intraperitoneal injection of tritium-labelled ENU is directly proportional
to injected dose over the range of 10 to 100 mg/kg. These findings
demonstrate that the decrease in mutation rate below linearity in
this range of doses, as described in the previous section, and the
reduced mutagenic effect when the 100 mg/kg dose is fractionated,
are not due to failure of the chemical to reach the testis in pro-
portionate amounts. The obvious conclusion is that, at a dose of
10 mg/kg, the spermatogonia have a capacity to repair a major part
of the genetic damage induced by ENU.

2.4. Effect of Cell Stage

For the estimation of genetic risk it is important to know the
relative mutational sensitivities of the various germ-cell stages.
The data now available for ENU show that, at doses of 100 mg/kg and
above, postspermatogonial stages and mature and maturing oocytes are
giving mutation rates approximately one order of magnitude lower
than those obtained from spermatogonia [13]. It should be noted,
however, that, even with the dose as low as 100 mg/kg to males, no
offspring were recovered from matings made during the sixth and
seventh weeks post-injection. So, for cells that were treated as
differentiating spermatogonia or early spermatocytes, the mutational
sensitivity to ENU is still not known. From exposure of arrested
oocytes, scored in conceptions occurring more than six weeks after
injection, only one mutation has been observed in over 10,000 offspring [13], a figure not significantly different from the control.
Thus, so far, it appears that the major genetic hazard from ENU
would lie in exposure of the stem-cell spermatogonia.

2.5. Nature of the Mutations

More than 25% of the mutations induced in spermatogonia by ENU
have a phenotypic expression intermediate between that of the viable
null allele and wild-type [13]. This indicates that at least this
proportion of the mutations are probably intragenic changes. That
the proportion could be much higher is supported by the following
two facts. First, there have been no occurrences of a d-se deficiency [13], a type of event that is not uncommon when the conditions of a radiation experiment are known to result in deletions.
Second, at five of the seven loci the proportion of mutations that
are lethal in homozygous conditon is very low, less than 5% [13].
From the above results, it is clear that there is no validity to
the speculation that our specific-locus method might be detecting
only deficiencies and not gene mutations.

3.   RESULTS WITH METHYLNITROSOUREA

Methylnitrosourea (MNU) is much more toxic than ENU, so we have
had to limit our injected amounts to 80 mg/kg or less. Because MNU
is a powerful mutagen in other biological systems, including Drosophila, and in some cases is much more potent than ENU, it was surprising to find that it has little or no mutagenic effect in mouse
spermatogonial stem cells [13]. Following injections of 70, 75, or
80 mg/kg (weighted mean dose 74 mg/kg), only 3 mutations have been
obtained, to date, in 21,079 offspring from cells treated in spermatogonial stem-cell stages. This is only slightly, and not significantly, above the control mutation frequency. A comparable dose
of ENU would have given about 41 mutations in an equivalent number
of offspring. From dominant-lethal studies by Generoso in our laboratory [18], from the results of Sega et al., on unscheduled DNA
synthesis, and with radioactively labelled MNU [19], and from the
data presented in the next paragraph, it is abundantly clear that
MNU reaches the germ-cells in the testis in active form.

Testing MNU for its mutagenic effect in post-stem-cell stages
produced another surprising result. With a dose of 75 mg/kg to the
male parents, the offspring conceived within the first five weeks
after injection showed only a low mutation rate. However, offspring conceived in the following week had an exceptionally high
mutation frequency that was totally unexpected. In 2832 offspring,
18 mutations have been scored. This, with a dose of only 75 mg/kg,
is similar to the mutation frequency obtained with 250 mg/kg of ENU
in spermatogonia. According to Oakberg's timing of spermatogenesis
[20], offspring conceived in the 6th week after injection probably
came from cells exposed as differentiating spermatogonia or prelep-
totene spermatocytes. As was pointed out earlier, the mutagenic
effect of ENU on these particular stages is not yet known.

Regardless of that outcome, it is apparent, from the low mutagenic response of stem-cell spermatogonia to MNU, that results with
ENU were not predictive of what would happen with MNU.

4.   RESULTS WITH ETHYLNITROSOURETHANE

After an injected dose of 312 mg/kg of ethylnitrosourethane,
3 specific-locus mutations have now been observed in 6001 offspring
derived from cells treated as stem-cell spermatogonia. In a similar
experiment with 400 mg/kg, 1 mutation has been obtained in 902 offspring. Thus, ethylnitrosourethane appears to be mutagenic in mouse
spermatogonia, but is clearly much less effective than ENU.

5.   COMPARISONS OF MOUSE SPECIFIC-LOCUS
TESTS TO OTHER TESTS

Our specific-locus test and that of Johnson and Lewis [21, 22] have now provided more information on the mutagenic effects of ENU than on those of any other chemical. It seems appropriate, there-fore, to combine these results with those obtained on other chem-icals for a fresh comparison with the results from other tests.

Some of the failures of other test systems to predict results in mammalian germ cells, e.g., the weak or zero mutagenic response of spermatogonia to chemicals that are potent mutagens in other tests, have already been discussed earlier in this paper. Some sweeping generalizations that I have regarded as misleading over-simplifications have been criticized elsewhere [23, 24]. These in-clude the ABCW hypothesis for radiation along with its extension to chemicals, and the view that the REC (rem-equivalent-chemical) value could be used to predict results from one organism to another. One of the bases for the criticism was the great variety of responses found among the different germ-cell stages of the mouse. The recent specific-locus studies with ENU and MNU dramatically emphasize this fact, and demonstrate the current difficulty of generalizing, even for related chemicals within one organism.

Another attempt at a simplifying concept that I have questioned only briefly elsewhere [13] is the proposal that mutation frequency induced by a chemical in mice can be effectively estimated by doing mutation and dosimetry studies in Drosophila, and then simply fol-lowing this with dosimetry in mouse germ cells or even just in soma-tic cells. The clear demonstration with ENU that mutation frequency with decreasing dose drops much faster than proportionately with dose in the testis, in other words, that repair is occurring at lower doses, indicates that this proposal would not work with ENU. If, as seems likely, repair is a common phenomenon with mutagenic chemicals in spermatogonia, then the concept faces a serious problem.

Many of the problems of extrapolating to mammalian germ cells from the results in simpler test systems, such as bacteria or mam-malian somatic cells, are obvious. For example, the variety of effects in different germ-cell stages in the mouse cannot be pre-dicted, and these range all the way from those with ethylmethane-sulfonate, which is mutagenic in postspermatogonial stages, but not in spermatogonia, to those with ENU, which is highly mutagenic in spermatogonia, but only weakly so in the postspermatogonial stages so far tested. MNU is highly mutagenic only in one restric-ted post-stem-cell stage.

Tests can, of course, be done on a variety of germ-cell stages in Drosophila, but differences between mammals and fruitflies in the development and maturation of these cells may affect their responses. For example, the mutagenic effect of ENU in Drosophila appears to be greatest in postspermatogonial stages, particularly spermatids, while in these stages in the mouse the mutational response is relatively weak.

Other obvious problems in extrapolating, such as metabolic differences, possible existence of repair mechanisms, etc., have been mentioned earlier. Because of the difficulties in extrapolating, there now seems to be widespread acceptance of the view that the simpler systems are of only limited value in predicting the risk of heritable damage. It is recognized that bathing cells in an essentially inexhaustible ocean of a chemical tells whether those particular cells mutate under those conditions, but may tell little about whether any of the mammalian germ-cell stages will respond similarly. Tests that do not directly involve mammalian germ cells are consequently now commonly regarded only as prescreens. Committees are engaged in tabulating the degrees of correlation between several of these tests and the tests on mammalian germ cells. It appears that the mouse spot test on embryonic somatic cells in vivo [25] will probably give the highest correlation with our specific-locus test on germ cells. No chemical that is positively mutagenic in germ cells has failed to produce mutations in the spot test, and the relative mutagenicity of different chemicals may be predicted more reliably by the spot test than by any other prescreen so far developed [26, 27].

6. USE OF THE MOUSE SPECIFIC-LOCUS TESTS
    IN RISK ESTIMATION

Because there is sometimes a misunderstanding of the purposes for using the specific-locus method, it should be explained at the outset that it was not designed for measuring total mutation rate in the mouse. In the past, in the absence of more relevant information on risk, some committees have, indeed, used the per-locus radiation-induced mutation rate for the seven loci, multiplied by an estimated total number of loci, to calculate a total per-genome rate. Some who have considered using this type of calculation for chemicals have regretted that the mutation frequencies in our test are based on only seven loci, although they have been willing to draw conclusions from Neurospora data based on only two closely associated loci and from Salmonella results based on only one part of one locus. Personally, I have never regarded such a use of the specific-locus mutation results as a satisfactory way of estimating total mutation rate per genome.

The specific-locus method in the mouse was designed for comparative purposes. Its first use was to compare radiation-induced mutation rates in the mouse with those in Drosophila. Following that, it has been used for extensive intraspecies comparisons: for comparing rates in different germ-cell stages in both sexes, for comparing the effects of different doses (dose-response curve), for comparing the effects of fractionated with single exposures, for comparison with the spontaneous mutation rate, for comparing qualitatively different radiation sources, for comparing the effects of different ages at time of treatment, for comparing the responses at different time intervals after treatment, for comparing the results from different routes of administration, and for comparing different chemicals. Because the same seven loci are used on both sides of each of these intraspecies comparisons, there is no bias, and the number of loci does not seem unduly restrictive. Additional information from other loci, particularly from those selected on another basis [2], is, of course, always desirable. All of the above comparisons are obviously useful for risk estimation.

The high mutation rate obtained with ENU, a compound that produces primarily gene mutations, rather than chromosomal aberrations, shows that the specific-locus test is a good detector of chemically-induced mutations. It is, therefore, suitable for risk estimation. With compounds as effective as ENU, mutagenicity can be detected with only a few hundred offspring. The number of offspring required for a satisfactory estimate of risk from chemicals that are weakly mutagenic depends on the multiple of the human exposure that will be tolerated in the mouse. This is illustrated later by examples.

The specific-locus test is not designed to measure total detriment resulting from mutations. At present this is best estimated from the test for dominant skeletal mutations [3] supplemented by the test for cataracts [4]. However, when more comparisons between these tests and the specific-locus test have been made, it should be possible to determine whether the specific-locus test is reasonably predictive of the frequency of these defects.

The specific-locus test is capable of measuring only a limited number of chromosomal aberrations, primarily small deletions. With radiation, gene mutations and small deletions are believed to constitute the major part of the genetic hazard [28]. This may not be true for chemicals, however. Although many chemicals are believed to be primarily inducers of gene mutations, others may induce mainly chromosomal aberrations, and give negative results in the specific-locus test. The skeletal-mutation test, which should always follow a positive specific-locus test for a fuller evaluation of risk, does detect chromosomal aberrations that have a dominant deleterious effect (some translocations, for example). The skeletal-mutation test should, therefore, also be done on any chemicals that are negative in the specific-locus test, but which have been shown to induce chromosomal aberrations.

Perhaps the most striking feature of the chemical mutagenesis results obtained with the mouse specific-locus method is that they have revealed the great complexity inherent in mammalian germ-cell

responses. For example, the mutation frequencies in different germ-
cell stages vary greatly, and the pattern of relative frequencies
changes markedly with different chemicals. Furthermore, these
drastic changes can be seen among closely related chemicals such as
ENU and MNU. Thus, many of the generalizations made in the past on
the basis of results from simpler organisms are shown to be untenable. The hope of reaching general principles that would have predictive value still seems far from our reach. For some time to
come, it would appear that each potentially hazardous chemical will
have to be tested, regardless of its similarity with other chemicals
already tested.

One encouraging fact is that performing the specific-locus test
is a lot less onerous when the dose that is tolerated by mice is
much greater than the human exposure. In that case, it may be possible to exclude the likelihood of an effect greater than one that
is considered an acceptable risk. This was done with the anti-
schistosomal drug, hycanthone [29], and also with 5-chlorouracil
[30], a chlorinated organic in drinking water. In the latter case,
the human exposure was calculated, with 95% confidence, not to induce a mutation rate higher than 2% of the spontaneous rate. This
conclusion was reachable with a tiny experiment that produced 0
mutations in 314 offspring. Justifications for the interpolations
necessary for the estimate of risk are described in detail in a
paper on risk estimation [24]. Other cases where the dose to the
mouse can greatly exceed the human exposure, and where the specific-
locus test can consequently be relatively inexpensive, are probably numerous. Food additives, pesticide residues, some pollutants,
and other chemicals ingested in only trace amounts, are likely examples.

On the basis of the results obtained so far with the specific-
locus method, what can be said about the likely magnitude of the
general mutagenic risk from chemicals? The series of findings that
many chemicals that were potent mutagens in other test systems were
only weak mutagens or nonmutagens in mouse spermatogonia encouraged
the view that mammals had protective mechanisms against transmissible
damage from chemical mutagens. For chemicals that were known to
reach the testis in active form, the weak or zero response in spermatogonia suggested that the protective mechanism was a capacity of
these cells to repair genetic damage. These results engendered the
somewhat comfortable feeling that perhaps no chemical could break
through the mammalian body's defense barriers to produce more than
a moderate mutagenic effect in spermatogonia. This view was shat-
tered by the discovery of the incredible mutagenic effectiveness of
ENU in mouse spermatogonia, an effectiveness 24 times greater than
that of any other chemical tested. Concern is considerably reduced,
however, by the recent finding with fractionated doses of ENU that
the spermatogonia can apparently repair at least a major portion of
the genetic damage induced by ENU when the repair system is not over-

whelmed by a high dose. Assuming that one tenth of the mutation
frequency obtained with the fractionated 100 mg/kg exposure correctly estimates the effect of a single dose of 10 mg/kg, then this
dose would induce a mutation frequency of only 75% of the spontaneous mutation rate.

Several speculations can be made on the basis of this result.
The existence of a repair mechanism against the most effective mutagen known in the mouse, a chemical that mice have presumably never
encountered in their evolutionary history, suggests that this mechanism may have a more general action, perhaps against other ethyl-
ating agents. The additional finding that the chemically related MNU
has only weak mutagenic action at high doses in spermatogonial stem
cells suggests that repair of methylations may be even more effective. Of course, it remains to be tested whether repair of ENU-
induced damage might also be more effective at still lower doses.
The possibility of a threshold dose is not excluded, although it
should be kept in mind that this apparently does not occur with
radiation-induced mutations in spermatogonia.

We can conclude that, so far, we have found no convincing evidence that would lead us to expect that the general mutagenic risk
from chemicals will turn out to be of panic proportions. However,
we have examined only a few chemicals with the definitive tests in
mammalian germ cells that are really needed to evaluate risk. With
60,000 man-made chemicals in our environment in 1981, and with new
ones being added at the rate of approximately 1000 per year, we can
hardly afford to relax.

REFERENCES

1. W. L. Russell, X-ray-induced mutations in mice, Cold Spring
    Harbor Symposia on Quant. Biol., 16:327-336 (1951).
2. F. M. Johnson and Susan E. Lewis, Mutation-rate determinations
    based on electrophoretic analysis of laboratory mice, Mutation
    Res., 82:125-135 (1981).
3. Paul B. Selby, Dominant skeletal mutations: applications in
    mutagenicity testing and risk estimation, in: "Mutagenicity,
    New Horizons in Genetic Toxicology," John A. Heddle, ed., pp.
    385-406, Academic Press, New York (1982).
4. U. H. Ehling, Risk estimates based on germ-cell mutations in
    mice, in: "Environmental Mutagens and Carcinogens, Proc. 3rd.
    Int. Conf. on Environ. Mutagens," T. Sugimura, S. Kondo, and
    H. Takebe, ed., pp. 709-719, University of Tokyo Press, Tokyo
    (1982).
5. W. L. Russell, Mutagenesis in the mouse and its application to
    the estimation of the genetic hazards of radiation, in: "Ad-
    vances in Radiation Research: Biology and Medicine," J. F.
    Duplan and A. Chapiro, eds., pp. 323-334, Gordon and Breach,
    New York, London, Paris (1973).

6.   U. H. Ehling and W. L. Russell, Induction of specific locus mutations by alkyl methanesulfonates in male mice, Genetics, 61:s14-s15 (1969).

7.   W. L. Russell, Sandra W. Huff, and Dorma J. Gottlieb, The insignificant rate of induction of specific-locus mutations by five alkylating agents that produce high incidences of dominant lethality, ORNL Biol. Div. Annu. Prog. Rep., 4535, pp. 122-123 (1969).

8.   L. B. Russell, P. B. Selby, E. von Halle, W. Sheridan, and L. Valcovic, The mouse specific-locus test with agents other than radiation: interpretation of data and recommendations for future work, Mutation Res., 86:329-354 (1981).

9.   R. B. Cumming and Marva F. Walton, Fate and metabolism of some mutagenic alkylating agents in the mouse, I. Ethyl methanesulfonate and methyl methanesulfonate at sublethal dose in hybrid males, Mutation Res., 10:365-377 (1970).

10. Gary A. Sega, Molecular dosimetry of chemical mutagens, Measurement of molecular dose and DNA repair in mammalian germ cells, Mutation Res., 38:317-326 (1976).

11. R. B. Cumming, Dose and metabolic fate of hycanthone in mice, ORNL Biol. Div. Annu. Prog. Rep., 4817, pp. 128-129 (1972).

12. W. L. Russell, E. M. Kelly, P. R. Hunsicker, J. W. Bangham, S. C. Maddux, and E. L. Phipps, Specific-locus test shows ethylnitrosourea to be the most potent mutagen in the mouse, Proc. Natl. Acad. Sci. USA, 76, 5818-5819 (1979).

13. W. L. Russell, Factors affecting mutagenicity of ethylnitrosourea in the mouse specific-locus test and their bearing on risk estimation, in: "Environmental Mutagens and Carcinogens, Proc. 3rd Int. Conf. on Environ. Mutagens," T. Sugimura, S. Kondo, and H. Takebe, eds., pp. 59-70, Univ. Tokyo Press, Tokyo (1982).

14. W. L. Russell, P. R. Hunsicker, G. D. Raymer, M. H. Steele, K. F. Stelzner, and H. M. Thompson, Dose-response curve for specific-locus mutations induced by ethylnitrosourea in mouse spermatogonia, Proc. Natl. Acad. Sci. USA, 79:3589-3591 (1982).

15. W. L. Russell, P. R. Hunsicker, D. A. Carpenter, C. V. Cornett, and G. M. Guinn, Effect of dose fractionation on the induction of specific-locus mutations by ethylnitrosourea in mouse spermatogonia, Proc. Natl. Acad. Sci. USA, 79:3592-3593 (1982).

16. V. Carricarte and G. A. Sega, personal communication.

17. G. A. Sega, personal communication.

18. W. M. Generoso, personal communication.

19. G. A. Sega, K. W. Wolfe, and J. G. Owens, A comparison of the molecular action of an S_N1-type methylating agent, methyl nitrosourea and an S_N2-type methylating agent, methyl methanesulfonate, in the germ cells of male mice, Chem.-Biol. Interact., 33:253-269 (1981).

20. E. F. Oakberg, Radiation response of the testis, Proceedings of the Third International Congress of Endocrinology, 1070-1076 (1968).

21. F. M. Johnson and Susan E. Lewis, Electrophoretically detected germinal mutations induced in the mouse by ethylnitrosourea, Proc. Natl. Acad. Sci. USA, 78:3138-3141 (1981).
22. Susan E. Lewis and F. M. Johnson, The nature of electrophoretically-expressed mutations induced by ethylnitrosourea in the mouse, Environmental Mutagenesis, 4:338 (1982).
23. W. L. Russell, The role of mammals in the future of chemical mutagenesis research, Arch. Toxicol., 38:141-147 (1977).
24. W. L. Russell, Comments on mutagenesis risk estimation, Genetics, 92:s187-s194 (1979).
25. Liane B. Russell, P. B. Selby, E. von Halle, W. Sheridan, and L. Valcovic, Use of the mouse spot test in chemical mutagenesis: interpretation of past data and recommendations for future work, Mutation Res., 86:355-379 (1981).
26. L. B. Russell, Relevance of the mouse spot test as a genotoxicity indicator, in: "Indicators of Genotoxic Exposure in Man and Animals," Banbury Report 13 (1982), in press.
27. L. B. Russell, The mouse spot test as a predictor of heritable genetic damage and other endpoints. in: "Chemical Mutagens," Vol. 8, F. J. de Serres, ed., Plenum Press, New York, in press.
28. BEIR III Committee (Advisory Committee on the Biological Effects of Ionizing Radiation of the United States National Academy of Sciences) (1980). The Effects on Populations of Exposure to Low Levels of Ionizing Radiation, pp. 71-134, Nat. Acad. Press, Washington, D. C.
29. W. L. Russell and Elizabeth M. Kelly, Absence of mutagenic effect of hycanthone in mice, ORNL Biol. Div. Annu. Prog. Rep., 4817, pp. 113-114 (1972).
30. W. L. Russell and R. B. Cumming, An example of conditions that make the mouse specific-locus test highly efficient at low expense, ORNL Biol. Div. Annu. Prog. Rep., 5072, pp. 126-127 (1975).

THE DETECTION OF INDUCED RECESSIVE LETHAL

MUTATIONS IN MICE

William Sheridan

National Institute of Environmental Health Sciences
P. O. Box 12233
Research Triangle Park, North Carolina 27709

1. INTRODUCTION

In order to best assess the genetic risks to humans of environmental factors, investigations of their effects in mammalian systems will be necessary. For the forseeable future most such studies will be conducted on laboratory animals, and in particular, because of the large body of existing information on its biology and genetics, the laboratory mouse.

One class of mutations which has been given particular attention in the detection of chemically induced genetic damage in Drosophila melanogaster is recessive lethals [1]. Since, theoretically, there is no reason to believe that recessive lethal mutations can not occur or be induced at any locus, it is of importance that investigations of the amount of induction by chemicals of this type of mutation be made in the mouse.

During the 1960's methods were developed for detecting recessive lethals in mice [2-5], however, they were used exclusively in radiation studies. In this paper we will first present the methodology in some detail for those who are unfamiliar with it, and then the results of an experiment using a chemical mutagen in order to demonstrate the utility of the method for use in estimating the risks from chemical agents.

2. METHODOLOGY FOR DETECTING LETHALS

In essence the method is based on a rather simple genetic concept, that is that if an animal is heterozygous for a recessive gene, then it should be expected that the gene will be transmitted

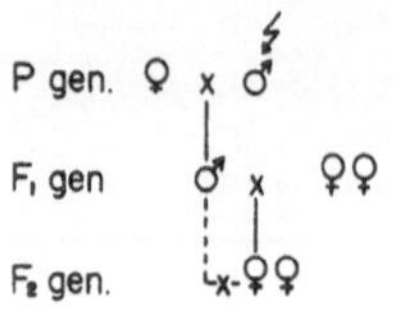

P gen. ♀ x ♂
         |
F₁ gen.   ♂ x ♀♀
              |
F₂ gen.   x-♀♀

Fig. 1.   Treatment and mating scheme for back-crosses to detect re-
          cessive lethals.

to half of the offspring, and in subsequent back-cross matings between parent and offspring half of the matings should allow the appearance of homozygous progeny who will express the recessive character. In the case of recessive lethal mutations, homozygosity will
lead to death of the animal and, therefore, increase in the frequency
of mortality is the primary indicator for the measurement of these
mutations.

        It is to be preferred that this type of study be conducted on
an inbred strain of mice. Important requirements for the strain are
that it have good fertility in both sexes, a relatively high litter
size, and a stable and moderate background level of spontaneous
intra-uterine fetal mortality. This latter consideration is particularly important since it is against this rate that increases in
mortality are judged for their statistical significance. The most
widely used strain in studies of recessive lethals has been the
CBA/Ca, or sublines thereof, since it was found to have the desirable characteristics mentioned above.

2.1.   Treatment and Breeding Schedule

        As illustrated in Fig. 1, male animals are usually the treated
sex in a recessive lethal induction experiment. There are several
reasons for such a choice: 1) males can be mated to several females
at a time to produce greater numbers of F_1 offspring per treated
animal, 2) particularly in irradiation studies treated females exhibit severe effects on reproductive capacity such as reduced litter-
size and early sterility [6] and physiological effects on reproduction caused by other mutagenic agents might be anticipated, 3) all
stages of germ-cell development are represented in adult male gonads
and by choosing an appropriate time interval between treatment and
mating these can be successively examined.

        Adult male mice (8-10 weeks old) are administered the suspect
mutagenic agent and then mated to 2 or 3 mature, untreated, virgin
females of the same strain. Thirty to forty treated males might be sufficient for a study if the rate of induction of dominant lethality
is not too high. In the latter case, adjustments in numbers must
be made. Matings occurring within the first weeks after treatment

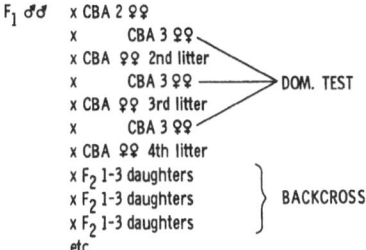

Fig. 2.   Interspersed outcross matings to detect dominant effects
prior to back-crosses.

will represent various post-meiotic germ cell stages, whereas matings
later than eight weeks after treatment, or following an induced tem-
porary sterile period, will correspond to germ cells treated at the
spermatogonial stem cell stage. If a sterile period is induced it
is of interest to measure its length as an indicator of the amount
of stem cell death which has been caused. This can conveniently be
done by caging the individual males with "sentinel" females and observing them for pregnancies [7].

At birth of the F_1 litters, date of birth, litter-size, sex of
young, and maternal and paternal origin are recorded. At weaning,
the litter-size and sex of young are again recorded, and male offspring are collected and identified by ear punching as to maternal
and paternal derivation. At least six to ten F_1 males from each
treated male should be kept for the study.

2.2.   Elimination of Translocation Bearers

At maturity, F_1 males are each given an individual code number
and mated to two females from the strain in order to produce F_2
daughters for the back-cross. However, the treatment of the parent
might also be suspected of causing chromosomal aberrations, as well
as gene mutations, which the F_1 animals may have inherited. In particular, inherited translocations will cause the F_1 male to give increased frequencies of death, which may range from 30-60%, among his
offspring in outcrosses [8]. Such high mortality rates, which will
stand during the entire reproductive life of the male [9], would of
course preclude determining any increase in the death rate attributable to recessive lethals in the back-cross matings. Such animals must be identified and removed from the study. A convenient
way of doing this is by utilizing the time between the matings to
produce F_2 litters and their birth, to mate each F_1 male in outcrosses to three strain females for one week. In the event that sufficient inbred females are not available, any good producing stock
of females, even random-bred, may be substituted. These outcross
females are then dissected in late pregnancy and the number of living
and dead implants are recorded (Fig. 2).

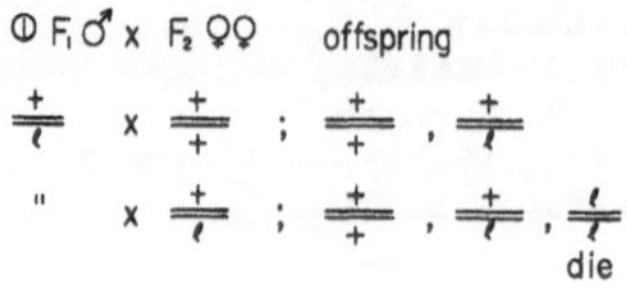

$$\textcircled{1}\ F_1 \male \times F_2 \female\female \quad \text{offspring}$$

$$\frac{+}{\ell} \times \frac{+}{+} \ ; \ \frac{+}{+} , \frac{+}{\ell}$$

$$" \quad \times \frac{+}{\ell} \ ; \ \frac{+}{+} , \frac{+}{\ell} , \frac{\ell}{\ell}$$
$$\underset{\text{die}}{}$$

Fig. 3.   Expected results from back-crosses involving an F_1 male
carrying a recessive lethal.

        While awaiting the outcome of the outcross mating, the males
are returned to their original mates to be present at the birth of
the F_2 litters. The females should be allowed to litter in separate
cages, the male being placed with the female deemed most likely to
litter first, in order that he might service her at the post-partum
estrus. If time permits the male can be moved to the cage of the
second female 24 hrs after the birth of the first female's litter.
Following this period of service, the male should once again be
outcrossed to three strain females which will be dissected in late
pregnancy. As before the male is returned to his mates for birth
of the second litter. Even a third outcross mating may be desirable.

        Any male showing high death rates in the outcrosses must be
suspected of being a translocation heterozygote and removed from the
test. If confirmation of his status is desired a cytogenetic analysis might be performed on the testes. Nevertheless, the male, his
mates and offspring are discarded. The outcross information gathered
on the other F_1 males is not wasted since it can be used at a later
stage to study possible dominant effects of any recessive lethals
uncovered.

2.3.   Back-Cross Matings

        The F_1 males are allowed to continue to produce F_2 litters un-
til a sufficient number of daughters (usually a minimum of 12) have
been obtained. The dates of birth, litter-size and sex of young at
birth and weaning are recorded. At weaning F_2 females are ear-
punched to identify paternal origin, and collected and held until
maturity. Male offspring are discarded.

        As the F_2 females mature (approx. 60 days old) they are back-
crossed to their sires in a succession of one week matings using up
to three daughters per week. The females are subsequently dissected
on the 17th day after the starting date of the mating and the uterine contents recorded under the code number of the male. As seen
in Fig. 3, an F_1 male carrying a lethal would produce two types of
daughters, those which have inherited the lethal and those without.
In the back-cross the females without the lethal should show the
intra-uterine mortality typical for the strain, whereas those with
the lethal should present an additional 25% death among the fetuses

due to homozygosity for the lethal. Individual litter sizes are usually too small for a determination of lethal heterozygote status to be made on single females, and therefore, a large number of implantations, comprising the young of many litters, are needed to find the total fetal death rate for the F_1 male.

2.3.1. Statistical Determination of Lethal Heterozygote Status of F_1 Males

Theoretically, the overall death-rate in back-crosses where the F_1 male is a lethal heterozygote should be 12.5% above the background frequency. Hamilton and Haseman [10] have developed a sophisticated statistic for analyzing recessive lethals which presents several advantages. The main advantage is that their procedure is sensitive to the patterns of fetal death rates across litters as well as to the total number of fetal deaths. Furthermore, they showed empirically that their technique was appropriate to a wide range of values for background frequencies of fetal death and total numbers of implants. This gives a distinct advantage in allowing a decision on lethal heterozygous status to be made in some cases on males which under earlier procedures would have yielded no decision. In addition to removing some males with more than 50 implants from an undecided classification, Hamilton and Haseman's procedure allows a decision to be made on many males which have given fewer than the 50 implants that Sheridan [5] and Lüning [11] considered to be the minimum number allowable. For those males, which even by the Hamilton and Haseman method cannot be classified as lethal free or lethal carrier, a probability can nevertheless be calculated as to which is the more likely state. This is of importance since it is easier, statistically speaking, to demonstrate a lethal carrier rather than an animal which is lethal free. Thus, knowing that the bulk of the unclassified males are probably lethal free will lend balance to the interpretation of the experimental results.

2.4. Controls

In any experiment it is desirable to have some sort of concomitant control. Certainly for a recessive lethal study it is necessary to have a sure knowledge of the current background rate of spontaneous fetal mortality in the inbred strain being used. Fortunately the rates are usually stable and generally well known from other studies for many of the major inbred strains. On the other hand, the rate of spontaneous mutation to recessive lethals or the numbers of heterozygotes for pre-existing recessive lethals within an inbred strain are most often unknown. In reviewing a large series of experimental investigations which included recessive lethal studies, Lüning and Searle [12] found that there was a marked variability in the control frequencies of recessive lethals between experiments despite the fact that they were conducted on the same inbred CBA strain. Ryman [13] examined this problem in a series of

computer simulations. He found that relatively large fluctuations
in the frequencies of recessive lethals present at any particular
generation of a strain, even with as large a number as 50 progenitor
pairs in the breeding nucleus, were to be expected.

This variability presents something of a dilemma. Often the
facilities and resources available to conduct recessive lethal experiments are limited, and, therefore, having to devote a large portion to maintain a control group would severely limit the space
which could be devoted to the experimental group. Lüning [11, 14]
proposed a solution to the dilemma. He suggested that by breeding
a large number of F_1 males from each treated P male one might detect the presence of pre-existing lethals by the appearance of several F_1 brothers carrying lethals. Allelism could be tested for by
matings of males to daughters of their brothers. Positive results
in such tests would not be proof of the spontaneous origins of the
mutation since there would remain the possibility, albeit an unlikely one, that it represented a cluster if derived from treated
spermatogonia. Nevertheless, these cases could be subtracted from
the results, and the remainder of the mutations could be considered
to represent a mix of new spontaneous or induced lethals. Thus, in
a way one would have an inbuilt control in the treated material.

3. CHEMICAL MUTAGEN INDUCTION OF RECESSIVE
   LETHAL MUTATIONS

Despite Haldane's plea in his 1956 paper [15] that experiments
on the mutagenic effects of chemicals such as food additives and
drugs be instituted, the emphasis in recessive lethal induction
studies continued for many years to be wholly on irradiation effects.
In order to rectify this lack of information regarding the potential
lethal inducing effects of chemicals, we initiated an investigation
with the chemical triethylenemelamine (TEM). Because this was to be
a first effort to induce and detect recessive lethals with the back-
cross method using a chemical, one was chosen which had already been
demonstrated to have strong effects in causing genetic damage such
as dominant lethality [16, 17] chromosomal abnormalities [18], and
increased frequencies of aberrant sperm [19].

In order to have some comparability with earlier irradiation
studies, adult male mice of the CBA/Ca strain were utilized, however, since the other genetic effects were primarily observed on
offspring derived from treatment of postmeiotic stages of germ cell
development, the males were mated to CBA females immediately after
treatment for the production of F_1 males. The mutagenic treatment
was administered as an intraperitoneal injection of 0.2 mg/kg body
weight of TEM dissolved in Hanks Balanced Salt Solution. F_1 males
were collected from matings during a two week period after treatment, thus representing sperm and late spermatid stages.

Table 1. Classification of F_1 Males from TEM-Treated Parents

CLASS	NUMBER OF MALES	FREQUENCY (%)
TRANSLOCATION	21	16.5
STERILE	8	6.3
LETHAL BEARER	29	22.8
NON-LETHAL	26	20.5
NO DECISION	18	14.2
PROBABLE LETHAL	2	
PROBABLE NON-LETHAL	15	
INSUFFICIENT OFFSPRING	25	19.7

At maturity the F_1 males were examined in back-cross studies as described above. The results are shown in Table 1. Following removal of translocation heterozygotes and sterile males (most of which were also translocation bearers) from the test, the remainder of the males were analyzed using the method of Hamilton and Haseman [10]. For 25 of the males the amount of back-cross data were too small for a variety of reasons to make a statistical analysis meaningful. The interesting feature is the high frequency of recessive lethal bearers detected. Of the 34 P males represented, lethals were found among the F_1 males of 22. Six of the P males had more than one lethal bearing son, however, allelism tests within these families could not demonstrate that the lethals were allelic. Therefore, all the lethals should be considered as individual events. At this time, we have demonstrated that 11 of the lethals were true mutations since they were transmitted to approximately half of the F_2 sons tested. The results suggest that in some of these families, for example No. 103 and 113, more than one lethal may have been present (Table 2). One may conclude from these experiments that recessive lethal mutations may be induced at relatively high levels by chemical mutagens, and detected by the back-cross method.

Table 2.   Test of F_2 Sons from F_1 Lethal Males

FAMILY NUMBER	SONS	LETHAL	NO DECISION	LETHAL FREE
10	10	6	1	3
35	10	4	2	4
39	10	5	1	4
44	10	5	1	4
73	10	5	-	5
78	8	4	1	3
95	8	6	1	1
103	9	7	2	-
113	7	6	-	1
116	4	1	1	2
117	13	7	1	5

4.   DISCUSSION AND CONCLUSIONS

The methodology presented here is unique in that it allows detection of induced mutations in practically the entire genome in
one test.   In addition, one also can determine the frequency of induced translocations and dominant mutations as well as study whether
or not the lethal mutations have dominant effects in heterozygotes.
(Since the latter topic is to be discussed later in this meeting,
we have not gone into detail on the subject.)   Slight modifications
of the procedure would allow, investigations of recessive visible
mutations.

Earlier we heard Dr. Johnson [20] (these proceedings) describe
briefly a recessive lethal study that he conducted which led him to
conclude that the number of loci capable of mutating to recessive
lethals might be very small; perhaps on the order of 100–200.   It
is generally estimated that there are 20,000–25,000 gene loci in
the mouse genome.   As mentioned earlier there is no reason to believe that recessive lethals cannot occur or be induced at a substantial number, if not all, of these loci.   It is conceivable that

his results were due to the rather complex hybrid cross involved, but more likely they can be attributed to the inadequate number of animals being studied leading to a spurious result.

Using a set of biochemical markers, Soares [21] studied the effects of doses of triethylenemelamine somewhat comparable to that in the presence experiment. In a total of 3902 progeny from treated post meiotic stages he found two heritable mutations which were presumably induced. Using the seven visible locus system, Cattanach [22] treated post meiotic stages with 0.2 mg/kg of TEM and detected 3 mutants among 1701 offspring. These studies may be compared to the present experiments where 29 confirmed lethal bearers were found among 127 F_1 males under study.

We can conclude that recessive lethal mutations represent an important class of genetic damage to be found following mutagenic treatment, whether this be by physical or chemical agents. The recognition of this fact by the United Nations Scientific Committee on the Effects of Atomic Radiation in their 1977 report to the General Assembly [23] led them to recommend that "it would appear preferable to use the direct estimate based on recessive lethals in hazard evaluations." The utility of the back-cross methodology in the detection of recessive lethals has been adequately demonstrated by the experiments described in this chapter. It is to be hoped that greater emphasis will be given in the future to the measurement of recessive lethal mutations.

REFERENCES

1.   E. Vogel and F. H. Sobels, The function of Drosophila in genetic toxicology testing, in: "Chemical Mutagens: Principles and Methods for their Detection," A. Hollaender, ed., Vol. 4, pp. 93-142, Plenum Press, New York (1976).

2.   T. C. Carter and Mary F. Lyon, An attempt to estimate the induction by x-rays of recessive lethal and visible mutations in mice, Genet. Res., 2:296-305 (1961).

3.   Mary F. Lyon, Rita J. S. Phillips, and A. G. Searle, The overall rates of dominant and recessive lethal and visible mutation induced by spermatogonial x-irradiation of mice, Genet. Res., 5:448-467 (1964).

4.   K. G. Lüning, William Sheridan, and H. Frölén, Genetic effects of supralethal x-ray treatment of male mice, Mutat. Res., 2: 60-66 (1965).

5.   William Sheridan, The effects of acute single or fractionated x-ray treatment on mouse spermatogonia, Mutat. Res., 5:163-172 (1968).

6.   William Sheridan, The radiosensitivity of offspring of an irradiated mouse population, I. 'Effects on the redproductive capacity of irradiated female offspring, Mutat. Res., 4:675-781 (1967).

7. William Sheridan, The effects of the time interval in fraction-
    ated x-ray treatment of mouse spermatogonia, Mutat. Res., 13:
    163-169 (1971).

8. H. W. Michelmann and W. Sheridan, Effects of transmitted trans-
    locations in mice, Abstracts of 9th annual meeting of Environ-
    mental Mutagen Society, p. 47 (1978).

9. William Sheridan, The induction by x-irradiation of dominant
    lethal mutations in spermatogonia of mice, Mutat. Res., 2:67-
    74 (1965).

10. M. A. Hamilton and J. K. Haseman, Statistical tests for re-
    cessive lethal-carriers, Mutat. Res., 64:269-278 (1979).

11. K. G. Lüning, Test of recessive lethals in the mouse, Mutat.
    Res., 27:357-366 (1975).

12. K. G. Lüning and A. G. Searle, Estimates of the genetic risks
    from ionizing irradiation, Mutat. Res., 12:291-304 (1971).

13. Nils Ryman, The frequency of recessive lethal heterozygotes
    among individuals obtained from inbred strains, A random-number
    simulation study, Mutat. Res., 42:363-372 (1977).

14. K. G. Lüning, Testing for recessive lethals in mice, Mutat.
    Res., 11:125-132 (1971).

15. J. B. S. Haldane, The detection of autosomal lethals in mice
    by mutagenic agents, J. Genet., 54:327-342 (1956).

16. E. R. Soares and W. Sheridan, Triethylenemelamine induced dom-
    inant lethals in mice - comparisons of oral versus intra-
    peritoneal injection, Mutat. Res., 43:247-254 (1977).

17. K. Bürki and William Sheridan, Expression of TEM-induced damage
    to postmeiotic stages of spermatogenesis of the mouse during
    early embryogenesis, I. Investigations with in vitro embryo
    culture, Mutat. Res., 49:259-268 (1978).

18. K. Bürki and William Sheridan, Expression of TEM-induced dam-
    age to postmeiotic stages of spermatogenesis of the mouse dur-
    ing early embryogenesis, II. Cytological investigations, Mutat.
    Res., 52:107-115 (1978).

19. E. R. Soares, William Sheridan, J. K. Haseman, and M. Segall,
    Increased frequencies of aberrant sperm as indicators of muta-
    genic damage in mice, Mutat. Res., 64:27-35 (1979).

20. F. M. Johnson and Susan Lewis, The detection of ENU-induced
    mutations in mice by electrophoresis and the problem of evaluat-
    ing the mutation rate increase (these proceedings).

21. E. R. Soares, TEM-induced gene mutations at enzyme loci in the
    mouse, Environ. Mutagen, 1:19-25 (1979).

22. B. M. Cattanach, Induction of paternal sex-chromosome losses
    and deletions and of autosomal gene mutations by the treatment
    of mouse postmeiotic germ cells with triethylenemelamine, Mutat.
    Res., 4:73-82 (1967).

23. UNSCEAR (United Nations Scientific Committee on the Effects
    of Atomic Radiation) report to the General Assembly: Sources
    and Effects of Ionizing Radiation, United Nations, New York
    (1977).

USING INVERSIONS TO DETECT AND STUDY RECESSIVE

LETHALS AND DETRIMENTALS IN MICE

Thomas H. Roderick

The Jackson Laboratory
Bar Harbor
Maine 04609

1. INTRODUCTION

Assays for induced mutations utilizing lower organisms, such as the Ames test, are most useful for rapid efficient screening of many potential mutagens. Test systems involving whole animals are necessary adjuncts to the simpler systems for at least four reasons.

Firstly, simpler tests do not always predict the mutagenic outcome in the whole animal. Secondly, there is a broad spectrum of mutagenic effects that needs to be assayed including point mutations, recessive lethals and detrimentals, deletions and other chromosomal rearrangements, non-disjunction, and traits of complex etiology either single or multifactorial. Not all mutagens have the same effect on this spectrum. Thirdly, as W. L. Russell and his colleagues have repeatedly demonstrated, whole animal tests are essential for assessing the mutagenic effects on different stages of oogenesis and spermatogenesis. These stages vary in their sensitivity and in repair and selection processes that follow them. Fourthly, whole animal systems that permit assessment of mutations are also needed for study of the burden of those mutagenic events so that a better estimate of the real cost of environmental mutagenesis can be made.

There are several well-validated whole-animal tests that cover the mutagenic spectrum as presently defined. These include the specific locus test, dominant lethal test, heritable translocation test, spot test, and X-chromosome loss test [1, 2]. Also Selby and Selby [3] have substantially developed an early method of Ehling to assess mutations that cause skeletal anomalies in mice. The importance of this test is that a high frequency of mutations can be assessed and

that these mutations may represent the category of complex irregularly inherited genetic traits encompassing the greatest portion of
genetic burden in man [4]. Kratochvilova and Ehling [5] have demonstrated the usefulness of a mutational assay for cataracts that may
be an important adjunct to the skeletal mutation assay.

The specific locus test is one of the most reliable methods for
measuring heritable mutations in mammals. The test developed by
Russell [6] has given us extensive information on mutational events
in mammals and is one to which we have looked for providing standards
and baselines for further mutagen testing in mammals. It has also
played the predominant role in predicting mutagenicity and burden in
humans. The test has limitations in that it is one of the more expensive whole-animal tests. Furthermore, as is true for most tests,
the information that can be derived from it about the total spectrum
of possible single gene mutational events is limited.

More efficient and complementary methodologies are greatly
needed. We need to know the kind and extent of genetic loads induced in specific regions of chromosomes where map distances can be
estimated precisely. We need a method which does not have possible
bias because of the specific loci studied and which does not have
the uncertainties involved in estimating lethals from an increase in
uterine deaths alone, the essence of the dominant and recessive
lethal tests. Finally, we need a method which will permit identification and easy maintenance of specific mammalian recessive lethals
and detrimentals so that they can be propagated and analyzed. Systems using chromosomal inversions in combination with each other and
in combination with Robertsonian translocations in mice provide some
of the necessary attributes for achieving an improved and complementary methodology.

2.   INVERSIONS IN THE MOUSE

2.1.   Inversions Induced and Their Major Properties

There are two major types of inversions as shown in Fig. 1.
Pericentric inversions are defined as those that involve the centromere. As such, they can often be recognized cytologically by a
shift in the centromeric position. To be detected in this manner
the inversion must be sufficiently large and have obvious asymmetry
in length on either side of the centromere. Chromosomal banding
techniques involving more effort permit identification of pericentric
inversions without an obvious shift in centromeric position. Reduced litter size can be expected in heterozygotes because of duplications and deficiencies in gametes resulting from crossing-over
within the inverted segment. Inducing pericentric inversions in the
mouse theoretically should be relatively difficult because of the
telomeric nature of all chromosomes of the standard Mus musculus
karyotype. The low probability of induction results from the essen-

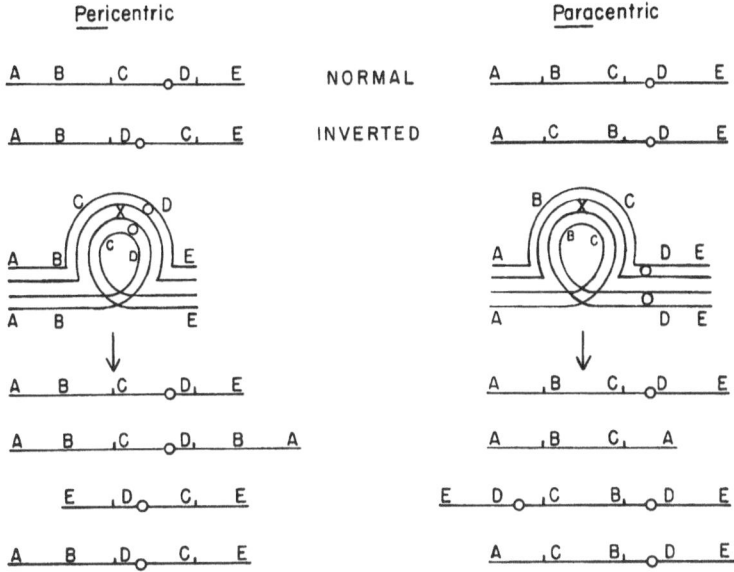

Fig. 1.   Pericentric and paracentric inversions. A two-strand crossover within the loop configuration prior to the first meiotic division leads to four possible gametic genotypes shown
below the arrows.

tially absent chromatin target on one side of the centromere. Producing pericentric inversions in Robertsonian chromosomeres of Mus
musculus was possible however and there have been naturally occurring cases as well [7, 8].

In contrast, paracentric inversions are defined as those involving chromosomal segments on either side but not including the
centromere (Fig. 1). This type of inversion lacks the cytologically
observed shift in centromeric position but has another property of
producing first meiotic anaphase bridges. Crossing-over within the
inverted segment produces a dicentric chromatid. A bridge results
when centromeres of the dicentric chromatid go to opposite poles in
first anaphase. The resulting deficiencies and duplications in gametes should impair fertility in heterozygotes for this type of inversion also. For either type of inversion, products of recombination within the inverted segment will not result in functioning gametes or viable offspring; in meiosis the product will be lost because of the anaphase bridge or in the zygote because of significant
aneuploidy of the gamete. The value for mutagenesis research is
that the only surviving or functioning gametes produced by inversion
heterozygotes will be those with no disturbance of genetic integrity
through meiotic recombination. The segment of the inversion can remain intact through many generations of propagation and manipulation

if appropriate genetic crosses are made. Another consequence is that
the inverted segment is operationally "allelic" with any marker on
the normal homolog and vice versa.

We have induced and detected paracentric inversions by looking
for first meiotic anaphase bridges in one testis of sons of males
treated with radiation or a chemical mutagen [9]. The frequency of
naturally occurring bridges at this division is about 4%, making it
necessary to search for animals which display a frequency significantly in excess of 4%. We also needed to induce and study pericentric inversions and thus established another method to use simultaneously. The chromosomes of the mouse are all telocentric providing little probability of inducing pericentric inversions, since
breaks would need to occur on both sides of the centromere. Therefore, we have given the mutagens triethylene melamine (TEM) or
cesium radiation to mice carrying at least three Robertsonian chromosomes and have examined the banded karyotypes of sons and daughters
for presence of pericentric inversions in these chromosomes. The use
of these stocks also permitted recovering paracentric inversions in
Robertsonian chromosomes. If fertile and viable, an ideal mutagen
testing stock would contain a Robertsonian chromosome with a pericentric inversion plus paracentric inversions in both arms thereby
permitting inhibition of crossing-over in as much of the genome as
possible in one segregating element.

Initially in our attempt to induce chromosomal inversions we
used between 800 and 900 R of X-irradiation. We found 15 inversions
in 2236 screened animals, a relative frequency of 0.0067. We
changed to triethylene melamine (TEM) because it was known to be an
effective breaker of chromosomes. Our initial experience with it
confirmed the experience of others, and, furthermore, suggested it
was more effective in producing inversions and other transmittable
chromosome aberrations in relation to its other toxic effects, such
as lowered reproductive performance in mutagen-treated males. Our
inversion induction rate with TEM was:

for the years 1971-1975: 17 inversions in 2608 screened ani-
                           mals or 0.0065;

for the years 1976-1980: 7 inversions in 3079 screened ani-
                           mals or 0.0023.

TEM has a finite and relatively short shelf life which made estimations of the given dose inaccurate and probably accounts for the
difference in rates above. Different batches of TEM from the manu-
facturer seemed to have different effectiveness. Although we used
litter size of mutagen-treated males as a relative indicator of the
biological effectiveness of TEM, this characteristic, too, seemed
to be highly variable under TEM treatment. Finally, the need to
continue the inversion induction and, more importantly, to induce

lethals in the inversion test system, forced us to turn to a more
reliable and measureable mutagen, 1000 R gamma rays from a cesium
source. The induction using this source was 2 inversions in 591
animals screened or 0.0034.

The general method was useful in inducing inversions. Table 1
shows all the inversions and presumptive inversions produced in our
laboratory to date. The 28 useful inversions are well spread around
the genome. In addition to these a large inversion on the X-chromosome was induced at Harwell, England [10-13]. Harwell also has a
smaller X-inversion [14]. With these, sex-linked lethals may be detected. The Harwell autosomal inversion on Chromosome 2 is very

Table 1. Useful Inversions Induced in the Mouse

Animal number	Inversion symbol[a]	Anaphase bridge frequency		Induced by[b]	Induced on[c]
618	In(1)1Rk	.340	.008	900 R (X)	DBA/2J
816	In(5)2Rk	.192	.010	850 R (X)	DBA/2J
2916	In(2)5Rk	.345	.014	900 R (X)	DBA/2J
3563	In(10)6Rk	.262	.018	900 R (X)	DBA/2J
5204	In(5)9Rk	.737	.020	TEM .2	C3D2F$_1$
6845	In(3)11Rk	.294	.055	TEM .2	C57BL/6J
6857	In(1)12Rk	.228	.047	TEM .2	C57BL/6J
7877	In(7)13Rk	.447	.040	TEM .2	C3D2F$_1$
8677	In(8)14Rk	.347	.043	TEM .2	DBA/2J
11323	In(10)17Rk	.648	.040	TEM .3	DBA/2J
11449	In(15)18Rk	.366	.024	TEM .3	DBA/2J
11675	In(2)19Rk	.609	.035	TEM .3	DBA/2J
11684	In(11)20Rk	.473	.037	TEM .3	DBA/2J
11865	In(15)21Rk	.469	.032	TEM .4	DBA/2J
11951	In(14)22Rk	.663	.036	TEM .4	DBA/2J
12174	In(1)24Rk	.728	.046	TEM .4	DBA/2J
12452	In(12)25Rk	.459	.048	TEM .4	DBA/2J
12953	In(9)26Rk	.237	.028	TEM .4	DBA/2J
16647	In(4)28Rk	.559	.085	TEM .3	POS (AxD)

Table 1 (continued)

Animal number	Inversion symbol[a]	Anaphase bridge frequency		Induced by[b]	Induced on[c]
21716	In(5)30Rk	.270	.047	TEM .4	POS (F)
21923	In(13)31Rk	.700	.102	TEM .3	POS (F)
25022	In(4)32Rk	.211	.029	TEM .5	POS (P)
25086	In(5)33Rk	.594	.059	Spontaneous	POS (M)
25094	In(8)34Rk	.374	.045	TEM .5	POS (P)
25382	In(15)35Rk	.107	.019	TEM .5	DBA/2J
29646	In(14)36Rk	.426	.042	1000 R (Cs)	DBA/2J
29726	In(19)37Rk	.591	.105	1000 R (Cs)	DBA/2J
31189	In(1)38Rk	.531	.088	1000 R (Cs)	DBA/2J

[a]Inversion symbols: In refers to inversion. The number in parenthesis is the chromosome in which the inversion occurs. The number following parenthesis or alone is the number of the inversion found in our laboratory. Rk is the symbol of our laboratory.

[b](X) indicates induced by X-rays delivered by a GE Maxitron 250 KVP at 20 MA, 1mm Cu, 1mm Al filtration. (Cs) indicates induced by a Shepherd Mark I Cesium 137 irradiator. The numbers by TEM and EMS indicate dosage in mg/kg body weight.

[c]Symbols represent strains, stocks, or hybrids. POS are stocks carrying Robertsonian chromosomes.

similar to our inversion In(2)5Rk [15]. Except for Chromosome 6, the chromosomes lacking induced inversions are among the smallest four. This is not surprising when one considers that the probability of inducing an inversion is probably correlated with chromosomal target size.

The cause of the bridges that occur naturally about 4% of the time is not known, but it is necessary to distinguish them quantitatively from a higher frequency of bridges that indicates a paracentric inversion [9]. Other indicators of inversions that we do not usually see in normal animals are broken anaphase bridges and telophase bridges, neither of which are included in our count for anaphase bridges. Although anaphase bridges are the main indicator, we used these other characteristics as well in searching for new inversions. These additional characteristics give us more information (Table 2). Numerous associated broken bridges suggest that the inversion is relatively near the centromere, whereas numerous telophase bridges suggest that the position of the inversion is much more distal. The frequency of anaphase bridges is a first indication of the length of the inversion itself. As we continue to define the inversions the other associated idiosyncrasies may become understandable and useful.

As we breed a prospective new paracentric inversion heterozygote we simultaneously backcross him to one of the parental strains. Our usual procedure is to give the mutagen to a DBA/2J male, cross him to a C57BL/6J female, screen for inversions in the F_1s and then backcross presumptive carriers to C57BL/6J. This cross has enabled us to look for linkage of a new inversion with many isozyme loci [9]. Originally we backcrossed to DBA/2J but the cross to C57BL/6J gives two to three times the litter size of the DBA/2J cross, and thus speeds the linkage testing process. The number of isozyme differences is the same and the two additional coat color markers in the DBA/2J cross, dilute and brown, are only 7 cM from Mod-1 and 6 cM from Mup-1, respectively. With the Robertsonian F_1s we also backcross to C57BL/6J. The coat color locus albino lies between Gpi-1 and Hbb, both of which are segregating in the F_1. With the number of loci we can utilize (Table 3) we can usually find the linkage of the inversion within three generations.

Our characterization of these loci necessitates use of several techniques including vertical starch and cellulose acetate electrophoresis, immunoelectrophoresis modified from the technique of Minna et al. [16] and isoelectric focusing [17]. These techniques permit characterization of many loci presently known and are sufficiently versatile to be used broadly for loci yet to be discovered. These loci have served to map in good detail the genetic segments of induced inversions. More importantly, they will be useful in determining the location of induced lethals as they are ascertained utilizing the inversion test system and, perhaps, in determining the loci contributing to lethality.

With M. T. Davisson we also looked at the karyotype cytologically, as soon as the presumptive inversion was successfully established, to see if the inverted segment could be observed. Occasionally it has also been useful to look at synaptonemal complexes of the inversion heterozygote to determine the approximate sizes of the inversion and the chromosome in which it is located. This work was done in collaboration with M. J. Moses and P. A. Poorman. When these cytological methods are successful (and they have been on several occasions), the genetic tests can be much more efficiently planned.

Table 2. Some Meiotic Characteristics of Paracentric Inversions

Inv.	Percent bridges	Telophase bridges	Bridge length[a]	Bridge descript.	Fragments present	Presence of broken bridges
471	38	Few	M	Thin, lumpy	Yes	Many
In-1	34	Many	ML	Lumpy	Yes	Few
In-2	19	None	S	Smooth	Few	Some
2414	34	Few	SM	Smooth	No	Some
In-3	14	None	S	Smooth	Few	Few
In-4	19	Few	ML	Thin	Yes	Few
In-5	34	Many	M	Thin, lumpy	Yes	Some
In-6	26	Few	SM	Lumpy	Yes	Many
In-7	15	None	SM	Solid	Few	None
In-8	18	None	SM	Thin	No	Few
In-9	73	Many	M	Solid, few lumps	Yes	Some
In-10	74	Many	SM	Lumpy	Yes	Few
In-11	29	Few	SM	Lumpy	No	None
In-12	22	Few	M	Lumpy	No	None
In-13	45	Few	SM	Lumpy	No	Few
In-14	35	Few	M	Few lumps	Yes	Some
In-15	69	Many	S	Lumpy	Few	Many
In-16	46	---	-	---	---	---
In-17	65	Many	S	Lumpy	Few	None
In-18	37	Few	M	Solid	No	None

Inv.	Percent bridges	Telophase bridges	Bridge length[a]	Bridge descript.	Fragments present	Presence of broken bridges
In-19	61	Many	SM	Thin	Yes	Few
In-20	47	Few	M	Thin, lumpy	Yes	Some
In-21	47	Many	S	Thin, lumpy	Few	Few
In-22	66	Many	S	Lumpy	Few	Some
In-23	36	Few	SM	Lumpy	Few	Few
In-24	73	Many	S	Thin, lumpy	Yes	Few
In-25	46	Few	SM	Few lumps	Yes	Some
In-26	24	Many	SM	Few lumps	No	None
In-27	50	Few	M	Thin, lumpy	Few	None
In-28	56	Many	M	Lumpy	Yes	Some
In-30	27	Few	SM	Lumpy	No	Some
In-31	70	Few	SM	Lumpy	Yes	Some
In-32	21	Few	S	Lumpy	Few	Some
In-33	59	Few	SM	Lumpy	Yes	Few
In-34	37	Few	SM	Few lumps	Few	Few
In-35	11	Few	very S	Smooth	Few	None
In-36	38	Few	S	Thin	Yes	Few
In-37	62	Many	SM	Thin	Yes	Few

[a] S = short; M = medium; L = long.

Table 3.   List of Genes Used in Locating
Inversions

Gene	Chromosome	Gene	Chromosome
a	2	* Gpt-1	15
* Acf-1	1	Gr-1	8
Alb-1	5	* Hba	11
Amy-1	3	* Hbb	7
Amy-2	3	* Idh-1	1
Apk	10	* Igh-4	12
Apl	17	* Lap-1	9
* b	4	Ldr-1	6
c	7	* Mod-1	9
C-3	17	Mod-2	7
Car-1	3	Mor-1	5
* Car-2	3	Mpi-1	9
* d	9	* Mup-1	4
* Es-1	8	Np-1	14
Es-2	8	* Pep-3	1
* Es-3	11	* Pgm-1	5
Es-8	7	Pgm-2	4
* Es-10	14	* Pre-1	12
Es-13	9	rd	5
Gdc-1	15	Sdh-1	2
Got-1	19	* Tam-1	7
Got-2	8		
Gpd-1	4		
* Gpi-1	7		

*Loci with allelic differences between C57BL/6J and DBA/2J.

An induced pericentric inversion, In(11.13LS)29Rk, involved about 34% of Robertsonian chromosome Rb(11.13)4Bnr as measured both on Giemsa-banded mitotic chromosomes and meiotic synaptonemal complexes [7]. Both carriers showed extremely reduced fertility. The original female carrier had only three offspring, passing the inversion on to one son. This son was sterile. From these limited data and the report by Nijhoff [8], it appears that pericentric inversions reduce fertility. This is not true from our observations on paracentric inversions. The difference may be related to the fact that crossing-over within the inverted segment of a pericentric inversion does not produce a chromosomal bridge at first meiotic anaphase as it does with a paracentric inversion.

Males heterozygous for paracentric inversions are usually fully viable and fertile permitting easy maintenance of inversions. Some interesting exceptions are:

In(2)5Rk is homozygous lethal, although heterozygotes breed well.

In(10)17Rk homozygotes are also homozygous for pygmy (pg/pg), indicating a simultaneous induction of the recessive mutation with the inversion.

In(11)20Rk homozygotes breed normally, but heterozygotes have reduced litter size. No translocation is apparent in these animals.

In(15)21Rk heterozygotes seem partially sterile much like translocation heterozygotes. This reduced fertility is under study. We are having difficulty making this inversion homozygous and suspect it may be homozyogous lethal.

In(1)24Rk was induced simultaneously with a translocation involving Chromosome 13. The Chr 1 translocation breakpoint is within the inverted segment.

In(12)25Rk is homozygous lethal.

In(4)28Rk is homozygous lethal.

Two inversions are located in Robertsonian chromosomes. In(4)32Rk was induced in the Chr arm of Robertsonian Rb(4.6)2Bnr. The inversion has crossed out of the Robertsonian and we have two separate stocks: one with the inversion in the Robertsonian and one with the inversion by itself. Heterozygotes with the inversion in the Robertsonian do not breed well. But, when the inversion is by itself, inversion heterozygotes do breed well.

In(5)33Rk arose spontaneously in the Chr 5 arm of Rb(5.15)3Bnr. Approximately one-half of the heterozygotes are sterile and one-half breed well.

In (8)34Rk may be homozygous lethal, as producing homozygotes has been impossible to date.

In(15)35Rk has a very low bridge count and is probably therefore a very small inversion. We normally would not have kept it but for the fact that from a synaptonemal complex of a heterozygous male, Moses and Poorman suggested it was on Chr 15. This provided the opportunity to determine its genetic linkage quickly and to maintain the stock.

It is surprising that except for the pygmy phenotype of animals homozygous for In(10)17Rk, and the few homozygous lethal inversions, males and females heterozygous or homozygous for most of the inversions show no outward phenotypic deviation from normal littermates. We do not seem to be observing position effects characteristic of inversions in Drosophila.

Both the In-32 and In-33 stocks in which the inversion is within a Robertsonian exhibit breeding problems. The four other stocks derived from Robertsonian-carrying males breed well although heterozygous Robertsonians are still segregating in both the In-28 and In-34 stocks. This observation and the fact that most of our heterozygous Robertsonian stocks breed well suggest that the fertility impairment is not the result of heterozygous Robertsonians alone. It may be due to the combination of an inversion and a Robertsonian, both heterozygous in the same chromosome, or to some other factor(s) not yet identified. In combining Robertsonians and inversions, we do find that some combinations breed better than others.

2.2. Inversion Systems

After inducing and characterizing inversions, we can construct balanced lethal test systems much like those that have had such practical advantages in the study of genetics in Drosophila. In(1)1Rk is the most suitable inversion so far produced for this purpose, because (1) it is long, (2) it covers a chromosomal segment that has good genetic markers, and (3) males and females carrying the inversion either as heterozygotes or homozygotes are fully viable and fertile. Idh-1 (isocitrate dehydrogenase-1), Sp (splotch), ln (leaden), and Pep-3 (peptidase-3) are the most useful loci in the inverted segment. Splotch is a viable dominant spotting marker that is lethal when homozygous. There also appears to be a slight reduction in viability of Sp/+ heterozygotes. The mutant allele is nearly fully penetrant but the expression is variable. Most animals exhibit a large white belly spot, although, in a few, only the tips of the feet are white. In only a few of the several hundred we have observed have we had trouble distinguishing a Sp/+ animals from the wild-type +/+. Leaden is a fully penetrant recessive coat-color marker with no known effect on fertility or viability. Loci Idh-1 and Pep-3 are polymorphic in feral populations and inbred strains

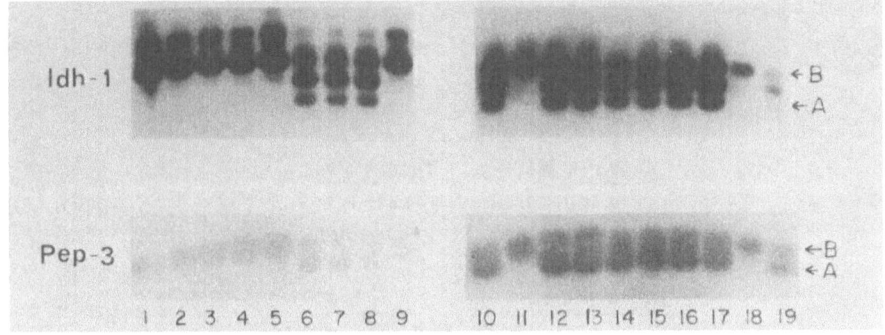

Fig. 2.   Two polymorphic loci located with In(1)1Rk. These two
          loci visualized by starch gel electrophoresis normally
          show 22% recombination. In these 19 offspring from the
          cross

$$\frac{In\ (Idh\text{-}1^b\ Pep\text{-}3^b)}{+\ (Idh\text{-}1^a\ Pep\text{-}3^a)} \times \frac{In\ (Idh\text{-}1^b\ Pep\text{-}3^b)}{In\ (Idh\text{-}1^b\ Pep\text{-}3^b)}$$

no recombination was found. Samples 1-17 are from kidney;
samples 18-19 are from testis.

and are fully penetrant codominant loci (Fig. 2). Both are determined by starch or cellulose acetate gel electrophoresis. Neither
has any detectable effect on fertility or viability.

Figure 3 describes the scheme in this test system. Mutagen is
given to homozygous inversion males who may be mated at any time
thereafter so that the appropriate germ cell stage is studied. Mates
are from a balanced splotch-leaden stock that is easily maintained.
All animals in generation-1 are presumptive carriers of one or more
unique mutations and each is the unique progenitor whose genotype is
being assayed for a mutation or mutations. Generation-1 males or
females heterozygous for splotch (Sp/+) are mated again to the
splotch-leaden stock. Animals noted as the alternate genotype can
also be used with only slight modification of the breeding scheme.
There are four genotypic classes in generation-2 and each is un-
ambiguously defined by its phenotype. Splotch animals are mated
with their splotch parent or mated among themselves to produce the
test generation. If a recessive lethal has been induced within the
inversion, then no wild type or test class individuals will appear.
If a deleterious recessive has been induced, then all the test class
animals should appear affected. If there is a partial reduction of
the test class or if part of the test class animals have a common
new phenotype, then a first surmise is that a recessive lethal or
recessive detrimental has been induced closely linked to the inverted segment.

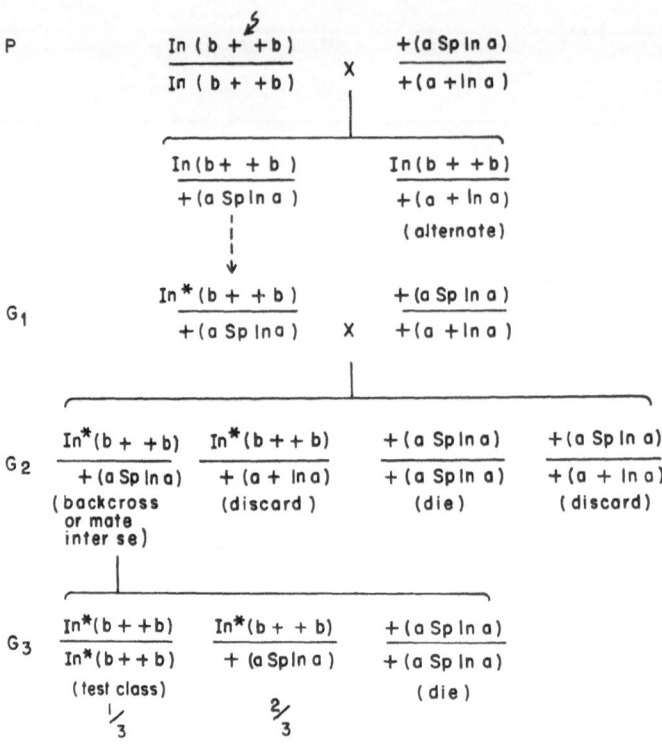

Fig. 3. The scheme for utilizing inversion In(1)1Rk to uncover recessive lethals and detrimentals. The marked loci within
the inverted segment are respectively isocitrate dehydro-
genase-1 (Idh-1), splotch (Sp), leaden (1n), and peptidase-3
(Pep-3). Chromosomes marked with an asterisk are potential
bearers of an induced mutation.

Irradiating postgonial stages has induced sterile animals as
well as lethals. The percentage of steriles is itself an important
indicator of a genetic effect (perhaps due mostly to trabslocations) but
it probably had no direct influence on the estimate of lethals. We
assume that the genetic cause of sterile animals is independent of
recessive lethals. The assumption, however, would not be true if,
for example, sterile animals in generation-1 were due in part to
recessive lethals. In any case, one has to consider the number of
steriles raised relative to the number of testable animals of gen-
eration-1.

Sterility was determined in two ways. In one experiment male
offspring of a mutagen-treated parent were defined as sterile if the
testes were devoid of spermatogenic activity. Often in such cases
the testes were small. In another experiment sons of mutagen-
treated parents were defined as sterile: (1) if they produced no

Table 4. Frequency of "Sterility" Found in Male Offspring of
          Mutagenized Sires

| | Stocks or hybrid where sterility observed | | Type of observed sterility |
Sire's treatment	B6D2F$_1$	B6 by POS stocks	
TEM .3 to .5 mg/kg	35/604 = .058	6/104 = .058	A
4 April 1978-			
30 August 1979			
TEM .3 to .8 mg/kg	19/253 = .075	6/348 = .017	A
2 January 1980-			
23 October 1980			
Cesium 1000 R	9/116 = .078	6/69 = .087	A
	Stocks combined		
X-Irradiation, 800-900 R	166 / 530 = .313		B
TEM .3 to .5 mg/kg	270+/1409 = .192		B*

A. Sterility determined by lack of spermatogenic activity in testis.

B. Sterility determined by mating to at least 2 females with (1) no offspring
    produced, (2) insufficient offspring to carry on the line, or (3) died
    after mating but before producing offspring.

B*. Study still in progress.

offspring when bred to two females, (2) if they produced insufficient offspring to carry on the line, or (3) if they died after
mating but before producing offspring (Table 4).

These experiments were established primarily for the purpose
of inducing inversions or recessive lethals so no further study of
the "sterile" animals was done. It is sufficient to note that up

to 1/3 of the male offspring derived from mutagen-treated sperm or
spermatids will not be able to continue the 3-generation test if a
significantly high dose (enough to reduce litter size by 1/2) is
given to the male parent.

In validating the inversion test system we have paid particular
attention to the test generation in order to prove that the inversion does indeed prevent the appearance of live born mice with re-
combinant genotypes. Therefore, another factor to consider in assessing the efficiency of this test system is the number of times
the test generation genotypes are, for one reason or another, in-
consistent with those expected. For example all animals of the test
generation should have the genotype:

$$\text{In } (\underline{Idh\text{-}1^b}, +, +, \underline{Pep\text{-}3^b}) \text{ / In } (\underline{Idh\text{-}1^b}, +, +, \underline{Pep\text{-}3^b})$$

or

$$\text{In } (\underline{Idh\text{-}1^b}, +, +, \underline{Pep\text{-}3^b}) \text{ / } + (\underline{Idh\text{-}1^a}, Sp, ln, Pep\text{-}3^a).$$

In other words all test class animals (In/In) should also be homozygous at Idh-1 and Pep-3, and all splotch animals should be heterozygous. Furthermore, no leaden homozygotes (ln/ln) should appear
in the test generation. To check these assumptions, we have monitored the Idh-1 and Pep-3 genotypes of one animal of the test class
and one splotch animal from each of the test lines. In the 1503
completed lethal tests, 81 test generation exceptions were noted,
and nearly all could be explained. In nine cases splotch heterozygotes were misidentified as normal (+/+), but further breeding
tests proved them to be Sp/+. Splotch heterozygotes are nearly always recognized by a belly spot or at least some white on the paws
or feet, but in these instances (9/1503 = 0.6% of the time) they
were not. This is a relatively minor problem since low frequency
misclassification in either direction in the test generation leads
to no consequence. If a lethal has been induced and one splotch
animal is misclassified as +/+, that animal would be in a sufficient
minority to necessitate checking its genotype by further matings.

In 31 of the 1503 tests, alleles at Sp or Idh-1 locus apparently recombined with the inversion somewhere in the 3-generation
test. This double crossing-over within an inversion of this length
was not expected at first but is now consistent with our experience
with other inversions. The consequence of this event is that at
most 2%, but probably much less of lethal tests are invalid. If the
double crossing-over event occurs in the last generation only one
animal would be misclassified which is not serious. If it occured
in generation-1, particularly if splotch were involved, the genotypic classes of the test generation would include a leaden (ln/ln)
class which is unexpected. If the inversion is lost in generation-1
due to double crossing-over, a leaden class should also occur in the

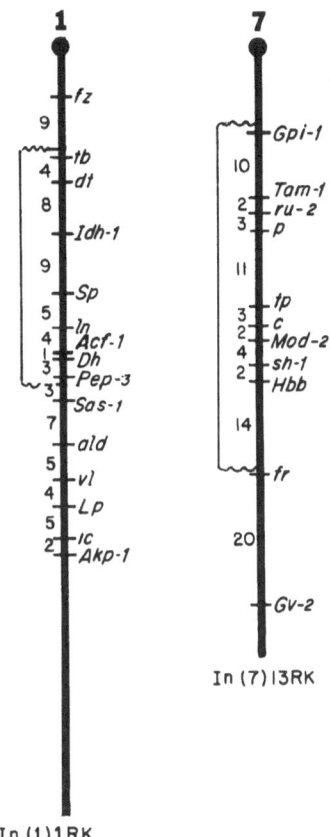

Fig. 4.   The two inversions making up the double-inversion test
          system.

test generation.   Such occurrences should be examined for their
causes, but the test of original generation-1 animal should be excluded.

In 5 cases the ln allele appeared to cross into the inversion.
Again this problem could be discovered and handled similarly as
above.   In 18 cases the inversion was lost during the 3-generation
test probably because of a double crossover event in the genera-
tion-1.   The consequence was the appearance of leaden (ln/ln) animals in the test generation.

In all cases, cessation of the test after an appearance of
leaden animals in the third generation is a conservative approach.

Another problem was occasional misreading of Pep-3 and Idh-1
gels.   In our experience this happened in 13 of the 1503 tests or

0.86% of the time. We could have avoided this class of error by always repeating gels if gel bands are not clearly discernible. Idh-1
and Pep-3 are not normally needed in this test system.

The last type of error was an interchange in the position of
samples lined up for insertion into the gels. This happened only once
in our experience. But again these types of errors will not affect
uncovering a lethal in the test generation, primarily because they
are not a normal part of identifying the test class.

From the genotypic exceptions we have found in the test generation, we can conclude that what double crossing-over with the inversion does occur will not cause problems in interpretation or assessment concerning lethals or detrimentals.

We have recently constructed a new inversion system utilizing
In(1)1Rk and In(7)13Rk which together will detect lethals induced in
over 7% of the genome (Fig. 4). In this mutation-test system, we
use Sp (splotch) and ln (leaden) in testing for lethals on Chr 1 and
isozyme markers Gpi-1 (glucose phosphate isomerase) and Hbb (hemoglobin beta chain) for assessing lethals induced in Chr 7. Gpi-1
and Hbb are good codominant electrophoretic markers, which along
with Sp and ln permit identification of all necessary genotypes and
appropriate assurance that inversion-inhibition is present. Only
Sp and Gpi-1, however, are necessary to classify animals for breeding, thus greatly simplifying the test.

Males and females doubly heterozygous for these two inversions
do not show obvious reduced fertility. Double inversion heterozygosity therefore does not impair this proposed mutagen test system.
Figure 5 shows a double bridge at first meiotic anaphase sometimes
seen in sections of testes of double inversion heterozygotes.

Use of the system is straightforward (Table 5). Generation-P
animals homozygous for both inversions are given a mutagen and, at
an appropriate time thereafter, mated to a stock not carrying either
inversion and carrying alternate alleles at all four loci of interest. Two classes of individuals appear in generation-1 of which only
the splotch (Sp/+) animals are kept. Each of these animals represents a possibility for unique mutations induced in or around either
In-1 or In-13. Asterisks signify the chromosomes in which mutations
could have occurred. Each of these splotch animals, then, is the
progenitor of the three-generation assay for induced recessives.
Each is mated to the original parental stock, and 6 phenotypic
classes of offspring are found in generation-2. The generation-2
animals that are splotch (Sp/+) are kept and all others discarded.
These represent 1/3 of the animals born alive in this generation.
Of the splotch animals, only those heterozygous at Gpi-1 or Hbb are
kept. These loci are characterized by gel-electrophoresis of blood
samples. These animals are 1/6 of those born in this generation and

Table 5. Mating Scheme for Double Inversion Test System

Generation

P

$$\frac{\text{In-1 (+ +)}}{\text{In-1 (+ +)}} , \frac{\text{In-13 (Gpi}^a, \text{Hbb}^d)}{\text{In-13 (Gpi}^a, \text{Hbb}^d)} \quad \times \quad \frac{+ \text{(Sp ln)}}{+ \text{(+ ln)}} , \frac{+ \text{(Gpi}^b, \text{Hbb}^s)}{+ \text{(Gpi}^b, \text{Hbb}^s)}$$

G1   Genotypes

$$\frac{\text{In-1 (+ +)}}{+ \text{(Sp ln)}} , \frac{\text{In-13 (Gpi}^a, \text{Hbb}^d)}{+ \text{(Gpi}^b, \text{Hbb}^s)} \quad \text{and} \quad \frac{\text{In-1 (+ +)}}{+ \text{(+ ln)}} , \frac{\text{In-13 (Gpi}^a, \text{Hbb}^d)}{+ \text{(Gpi}^b, \text{Hbb}^s)}$$

G1a   Matings

$$\frac{\text{In-1 (+ +)}^*}{+ \text{(Sp ln)}} , \frac{\text{In-13 (Gpi}^a, \text{Hbb}^d)^*}{+ \text{(Gpi}^b, \text{Hbb}^s)} \quad \times \quad \frac{+ \text{(Sp ln)}}{+ \text{(+ ln)}} , \frac{+ \text{(Gpi}^b, \text{Hbb}^s)}{+ \text{(Gpi}^b, \text{Hbb}^s)}$$

G2   Genotypes

(1) $$\frac{\text{In-1 (+ +)}^*}{+ \text{(Sp ln)}} , \frac{\text{In-13 (Gpi}^a, \text{Hbb}^d)^*}{+ \text{(Gpi}^b, \text{Hbb}^s)}$$

(2) $$\frac{\text{In-1 (+ +)}^*}{+ \text{(+ ln)}} , \frac{\text{In-13 (Gpi}^a, \text{Hbb}^d)^*}{+ \text{(Gpi}^b, \text{Hbb}^s)}$$

| | Ratio | Phenotype | | | | Decision |
		Sp	ln	Gpi-1	Hbb	
(1)	1/6	Sp	+	AB	DS	Keep, mate inter se or with parent
(2)	1/6	+	+	AB	DS	Discard

(Continued)

Table 5 (continued)

G2	Genotypes (continued)	Ratio	Phenotype				Decision
			Sp	ln	Gpi-1	Hbb	
(3)	$\dfrac{\text{In-1 } (+\ +)^* \quad + (Gpi^b,\ Hbb^s)}{+ (Sp\ ln), \quad + (Gpi^b,\ Hbb^s)}$	1/6	Sp	+	B	S	Discard
(4)	$\dfrac{\text{In-1 } (+\ +)^* \quad + (Gpi^b,\ Hbb^s)}{+ (+\ ln), \quad + (Gpi^b,\ Hbb^s)}$	1/6	+	+	B	S	Discard
(5)	$\dfrac{+ (Sp\ ln) \quad \text{In-13 } (Gpi^a,\ Hbb^d)^*}{+ (Sp\ ln), \quad + (Gpi^b,\ Hbb^s)}$	0	Die				
(6)	$\dfrac{+ (Sp\ ln) \quad \text{In-13 } (Gpi^a,\ Hbb^d)^*}{+ (+\ ln), \quad + (Gpi^b,\ Hbb^s)}$	1/6	Sp	ln	AB	DS	Discard
(7)	$\dfrac{+ (Sp\ ln) \quad + (Gpi^b,\ Hbb^s)}{+ (Sp\ ln), \quad + (Gpi^b,\ Hbb^s)}$	0	Die				
(8)	$\dfrac{+ (Sp\ ln) \quad + (Gpi^b,\ Hbb^s)}{+ (+\ ln), \quad + (Gpi^b,\ Hbb^s)}$	1/6	Sp	ln	B	S	Discard

G2a Matings

$$\dfrac{\text{In-1 } (+\ +)^* \quad \text{In-13 } (Gpi^a,\ Hbb^d)^*}{+ (Sp\ ln) \quad + (Gpi^b,\ Hbb^s)} \times \dfrac{\text{In-1 } (+\ +)^* \quad \text{In-13 } (Gpi^a,\ Hbb^d)^*}{+ (Sp\ ln) \quad + (Gpi^b,\ Hbb^s)}$$

G3	Genotypes	Ratio	Phenotype Sp	In	Gpi-1	Hbb	Decision
(1)	In-1 (+ +)* In-13 (Gpi^a, Hbb^d)* / In-1 (+ +)*' In-13 (Gpi^a, Hbb^d)*	1/12	+	+	A	D	Test class for presence of recessive lethal or detrimental in both inversions
(2)	In-1 (+ +)* In-13 (Gpi^a, Hbb^d)* / In-1 (+ +)*' + (Gpi^b, Hbb^s)	1/6	+	+	AB	DS	Test class for In-1 only and usuable to maintain lethals induced in In-13 for further study
(3)	In-1 (+ +)* + (Gpi^b, Hbb^s) / In-1 (+ +)*' + (Gpi^b, Hbb^s)	1/12	+	+	B	S	Test class for In-1 only
(4)	In-1 (+ +)* In-13 (Gpi^a, Hbb^d)* / + (Sp In) In-13 (Gpi^a, Hbb^d)*	1/6	Sp	+	A	D	Test class for In-13 only and usable to maintain lethals induced in In-1 for further study
(5)	In-1 (+ +)* In-13 (Gpi^a, Hbb^d)* / + (Sp In) + (Gpi^b, Hbb^s)	1/3	Sp	+	AB	DS	Usable to maintain lethals induced in either In-1 or In-13 or both
(6)	In-1 (+ +)* + (Gpi^b, Hbb^s) / + (Sp In) + (Gpi^b, Hbb^s)	1/6	Sp	+	B	S	Usable to maintain lethals induced in In-1
(7)	+ (Sp In) In-13 (Gpi^a, Hbb^d)* / + (Sp In) In-13 (Gpi^a, Hbb^d)*	0	Die				

(continued)

Table 5 (conclusion)

G3	Genotypes, continued	Ratio	Phenotype				Decision
			Sp	In	Gpi-1	Hbb	
(8)	$\dfrac{+ (Sp\ In)}{+ (Sp\ In)}, \dfrac{In\text{-}13\ (Gpi^a,\ Hbb^d)^*}{+ \quad (Gpi^b,\ Hbb^s)}$	0	Die				
(9)	$\dfrac{+ (Sp\ In)}{+ (Sp\ In)}, \dfrac{+ (Gpi^b,\ Hbb^s)}{+ (Gpi^b,\ Hbb^s)}$	0	Die				

*Chromosome in which a recessive mutation could have been induced.

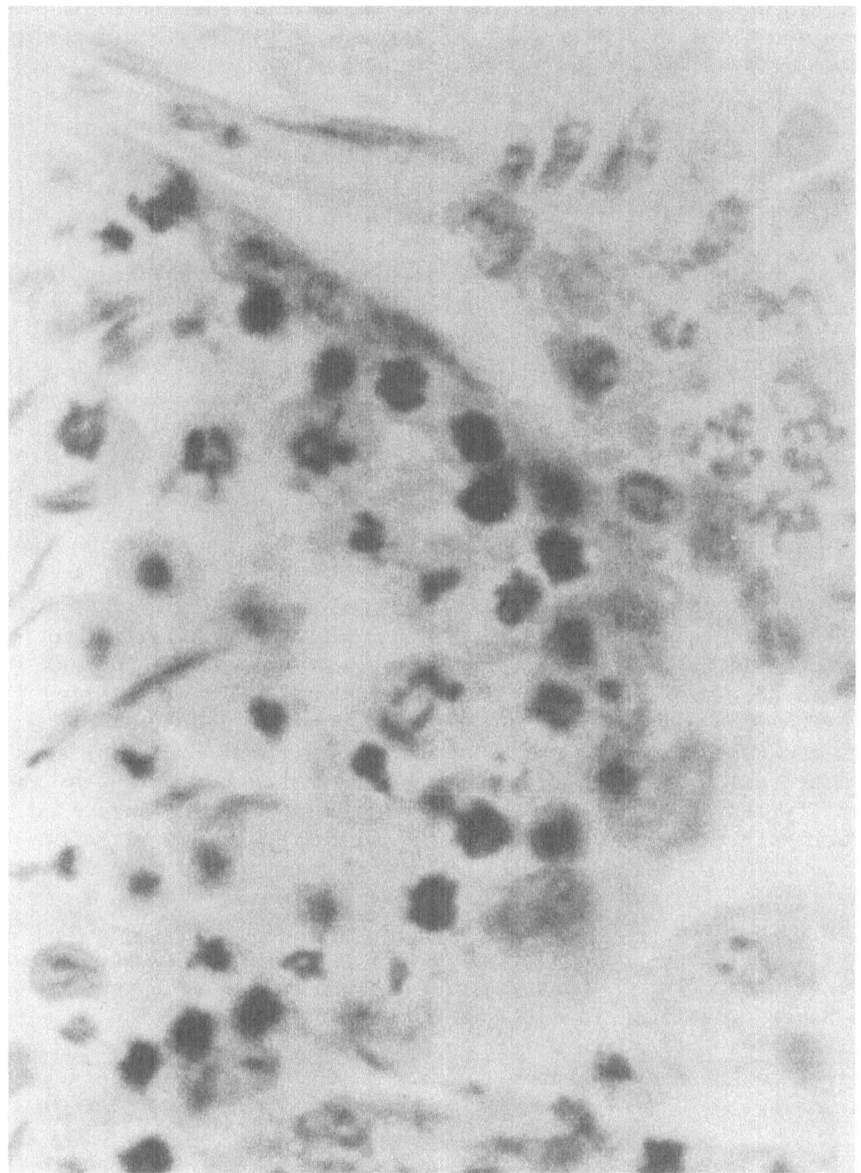

Fig. 5. A double chromatin bridge at first meiotic anaphase ob-
         served in a section of testis from a double inversion
         heterozygote.

represent double inversion heterozygotes like one parent. They are mated among themselves or with their identical parent to produce generation-3.

Generation-3, the test generation, contains six viable pheno-typic classes, again, distinguishable from each other by typing at the Sp and Gpi-1 or Hbb loci. The absence of genotypic classes 1, 2, and 3 (1/3 of the animals) indicates a recessive lethal was in-duced in or around In-1. Absence of classes 1 and 4 (1/4 of the animals) indicates a lethal was induced in In-13. Classes 4, 5, and 6 (2/3 of the animals) can be used to maintain any recessive lethals induced in In-1, and classes 4 and 5 (1/2 of the animals) can be used to maintain recessives induced in In-13. The test has many attributes of that utilizing In-1 alone. For example, in the test generation the absence or reduced frequency of any nonsplotch animals implies a recessive lethal in or very near the inverted segment of In-1. If the unlikely event of lethals in both segments occur, one would expect only the presence of classes 5 and 6. If this were to occur classes 5 and 6 will be used to maintain the lethals. A lethal or detrimental in In-1 could be studied alone. The lethal or detrimental in In-13 would need to be separated from that in In-1 by outcrossing animals of class 5 to a stock homozygous for Gpi^b and Hbb^s. Only splotch (Sp/+) animals would be selected of which only heterozygotes for Gpi or Hbb would be saved for mating.

We have also begun to utilize the X-chromosome inversion from the MRC Radiobiology Unit at Harwell, England. This inversion designated In(X)1H has potential value in a two-generation test for X-linked recessive lethals and detrimentals [11, 12]. Phillips [18] had begun to use it to induce lethals before her death. Table 6 shows the system we will use. We have chosen bent tail (Bn) as a dominant tail marker and Pgk-1 as an electrophoretic marker as aids to verify that lethals have been induced when they are suspected. Particularly these markers are needed to guard against misinter-preting lethals in the inverted region that may cross over with the inversion. This inversion is long enough to permit 2- and 3-strand double crossovers to occur [12]. We will try to place Bn inside the inversion. This should be possible since it is near the center of the inversion. Pgk-1, however, may not cross over with the inver-sion because it lies close to the distal break point. The rare 'b' allele of Pgk-1 found in a feral population of mice in Denmark was imported to the Jackson Laboratory by E. M. Eicher who kindly gave us animals.

The test is similar to those described above for autosomal re-cessives but has a distinct advantage in requiring one less genera-tion. The mutagen is given to males carrying the $Pgk-1^b$ allele. When mated to the female of the homozygous inversion stock, they produce two types of offspring that are mated together to obtain the test generation. In the test generation four classes of animals

Table 6. Proposed X-Linked Recessive Mutation Test System Using $\underline{In(X)1H}$. The mutation is given to the male parent in generation 1

Generation

P

$$\frac{+\ (+\ Pgk^b)}{Y} \times \frac{In\ (Bn\ Pgk^a)}{In\ (Bn\ Pgk^a)}$$

G1

$$\frac{In\ (Bn\ Pgk^a)}{+\ (\ +\ Pgk^b)^*} \times \frac{In\ (Bn\ Pgk^a)}{Y}$$

G2	Genotypes	Ratio	Phenotypes Bn	Phenotypes Pgk-1	Decision
(1) ♀	$\dfrac{In\ (Bn\ Pgk^a)}{In\ (Bn\ Pgk^a)}$	1/4	Bn	A	Discard
(2) ♀	$\dfrac{In\ (Bn\ Pgk^a)}{+\ (\ +\ Pgk^b)^*}$	1/4	Bn	AB	Can be mated to type (3) ♂♂ to maintain any induced lethals
(3) ♂	$\dfrac{In\ (Bn\ Pgk^a)}{Y}$	1/4	Bn	A	Can be mated to type (2) ♀♀ to maintain any in- duced lethals
(4) ♂	$\dfrac{+\ (+\ Pgk^b)^*}{Y}$	1/4	+	B	Test class

*Chromosome with possible induced mutation.

occur. Class 4 males will have normal tails and can be distinguished from those of class 3 with bent tails. If class 4 males are absent in the progeny of a generation-1 mating, then there is evidence for an induced X-linked lethal. If there are only a few animals appearing like class 4 males, then one might suspect that the $\underline{Bn}$ allele was not penetrant or that the lethal $\underline{Bn}$ had recombined with the inversion. Those exceptional animals can be typed for $\underline{Pgk-1}$ and if found to be $\underline{Pgk-1A}$, either explanation would be probable.

If we are unsuccessful in placing $\underline{Bn}$ within the inversion, an alternative scheme is possible. We could link $\underline{Bn}$ with the rare $\underline{Pgk-1^b}$ allele and then give the mutagen to animals of the inversion stock. We prefer the former system because it permits assessing

naturally occurring lethals in laboratory or feral populations. One
can always use the reduction of males by 1/2 in the test generation
as the sole criterion for a presumed lethal, but this is less reliable and insensitive in ascertaining lethals outside the inversion.
Furthermore, it is less convenient to maintain and confirm lethals
without the markers.

In 1977 M. T. Davisson and I imported 41 Robertsonian chromosomes in 8 independent stocks from Prof. Alfred Gropp in Germany.
The importation was undertaken in part to use Roberstonians in combination with inversions to tie up additional elements of the genome
in a mutation-testing system. It has been shown by us and others
that many Robertsonian chromosomes of the mouse inhibit crossing-
over in the proximal regions of the chromosome arms.

We have successfully combined inversion In(1)1Rk and Robertsonian Rb(1.3)1Bnr so that this chromosome has the middle portion of
the long arm inverted. The stock is homozygous and breeds well with
a litter size of 6 to 9. The stock is also homozygous for Rb(8.12)5Bnr
and Rb(9.14)6Bnr. Thus it is a good stock to use for placing inversions in other Robertsonians to create a stock with multiple inver-
sion-bearing Robertsonian chromosomes. We already have available inversions in Chrs 3, 8, 9, 12, and 14. In(1)12Rk has also been in-
serted by recombination into the long arm of Rb1Bnr. This stock is
also homozygous.

An objective is then to produce a pericentric inversion large
enough to tie up genetic segments proximal to the inversions in both
arms. Alternatively, the inhibition of crossing-over in the proximal
segments by the Roberstonian centromeres may be sufficient to achieve
this objective. We have already shown that in the presence of
Rb(1.3)1Bnr recombination between the most proximal marker fuzzy
(fz) is reduced from 6% to 1% [19]. We know from observing our
colony that Robertsonians by themselves seldom affect fertility
severely. In a study using Robertsonians as centromere markers,
D. M. Juriloff, a postdoctoral associate in our laboratory, noted
that in males or females heterozygous for 1, 2, 3, or 4 Robertsonians the fertility did not fall significantly below that of controls
until 4 Robertsonians were present. As mentioned earlier, preliminary evidence suggests that pericentric inversions do impair fertility significantly. If this observation is confirmed Roberstonian
inhibition of recombination may be the method of choice for tying
up the central part of two-armed chromosomes.

It is desirable to combine mutation systems for several reasons.
One reason is to improve the efficiency of a regimen that involves
relatively costly whole-animal assays. Another reason is to gain
understanding of relationships and corroborations of various test
systems under the same environmental and genetic conditions. A
third reason, related to the second, is to provide validation of

systems as they are developed comparing them to older established
protocols. We have already incorporated a specific locus test in
conjunction with the double inversion test system assaying for mutations at loci ln, Pep-3, Idh-1, Gpi-1, and Hbb. We are also trying
to devise a workable system whereby the X-chromosome inversion might
be put together with the two autosomal inversions, thus combining
the two- and three-generation assays in a single system.

3. VALIDATION

3.1. Rate of Lethal Induction

To date, using inversion In(1)1Rk we have induced seven lethals
in 1663 complete three-generation tests. We estimate that this particular inversion covers approximately 3% of the genome and since
we also will have picked up some lethals outside the inversion, we
are screening approximately 3.5% of the genome which contains approximately 50,000 genes. So far in the validation using X-irradiation we have tested 364 potential carriers of newly induced recessives through the three-generation test. The treated males received
an average of 892 R to post spermatogonial stages. We have uncovered two recessives giving an estimate of induced mutations of ap-
proximately $2/(1750 \text{ loci} \times 892 \text{ R} \times 364 \text{ gametes}) = 0.35 \times 10^{-8}$ per R
per locus. Believing that we were inducing at least as many chro-
mosomal aberrations using triethylene melamine (TEM) for the same
amount of induced sterility, we have since used a dose of 0.3 mg/
kg TEM to validate the lethal test system. This dose is approximately equal to 1000 R in inducing sterility in the offspring, so
we have estimated 1000 R as an equivalent R-dose. In this experiment five lethals have been uncovered in 1299 offspring tested
through the three-generation system. This provides an estimate of
recessive lethal induction of $5/(1750 \text{ loci} \times 1000 \text{ R} \times 1299 \text{ gametes}) =
0.22 \times 10^{-8}$ per equivalent R per locus. Because of the problem with
TEM mentioned above we think this second estimate is low. But together both estimates are considerably below those given by Russell
for post-spermatogonial cells of 16 recessive$/(7 \text{ loci} \times 16813 \text{ ga-}$
metes$) = 45.32 \times 10^{-8}$ per R per locus, and for spermatogonial cells
9 recessives$/(7 \text{ loci} \times 200 \text{ R} \times 31253 \text{ gametes}) = 20.57 \times 10^{-8}$ per R
per locus [20]. This discrepancy of one hundred fold is perplexing
and deserves comment.

The difference could be explained if our lethal response is on
the "down side" of the dose-response curve, that is if our dose of
both mutagens is sufficiently high that there is selection against
the carriers of induced lethals.

The difference might be due to a difference in induced rates between specific loci which have viable allelic alternatives and non-
specific loci tested by the inversion system. There is little doubt
about the genetic and cytological length of the inversion, so its

estimated portion of the genome of 3% seems sufficiently accurate. The number of genes in the inversion could be less than 1,750, but if so could not be of 100-fold less. This would necessitate a significant revision in our estimate of the approximate number of loci in mammals or suggest a significant departure from random distribution of genes along this chromosome.

The difference could lie in the different character of the two systems. In the seven-specific-locus system, mutations are scored in the offspring of mutagen-treated animals. In the inversion system, the heterozygous offspring must be able to breed to be tested for carrying an induced recessive. The radiation dose we gave produced sterility or non-productivity in 31% of the offspring. The dose of TEM produced 19% infertility or non-productivity. It would not be surprising if some of these offspring were sterile because of induced recessive lethals (perhaps the dominant effect of deletions) but it is difficult to imagine that the sterility of so many would occur because of deletions on Chr 1 alone, where the inversion is located. Nearly 2/3 of the steriles of the radiation study and all of the steriles of the TEM study would need to carry induced recessives on Chr 1 to account for the difference.

Is it possible that the non-specific loci of the inversion test are somehow different from the specified loci of the specific locus test? One explanation could be that recessive lethals induced at non-specific loci (where viable allelic alternatives are unknown or impossible) are actually dominant lethals in most cases. If this were true, we might then speculate that only 1% of those 1,750 loci within the inversion (= 18 loci) would be capable of providing viable allelic alternatives and there is little doubt that this is by far an underestimate. The linkage map of the mouse already shows 15 known loci within the inverted segment [21].

For an estimate of the specific-locus induced rate of mutation from this study, we can use data for lethals induced at loci $\underline{ln}$, $\underline{Dip-1}$, and $\underline{Idh-1}$. All of these would be recognizable. The $\underline{ln}$ mutations would be recognized in the first generation, the other two recognized by the absence of one of the electrophoretic bands in the splotch-leaden heterozygotes of the test-generation. We found only a $\underline{ln}$ mutation. A rough estimate of the induced rate, again assuming the mutational equivalency of 1000 R and 0.3 mg/kg TEM, would be 1 recessive/(3 loci × 1000 R × 1299 gametes) = 25.7×10^{-8} per R per locus, which is close to the Russell estimate. In the further validation of this method we are continuing to note the mutations induced at these three loci and to compare their induced rates with those at the other non-specified loci within the inversion.

Kohn [22] showed that mutation rates at histocompatibility loci are nearly 100-fold less than those of the mean of the specific loci. He attributed this difference to selection or repair or both.

Table 7. Lethals Induced Using In(1)1Rk

Lethal No.	Animal No.	Mutagen	+/+ :	Sp/+	Comments
1	2862	900 R X-ray	64	753	
2	319	900 R X-ray	3	436	
3	2883	TEM .3 mg/kg			Found as a 2nd generation mutant at the leaden (ln) locus
4	765	TEM .3 mg/kg	0 :	79	
5	959	TEM .3 mg/kg	6 :	509	
6	1296	TEM .3 mg/kg	0 :	185	
7	1312	TEM .3 mg/kg	25 :	231	

Lüning [23] also noted a much lower rate of induced recessive lethals per gamete than would be expected from results of the specific locus tests. His estimate is 4.3×10^{-5} recessives per rad per gamete. Making comparable assumptions concerning the number of loci in the mouse (50,000), an adjusted estimate for his induced recessive lethals per rad per locus is 0.09×10^{-8} which is similar to our estimate.

3.2. Lethals Induced

Seven lethals have been induced using this system (Table 7). Lethals 4 and 6 so far have yielded no test-class animals suggesting to us they are within the inversion and probably near one of the inversion break points. Lethals 2 and 5, however, have yielded about 1% test-class animals making it difficult to determine whether they are within the inverted segment or not. If within they must be located very close to the middle of the inverted segment where crossing-over on either side would have the highest opportunity. Even the splotch locus which is in the center of the inversion has recombined with the inversion only 1/10 as often. The other possibility is that these lethals are immediately outside the inverted segment. Lethals 1 and 7 are certainly outside the inverted segment but probably not farther than 5 cM from the break points. We have consistently found a higher recombination of loci just outside the segment in inversion heterozygotes and thus the recombination fraction overestimates often by a factor of 2 the actual genetic distance from the inversion. No complementation tests of these lethals have been done yet.

One of the important gaps in our knowledge is the extent of burden that induced recessive lethals produce in mammalian systems. So far most of the studies on recessive lethals have concerned lethal alleles at known specific loci where previously discovered viable allelic alternatives permitted identification and naming of these loci. The lethals induced in this study except for lethal 3 are apparently in presently unmarked areas involving no known loci on Chromosome 1. In collaboration with D. M. Juriloff, we completed an initial examination of these lethals with the specific objective of determining the time of death in embryogenesis. Two additional lethals were also studied. These are the lethal effects of the homozygous condition of inversions In(2)5Rk and In(12)25Rk, two of the three inversions displaying homozygous lethality.

Lethal 1 acts after implantation, as was evidenced by deciduomas (moles) of high frequency. Lethal 2 may act at birth and possibly at preimplantation as well. Lethal 4 acts at birth or preimplantation. Lethal 6 acts at day 7 to 10 in embryogenesis, but also some at preimplantation. The lethal of In-5 acts at preimplantation, and the lethal of In-25 at day 7 to 10 in embryogenesis. No embryological studies have yet been done on Lethal 7 or the recessive lethal involved with In(4)28Rk.

The wide variety of effects of these lethals was not unexpected. The impact of the burden of these lethals in heterozygous condition (the dominant effect of recessive lethals) is not known but preliminary evidence suggests possible slight effects on fertility.

4. PROSPECTS AND CONCLUSIONS

The present study is the first where mammalian chromosomal inversions have been induced and genetically defined. We have induced and are maintaining 28 different autosomal inversions each of which can be utilized by itself or in combination with others in a mutation assay system. The length of the inversion, and what known genetic markers it overlaps determine the value of each in such a test. What is particularly important is that bearers of two or three paracentric inversions in heterozygous state are generally not affected in their fertility or viability.

One system we have developed and have begun to validate uses one inversion In(1)1Rk. It is a simple system only requiring recognition of a coat color mutation, splotch. Any technician with a little training could conduct the test. Further analysis of exceptions when they occur would necessitate more sophisticated expertise. So far we have induced seven recessive lethals using the test. There are important aspects of the burden of these recessive lethals that need to be assessed.

The system by itself is comparable to and perhaps better in efficiency and cost per locus than the specific locus system [24]. Also important is the fact that loci with no other known phenotypic expression are being assessed for their mutagenic changes.

It should be mentioned here that any system utilizing inversions will detect pre-implantation lethals. They will be observed by an absence of the test class but the dams should show no embryonic or fetal wastage in the form of deciduomas or dead embryos. This is a class of recessive lethals that would not be detected by a search for fetal wastage.

The double inversion test system covers twice the length of the genome of that utilizing In(1)1Rk alone and is therefore nearly twice as efficient per locus. For the second inversion in this system, In(7)13Rk, we have not yet incorporated easily detectable phenotypic markers other than electrophoretic variants, but we believe this can be done with markers at the albino (c) and pink eye (p) loci.

We are testing the value of the system for assessing mutagens by inducing lethals using TEM and gamma-rays from a cesium source. Collaborating with M. L. Petras we are using it to assess recessive lethals in wild populations of mice. We feel it is a very important advantage to be able to ascertain and retain for study recessive lethals and detrimentals in wild populations of a mammal. There is no reason why at least three inversions on nonhomologous chromosomes cannot be placed together in a single stock so that a larger portion of the genome can be tested. The major problem with combining three or more inversions in a single test system is that more animals need to be raised in each generation to get the gene combinations necessary to continue the test.

If the inversion system in mice can ultimately cover from 10 to 15% of the genome, then, assuming 50,000 loci, mutations will be examined simultaneously at 5,000 to 7,500 loci. Since the inversion system requires three generations, it could be equivalent to 1,667 to 2,500 loci per generation, or about 238 to 357 times as powerful as the specific locus system. To counter these advantages, one must consider the possibility from our findings that further validation will show the inversion system to be 1/100 as powerful in detecting lethal mutations per locus in which case the two systems would be nearly equal in overall efficiency. What must then be decided is whether the inversion system assesses mutations at different kinds of loci and therefore is a useful adjunct to the specific locus system. In any event, we see the inversion system as complementing rather than replacing the specific locus system, which has provided basic data for comparisons of mutagens, dose rates, and sensitivities of various cell stages.

ACKNOWLEDGMENTS

This paper is dedicated to the memory of Rita J. S. Phillips of the MCR Radiobiology Unit at Harwell, England who was my friend and colleague in work on mammalian inversions. I wish to acknowledge the assistance and collaboration of Norman L. Hawes, Mary N. Murphy, Dr. Muriel T. Davisson, Dr. Diana M. Juriloff, Dr. Montrose J. Moses, Patricia A. Poorman, and Dr. Michael L. Petras. This work was supported by Contract EV-76-S-02-3267 with the U.S. Department of Energy and NIH research grants GM-19656 with the National Institute of General Medical Sciences and N01-ES4-2156 with the National Institute of Environmental Health Sciences. The Jackson Laboratory is fully accredited by the American Association for Accreditation of Laboratory Animal Care.

REFERENCES

1. L. B. Russell and B. E. Matter, Whole-mammal mutagenicity tests: evaluation of five methods, Mutat. Res., 75:279-302 (1980).
2. W. M. Generoso, J. B. Bishop, D. G. Goslee, G. W. Newell, C.-J. Sheu, and E. von Halle, Heritable translocation test in mice, Mutat. Res., 76:191-215 (1980).
3. P. B. Selby and P. R. Selby, Gamma-ray-induced dominant mutations that cause skeletal abnormalities in mice. II. Description of proved mutations, Mutat. Res., 51:199-236 (1978).
4. H. B. Newcombe, Problems of assessing risks versus mutations, Genetics, 92 (Suppl. 1, Part 1), s199-s210 (1979).
5. J. Kratochvilova and U. H. Ehling, Dominant cataract mutations produced by gamma-irradiation of male mice, Mutat. Res., 62: 221-223 (1979).
6. W. L. Russell, X-ray induced mutations in mice, Cold Spring Harbor Symp. Quant. Biol., 16:327-336 (1951).
7. M. T. Davisson, P. A. Poorman, T. H. Roderick, and M. J. Moses, a pericentric inversion in the mouse, Cytogenet. Cell Genet., 30:70-76 (1981).
8. T. H. Nijhoff, Pericentric inversion in Rb(6.5)1Ald, Mouse News Letter, 60:77 (1979).
9. T. H. Roderick, Producing and detecting paracentric chromosomal inversions in mice, Genetics, 76:109-117 (1971).
10. R. J. S. Phillips and M. H. Kaufman, Bare-patches, a new sex-linked gene in the mouse, associated with a high production of XO females. II. Investigations into the nature and mechanism of the XO production, Genet. Res., 24:27-41 (1974).
11. R. J. S. Phillips, Factor for XO production - In(X)1H, Mouse News Letter, 52:35 (1975).
12. E. P. Evans and R. J. S. Phillips, Inversion heterozygosity and the origin of XO daughters of Bpa/+ female mice, Nature, 256: 40-41 (1975).
13. G. Fisher and M. F. Lyon, Crm outside In(X)1H, Mouse News Letter, 64:58 (1981).

14. R. J. S. Phillips and E. P. Evans, A new X-linked inversion In(X)3H, Mouse News Letter, 61:39 (1979).

15. E. P. Evans and R. J. S. Phillips, A phenotypically marked inversion, In(2)2H, Mouse News Letter, 58:44-45 (1978).

16. J. D. Minna, G. M. Iverson, and L. A. Herzenberg, Identification of a gene locus for immunoglobulin H chains and its linkage to to the H chain chromosome region in the mouse, Proc. Nat. Acad. Sci. U.S., 58:188-194 (1967).

17. J. B. Whitney III, G. T. Copland, L. C. Skow, and E. S. Russell, Resolution of products of the duplicated hemoglobin alpha-chain loci by isoelectric focusing, Proc. Nat. Acad. Sci. U.S., 76: 867-871 (1979).

18. R. J. S. Phillips, Sex-linked lethal tests with In(X)1H, Mouse News Letter, 60:45-46 (1979).

19. M. T. Davisson and T. H. Roderick, Linkage data, Mouse News Letter, 53:54 (1975).

20. E. L. Green and T. H. Roderick, in: Radiation Genetics in Biology of the Laboratory Mouse, E. L. Green, ed., pp. 165-185, 2nd ed., McGraw-Hill, New York (1966).

21. T. H. Roderick and M. T. Davisson, Linkage map of the mouse (Mus musculus), in: Genetic Maps, S. J. O'Brien, ed., Vol. 2, pp. 277-285, National Cancer Institute, Frederick, Maryland (1982).

22. H. I. Kohn, X-ray mutagenesis: results with the H-test compared with others and the importance of selection and/or repair, Genetics, 92 (Suppl. No. 1, Part 1), s63-s66 (1979).

23. K. G. Lüning, Some problems in the assessment of risks, Genetics, 92 (Suppl. No. 1, Part 1), s203-s209 (1979).

24. T. H. Roderick, Chromosomal inversions in studies of mammalian mutagenesis, Genetics, 92 (Suppl. No. 1, Part 1), s121-126 (1979).

CATARACTS - INDICATORS FOR DOMINANT MUTATIONS

IN MICE AND MAN

Udo H. Ehling

Institut für Genetik
Gesellschaft für Strahlen- und Umweltforschung (GSF)
D-8042 Neuherberg
Federal Republic of Germany

1. INTRODUCTION

A cataract is an opacity of the lens causing a reduction of visual function [1]. The organogenesis of the lens in various mammals is similar [2]. Therefore, a gene, which disturbs in different species the same process of the normal development of the lens, leads to the manifestation of the same cataract type in these different species. Ehling [3] pointed out that morphologically comparable cataracts in humans and mammals have very often the same mode of inheritance. Marner [4] described in humans the transmission of a zonular cataract through 8 generations. A similar cataract with dominant inheritance was described for dogs [5, 6], rabbits [7], and rats [8]. These examples may be sufficient to emphasize that cataract mutations observed in the mouse can be directly compared with the manifestation of cataracts in man.

Progress in the estimation of the genetic risk of radiation-induced mutations was made by the systematic investigation of dominant mutations affecting major body systems of the mouse. We developed methods to determine the dominant mutations which affect the skeletal system [9, 10] and the lens [11, 12] of the mouse. Based on the induction of dominant mutations in mice, a concept for the direct estimation of the risk of radiation-induced genetic damage to the human population expressed in the first generation was developed [13, 14]. The same approach can be used to estimate the chemically-induced genetic damage expressed in the first generation in man [15].

The detection of dominant cataract mutations was combined with
the scoring of specific locus mutations. The essential feature of
the specific locus method is the choice of parents that differ with
respect to alleles at 7 selected loci. In contrast to the specific
locus experiments, the determination of the mutation rate of dominant cataracts in mice allows the investigation of mutations at unselected loci coding for one organ system. Taking into account that
in humans 20 well established dominant cataracts are known [16],
then it is likely that in the combined experiment mutations at a
total of at least 30 loci are scored.

2.   DOMINANT CATARACT MUTATIONS IN MICE

The systematic investigation of induced dominant cataracts in
mice was initiated by Ehling in 1977. In a series of papers the induction of radiation-induced dominant cataracts [11, 17, 18] and
ethylnitrosourea-induced dominant cataracts [12, 19] were reported.

To detect dominant cataract mutations, the F_1-offspring were
examined with a slit lamp at 4-6 weeks of age. Mydriasis was
achieved with 1% atropin applied at least 10 minutes before examination. The lenses were scanned with a narrow beam of light with an
angle of 20°-30°. Presumed mutant individuals exhibiting a lens
opacity were outcrossed to normal mice, and at least 20 offspring
were examined to confirm the genetic nature of the cataract. When,
among 20 offspring, no individual exhibited the phenotype, it was
concluded that the lens opacity was not due to a dominant mutation
with a penetrance value equal or greater than 0.32 [20]. For those
F_1 individuals that produced offspring with the lens opacity phenotype, the presumed mutant was considered confirmed, and mutant lines
were established to determine the penetrance and the expressivity
of the gene. Furthermore, the viability of the mutant and the effect
of the mutation in the homozygous condition was studied [11, 12, 18].

The statistical comparison of the experimental and control mutation rates can be based on Fisher's exact treatment of a 4-fold
contingency table [21]. For the calculation of the confidence limits the tables of Crow and Gardner [22] were used.

In a total of 10,691 offspring no spontaneous cataract mutation was detected. In the literature, the following dominant cat-
aracts·were described for the mouse: blind, Bld, chromosome 15 [23],
congenital cataract, Cad [24], dominant cataract, Cat^{Fr} [25-27],
cataract and small eyes, Cts [28], Dickie's small eye, Dey, chromosome 2 [29], eye lens obsolence, Elo, chromosome 1 [30], eye opacity,
Eo, chromosome 2 [31], and lens opacity, Lop, chromosome 10 [32].
The location of the genes is based on Mouse News Letter [33]. Additionally ten dominant cataract mutations were described by Kratochvilova [18]: nuclear and zonular cataract (Nzc), nuclear cataract
(Nuc), anterior pyramidal cataract (Apyc), anterior polar cataract

(Apoc), anterior capsular cataract (Acc), anterior lenticonus with microphthalmia (Alm), vacuolated lens with microphthalmia (Vlm), iris anomaly with cataract (Iac), iris dysplasia with cataract (Idc), and anisocoria (Anc). For the unlocated genes the possibility of allelism must be considered.

Of the five located cataract genes two mutations are on chromosomes, which carry also specific locus mutations. Non-agouti (a) is on chromosome 2, while fuzzy (fz) and leaden (ln) are on chromosome 1. Non-agouti is used as marker in the Harwell, Moscow, and Oak Ridge test stocks. Leaden is a marker of the Harwell and Moscow test stocks and fuzzy is a marker exclusively of the Harwell test stock [34].

3. SPECIFIC LOCUS MUTATIONS IN MICE

With the specific locus method, mutations to recessive alleles of a small number of selected loci can be detected in the first generation. The multiple recessive tester stock that has been extensively used in radiation and chemical mutagenesis studies has the following 7 markers: a/a (non-agouti); b/b (brown); $c^{ch}p/c^{ch}p$ (chinchilla, pink-eyed dilution); d se/d se (dilute, short-ear); s/s (piebald). The 7 loci are distributed among five autosomes. The c^{ch}- and p-loci are linked on chromosome 7, with an average recombination of about 14%, and the d- and se-loci are closely linked, being only 0.16 centimorgans apart on chromosome 9 [35].

A detailed description of the other multiple tester stocks was published by Ehling [34]. A mutation in the germ cells of wild-type animals at any of the 7 loci represented by the recessives in the test stock will be detected in the first generation offspring. Different types of intragenic mutations, small deficiences involving the marked loci and, rarely, events of a grosser nature, like nondisjunction can be detected with the specific locus mutation assay. In addition, with the specific locus method it is possible to detect mosaics and clusters. One of the important assets of the specific locus method is that the estimates of spontaneous mutation frequencies obtained by different workers and laboratories are in good agreement [35].

The statistical analysis of the results is identical with the methods mentioned for the dominant cataract data.

Specific locus mutations in mice were used to demonstrate that the mutation frequency induced by chemical mutagens depends on the spermatogenic stages and the different treatment conditions [15, 34]. For mitomycin C and procarbazine the oocytes are less sensitive than spermatogonia for the induction of specific locus mutations [15, 36]. The induction of specific locus mutations by radiation is the basis for the estimation of the radiation genetic risk in man [13, 17].

4. COMBINED DETECTION OF DOMINANT CATARACT
    AND SPECIFIC LOCUS MUTATIONS

The advantage of a combined investigation of dominant cataract mutations and specific locus mutations in mice is at least three-fold:

1. The number of scorable mutations is increased in comparison with a simple specific locus experiment by a factor of four.

2. The combined investigation allows the comparison of the mutation frequency of selected and unselected loci.

3. In the same experiment the frequency of mutations with a dominant and a recessive mode of inheritance can be compared.

4.1. Experimental Procedures

Groups of $(101xC3H)F_1$ hybrid male mice, 11 weeks old, were exposed to 3 doses of ^{137}Cs γ-rays. In one experiment, a dose of 455 + 455 R (55 R/min) with a 24-h fractionation interval, and in the others, a single dose of 534 or 600 R (53 R/min) were used. During irradiation the males were placed in a lucite container and their heads were shielded with lead. The control males were handled in the same way but sham-irradiated.

Ethylnitrosourea (ENU), Serva, Heidelberg, was dissolved in 0.07 M phosphate buffer (pH 6.0) immediately before use. The mutagenicity of ENU was tested in hybrid male mice $(101xC3H)F_1$, 10-12 weeks old. Single doses of 160 mg/kg and 250 mg/kg of the test compound in a volume of 0.5 ml were injected intraperitoneally. All injections were completed within 40 min after the chemical was dissolved. The weights of the hybrid males ranged from 24 to 29 g. The amount of mutagen injected did not vary from the nominal value for the particular animal by more than 5%. The highest dose of ENU had no effect on the survival of the treated animals. The control males received an equal volume of the solvent.

Immediately after treatment, each male was caged with an untreated test-stock female, 10-13 weeks old. The offspring were counted, sexed and carefully examined externally at birth for any variant phenotype. The litters were examined again when cages were changed, the final examination for specific locus mutations being at 19-21 days of age. At that time the offspring were ledgered and ear punched. Specific locus mutations were confirmed by an allelism test.

After they had been scored for specific locus mutations, all F_1 offspring were examined biomicroscopically, with the aid of a slit lamp,

Table 1. Recessive and Dominant Mutations Induced in Mice by γ-Rays

Dose (R)	Dose rate (R/min)	Germ cell stage treated	Number of F1 offspring	Number of mutations at 7 specific loci	Mutation rate per locus $\times 10^5$	Number of dominant cataract mutations	Mutation rate per gamete $\times 10^5$
0	–	–	103 218	6[a]	0.8	–	–
0	–	–	8 174	2	3.5	0	0
534	53	postspermatogonia	1 721	3	24.9	1	58.1
600	53	postspermatogonia	865	3	49.5	1	115.6
455+455	55	postspermatogonia	272	2	105.0	1	367.6
534	53	spermatogonia	10 212	7	9.8	3	29.4
600	53	spermatogonia	11 095	14	18.0	3	27.0
455+455	55	spermatogonia	5 231	9[b]	24.6	6	114.7

[a] Untreated historical control of the laboratory.
[b] A simultaneous d-se mutation included, which is caused by double non-disjunction.

for lens opacities between the ages of 4 and 6 weeks. Pupils were
dilated by treatment with atropin. Presumed mutant individuals exhibiting a lens opacity were outcrossed to strain-101 mice. The
phenotype was confirmed by breeding tests.

4.2. Results

    The results of the radiation experiments were discussed re-
cently in detail [12]. A summary of the data is given in Table 1.
A total of 15 dominant cataract mutations were observed in 29,396
offspring. Comparing the overall frequency of induced cataract mutations and specific locus mutations in postspermatogonia and spermatogonia, there were 2.5-2.7 times more recessive mutations than dominant mutations induced by γ-radiation. Taking into account that in
humans 20 well established dominant cataracts are known, according
to McKusick [16], then it is very likely that approximately three
times as many loci coding for dominant cataracts are scored in this
experiment than for recessive mutations. Therefore, on a per locus
rate, radiation induced approximately 8 times more recessive mutations than cataract mutations. In this regard it is interesting to
note that H. J. Muller in 1950 reported that the ratio of radiation-
induced recessive visibles to dominant mutations in spermatozoa of
Drosophila is 5:1 [37].

    The two characteristic features of radiation-induced specific
locus mutations and of dominant skeletal mutations can also be demonstrated for the induction of dominant cataracts. The mutation frequency per gamete for dominant cataracts due to the fractionation
of the dose is 2.6 times higher than the combined frequency of the
mutation rate in spermatogonia after single exposure. Similarly the
frequency of dominant mutations induced in postspermatogonial stages,
is 2-4 times higher than in spermatogonia [9].

    The radiation-induced frequency of specific locus mutations
agrees well with the results observed in Harwell [38] and Oak Ridge
[39].

    The frequency of chemically-induced dominant and recessive mu-
tations is summarized in Table 2. These results were partly discussed in recent publications [12, 19, 40]. The testing of some
presumed cataract mutations is still incomplete. In the 250 mg/kg
dose group a total of 6 specific locus mutations appeared to be
phenotypically intermediate between the test allele and the wild-
type, and for some of them it was not possible to tell, solely from
the phenotype, which locus was involved. These mutations were excluded from an earlier publication [12]. After completing the
allelism test, these mutations are now included in Tables 2 and 3.
In addition, 2 lens opacities of the 250 mg/kg ENU-group (Table 2)
originally not classified as presumed mutations [12, 19] proved to

Table 2. Ethylnitrosourea-Induced Recessive and Dominant Mutations in Mice

Treatment	Dose (mg/kg)	Germ cell stage treated	Number of F1 offspring	Number of mutations at 7 specific loci	Mutation rate per locus x 10⁵	Number of dominant cataract mutations	Mutation rate per gamete x 10⁵
Solvent control	0	—	67 243	3[a]	0.6	—	—
	0	—	2 517	0	0	0	0
ENU	160	postspermatogonia	2 630	0	0	0	0
		spermatogonia	6 435	27[b]	59.9	11	170.9
	250	postspermatogonia	3 360	0	0	2	59.5
		spermatogonia	9 352	60[c]	91.7	12	128.3

[a] Treated historical control of the laboratory.

[b] Includes 2 clusters of 2 mutations.

[c] Includes 5 clusters of 2 mutations and 1 cluster of 3 mutations.

Table 3. Distribution Among Seven Loci of Mutations in Spermatogonia of the Mouse After Treatment with Ethylnitrosourea

Dose (mg/kg)	Locus								Total
	a	b	c	d	d/se	se	p	s	
0	0	5	2	1	0	0	3	0	11
160	2	5	7^a	3	0	3	4	3	27
250	2	6	6	15^b	0	6	21^a	4^c	60

[a] Includes 2 clusters of 2 mutations.
[b] Includes 2 clusters of 2 mutations and 1 cluster of 3 mutations.
[c] Includes 1 cluster of 2 mutations.

be of dominant inheritance. A total of 25 dominant and 87 recessive mutations were observed in 21,777 offspring. Clusters of mutations, more than one mutant offspring occurring in the progeny of a single treated animal occurred at the c-, d-, p-, and s-locus (Table 3). Excluding the highly unlikely occurrence of 2 or 3 independent mutations at the same locus in offspring of a treated male, the number of independent specific locus mutations in the low dose group is very likely 25 and in the high dose group 53. Comparing the frequency of dominant (23) and recessive mutations (78) in spermatogonia, 3.4 times more specific locus mutations were induced than dominant mutations. On a per locus basis the ratio is approximately 1:10.

The dose-effect-curve of ENU-induced specific locus mutations is linear (Fig. 1). The regression analysis of the specific locus mutations gives a linear equation, $Y = 1.1 + 0.364 \times D$ ($r^2 = 0.99$, goodness of fit $\chi^2 = 0.83$). A similar dose-response relationship was observed for procarbazine in the dose range of 200–600 mg/kg. The point estimate of the mutation rate in spermatogonia for procarbazine at the 800 mg/kg level was one-third of that expected from a linear extrapolation [41].

One possible way of accounting for the departure from linearity is to assume that there is heterogeneity among A_S-spermatogonia with regard to survival and that sensitivity to damage resulting in cell death is positively correlated to sensitivity to mutation. Thus, at the higher dose, more sensitive A_S-spermatogonia would be killed and the mutation rate in the more resistant cells would be lower [42]. This simple explanation did not fit the data of the fractionation experiments with procarbazine [43]. The results of the ENU experiments contradict this explanation. The temporarily sterile phase is more pronounced after injection of 250 mg/kg of ENU than after

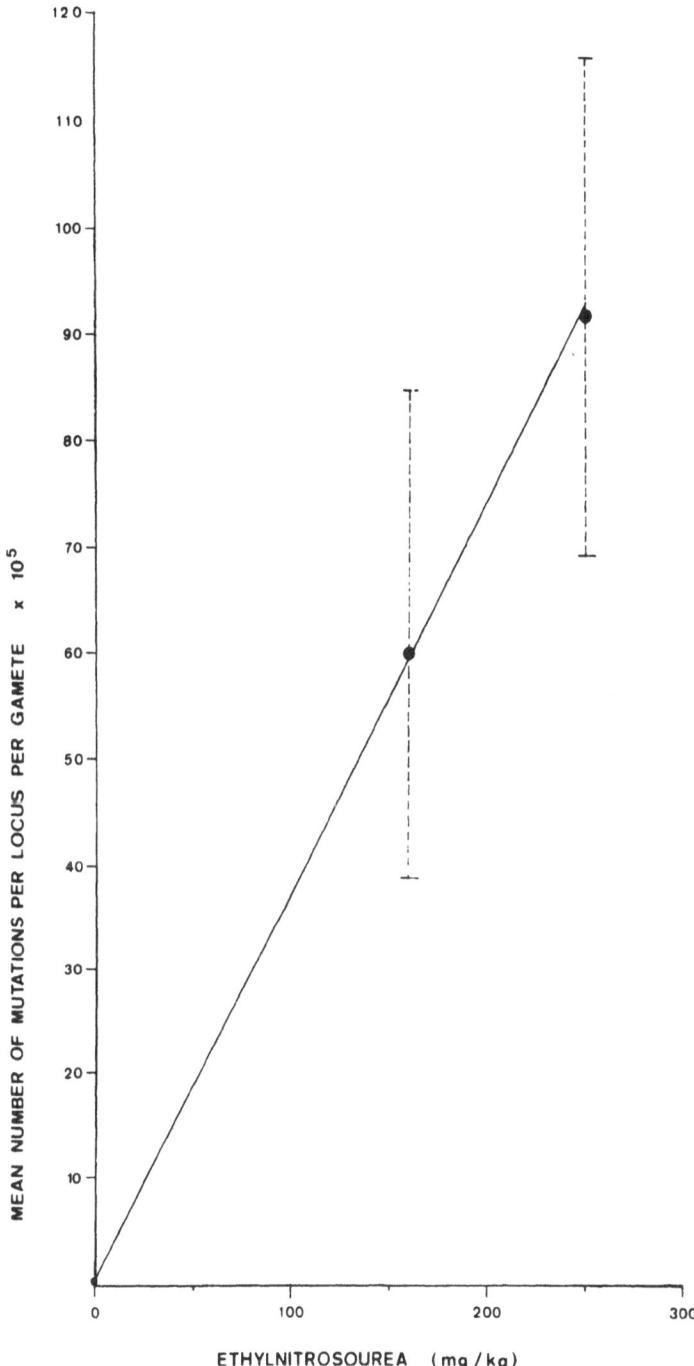

Fig. 1.   Specific locus mutation rates in spermatogonia of mice,
plotted with 95% confidence intervals for various doses of
ENU.   The straight line is fitted to the control, the 160
and 250 mg/kg single dose points.

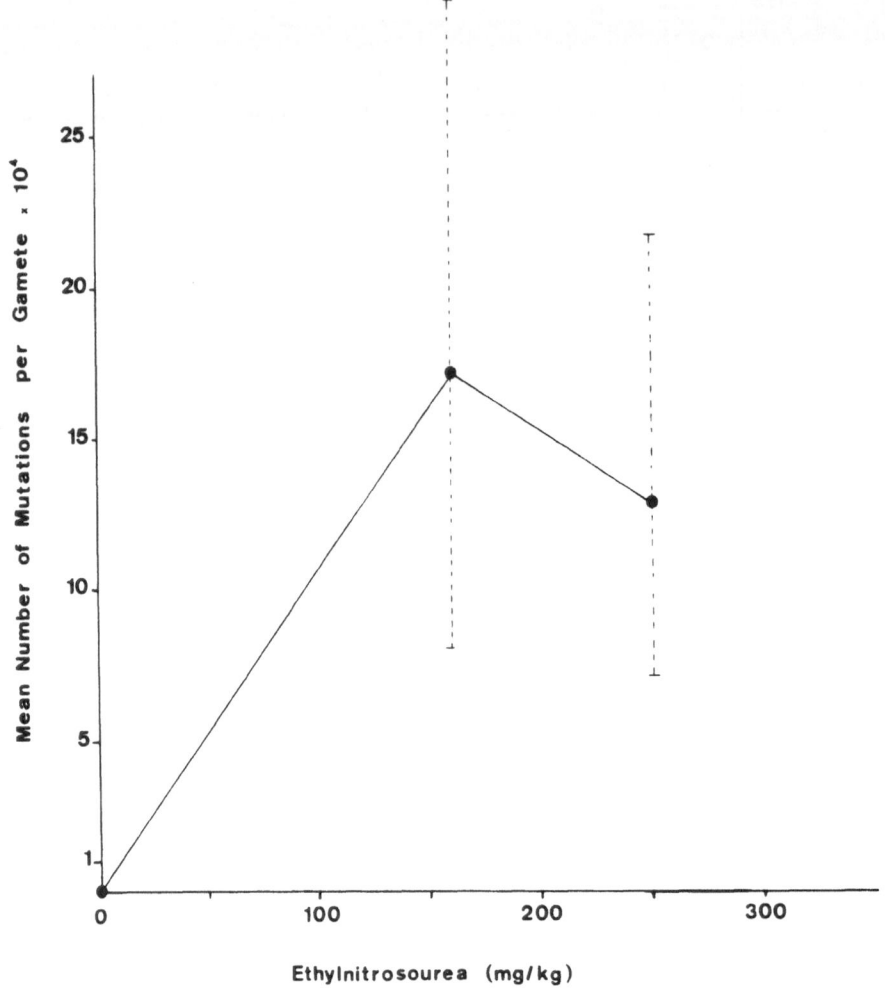

Fig. 2.   Dose-effect relationship for ENU-induced dominant cataracts
in spermatogonia of mice, plotted with 95% confidence intervals.

injection of 800 mg/kg of procarbazine, but the induced mutation frequency is 53 times higher after i.p. injection of 250 mg/kg of ENU
than after i.p. injection of 800 mg/kg of procarbazine.

The dose-effect relationship of ENU-induced dominant cataracts
is not linear [40], but similar to the dose-response-curve for the
induction of specific locus mutations by procarbazine (Fig. 2).   The
mutation rate per gamete per mg/kg for dominant cataract mutations
for the low dose group is 10.7×10^{-6} and for the high dose group
5.1×10^{-6}.   The different shapes of the dose-effect-curves for recessive and dominant mutations are not due to the distinct genetic
endpoints scored, but to the difference in the mutation spectrum of

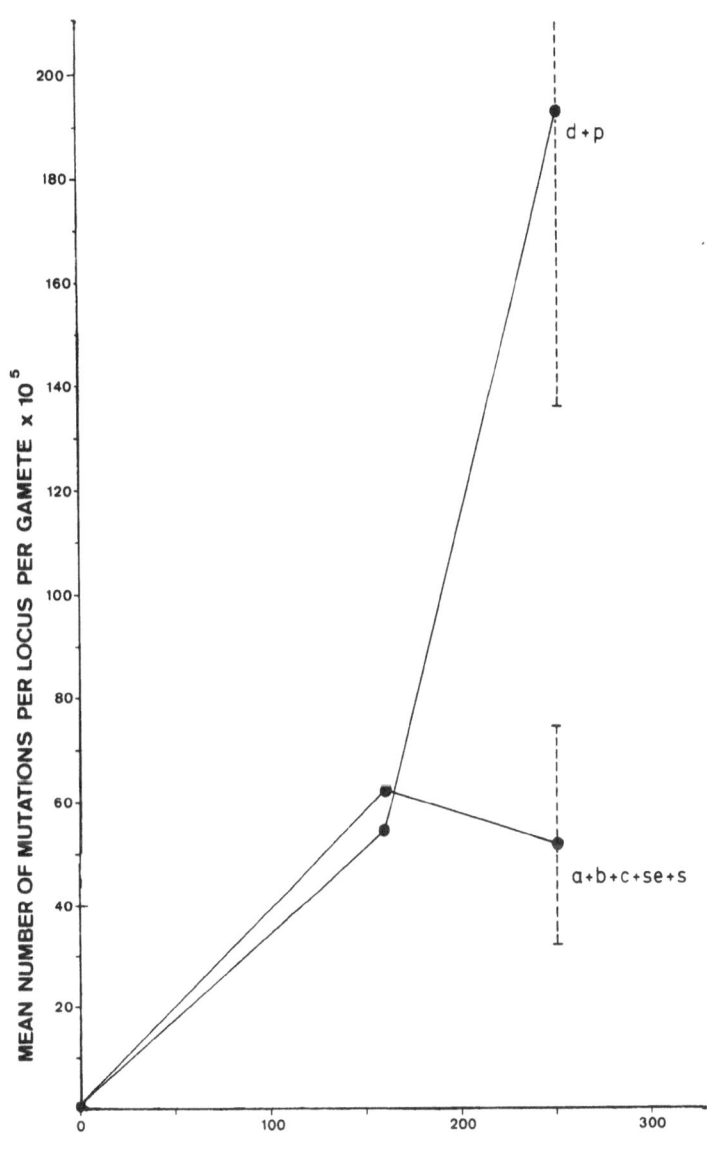

Fig. 3. Differential mutability of specific locus mutations in
         spermatogonia of mice after ENU treatment. The 250 mg/kg
         points are plotted with 95% confidence intervals.

         Group 1: d = dilute; p = pink-eyed dilution;

         Group 2: a = non-agouti; b = brown; c = chinchilla;
                   se = short-ear; s = piebald.

ENU-induced specific locus mutations (Table 3). The d-locus mutates
2 times more often in the high dose group than in the low dose group.
A still more pronounced difference exists for the p-locus. The p-
locus had also the highest mutation rate in our extensive procarbazine experiment [41].

It is interesting to compare, under the same laboratory conditions, the mutation spectra of chemically- and radiation-induced
specific locus mutations. Ionizing radiation induced more mutations
at the s-locus (36%) than procarbazine (13%) and ENU (8%). In addition, in 87 ENU- and 39 procarbazine-induced specific locus mutations in spermatogonia no double mutants were observed of the closely
linked d/se-loci. In a sample of 50 γ-ray-induced specific locus
mutations in spermatogonia of mice we observed 5 double mutants [41].
From a more extensive investigation of procarbazine-induced specific
locus mutations, including a viability test of the induced mutations,
it was concluded that procarbazine-induced mutations may be mainly
due to base-pair changes [41]. A similar conclusion may be reached
after completion of the viability test for ENU-induced specific locus
mutations.

The mutation spectrum changes with the applied dose of ENU.
This point is also illustrated by Fig. 3. The figure demonstrates
that the exceptionally high mutation frequency of ENU in the higher
dose group is mainly due to mutations at the d- and p-locus. The
difference in the mutation frequency per locus between the two groups
(d, p vs. a, b, c, se, s) is for 250 mg/kg ENU highly significant
(P < 0.0001). The dose-effect relationship for the remaining 5 loci
(a, b, c, se, s) is similar to the dose-effect-curve for ENU-induced
dominant cataract mutations (Fig. 2). In this context it is interesting to note that experiments in progress indicate that the point
estimate of the mutation rate of 40 mg/kg of ENU is below the frequency expected from a linear extrapolation (Fig. 1). These results
support the differences in the mutation spectrum already observed
for the 160 and 250 mg/kg group of ENU.

5.   ESTIMATION OF THE GENETIC RISKS

For the determination of the amount of human ill health arising from an increase in the mutation rate due to exposure to mutagenic agents, it is essential to recognize that in spite of extensive efforts, the attempts to detect radiation-induced genetic
damage in man, so far, have proved inconclusive. Therefore, the
UNSCEAR-Report [44] stated that there is no alternative for "using
results obtained with experimental animals in estimating rates of
induction in man. The limitations of such a procedure are obvious
when it is realized that animal species differ from each other in
their susceptibility to the induction of genetic changes by radiation and that there is no evidence indicating whether the genetic
material of man is more or less sensitive to radiation than that of

other animal species. The only mammal which has been studied in some detail with respect to radiation genetics is the mouse. Results of mouse experiments must therefore form the main basis for the assessment of genetic risks in man." This conclusion is likewise important for the attempt to estimate the risk of chemical mutagens to man. Using the mouse data for the risk estimation of chemical mutagens, it is necessary to adopt a principle enumerated in the BEIR-Report: "Use simple linear interpolation between the lowest reliable dose data... In order to get any kind of precision from experiments of manageable size, it is necessary to use dosages much higher than are expected for the human population. Some mathematical assumption is necessary and the linear model, if not always correct, is likely to err on the safe side" [45].

One way of expressing the genetic risk is the determination of the doubling dose. This approach has been used for data obtained with the specific locus method in mice [34, 44, 45, 46]. With the doubling dose method the relative individual and population risk can be determined.

5.1. Relative Individual Risk

A useful approach to determine the genetic risk of a patient is the comparison of the therapeutic dose with the doubling dose. The doubling dose can be defined as the dose necessary to induce as many mutations as occur spontaneously in one generation. One underlying assumption for the calculation of the doubling dose is a linear dose-response relationship. Another assumption is the similarity between spontaneous and induced mutations.

Detailed discussions of the individual risk of patients due to treatment with cyclophosphamide [36], isoniazid [15], and procarbazine [43] have recently been published. The calculation of the individual risk is of importance for patients treated with a drug. The individual risk can serve as a guideline for counseling a patient if, for instance, an abortion is advisable. In addition, from the individual risk in combination with the frequency of the disease, one can calculate the population risk.

5.2. Relative Population Risk

The ratio between the population dose and the doubling dose times 100 gives the expected increase of mutations due to the treatment with a mutagen in percent of the spontaneous mutation rate. Procarbazine-induced specific locus mutations fulfill the criteria for the estimation of the doubling dose [41]. Due to the detailed knowledge of the influence of the fractionation of the dose on the mutation rate and the different mutation frequencies of spermatogonia and oocytes of the mouse a realistic estimation of the population risk is possible [15].

Table 4. Estimated Effect of 0.01 Sv per Generation of High Intensity
          Exposure of Spermatogonia for 1 Million Liveborn Individuals

	Cataracts	Skeletal defects
Mutations/gamete/R	$0.45 - 0.55 \times 10^{-6}$	10.1×10^{-6}
Multiplication factor for the overall dominant mutation rate	36.8	4.6
Expected cases of dominant diseases	17 - 20	46

In contrast to the estimation of the relative risk, it is possible to determine directly the number of mutations expected in the
first generation due to the exposure with a mutagen.

5.3. Absolute Population Risk

Based on the induction of dominant mutations in mice, I developed in 1974/76 a concept for the direct estimation of the risk
of radiation-induced genetic damage to the human population expressed
in the first generation [13, 14]. The quantification of the genetic
risk is based on the following assumptions:

1. The dose-effect-curve for the induction of dominant cataract
    mutations is linear.

2. Dominant cataract mutation rates are representative for all
    dominant mutations.

3. The ratio of dominant cataract mutations (20) to the total
    number of well established dominant mutations (736) in man
    [16] is the same as in the mouse. This ratio gives a multi-
    plication factor of 36.8. The multiplication factor is used
    to convert the induced mutation rate of dominant cataracts to
    the estimation of the overall dominant mutation rate.

The radiation-induced frequency of dominant cataracts for single
exposure (Table 1) is $0.45-0.55 \times 10^{-6}$ mutations/gamete/R. The
mutation rate has to be multiplied by 36.8 for the calculation of
the overall frequency for dominant mutations [$17-20 \times 10^{-6}$ mutations/
gamete/R]. For an acute exposure with high-intensity radiation of
spermatogonia of man with 0.01 Sv, one can expect 17-20 induced dominant mutations in the first generation in 1 million liveborn individuals (Table 4).

No experimental data are available for the induction of dominant cataract mutations by low-intensity radiation in male mice or for the mutation rate in female mice. Therefore, an estimation of the population risk is only possible with reservations. Such an estimation can only be based on the generalization of results obtained with the specific locus method [17]. In general, for exposure with low-intensity we expect only one-third of the frequency with high-intensity exposure [44, 45].

For the risk due to total population exposure, a sensitive factor for female mice of 1.4 [47] or 2 [17] is used. Because of the higher DNA content in the germ cells of man in comparison to those of mice [48] some authors use a factor of 1.2 for the extrapolation from mice to man [17, 44]. Other authors claim that mice and man are equally sensitive to the induction of mutations by radiation [46, 49]. Using these additional assumptions, it is possible to calculate the genetic risk of a population after exposure with 0.01 Sv per 1 million liveborn individuals to be 5 to 15 expected cases of dominant diseases in the first generation [11, 17, 50].

One advantage of the direct estimation of the genetic risk is that the results based on the induction of dominant cataracts can be compared with the data based on dominant hereditary disorders of another system, for example, on the induction of dominant skeletal mutations. For the quantification of the genetic risk based on the induction of dominant skeletal mutations similar assumptions are made as for the estimation based on dominant cataract mutations in mice:

1. The dose-effect-curve for the induction of dominant mutations affecting the skeleton of mice is linear.

2. Dominant skeletal mutation rates are representative for all dominant mutations.

3. The multiplication factor of 4.6 is used to convert the induced mutation rate of dominant skeletal defects to the estimation of the overall dominant mutation rate [17].

The radiation-induced frequency of dominant skeletal mutations for single exposure is 10.1×10^{-6} mutations/gamete/R [9]. This rate can be converted to the overall frequency for dominant mutation by multiplication with 4.6 [17]. It follows that for acute exposure with high-intensity radiation of spermatogonia of man with 0.01 Sv, we can expect 46 induced dominant mutations in the first generation in 1 million liveborn individuals (Table 4).

In contrast to the determination of the multiplication factor for cataracts, it is difficult to determine the multiplication factor for the conversion of the mutation rate for dominant skeletal

defects into the estimation of the overall dominant mutation rate
[17, 50]. These difficulties are also well documented by the UNSCEAR-
Report [49] and the BEIR-III-Report [45]. In addition, in the experiments to determine the mutation rate of dominant skeletal defects by
Ehling [9] there was for the great majority of cases no proof of inheritance of the defect. The evaluation of the mutations was mainly
based on statistical arguments. Three dominant mutations affecting
the skeleton were found to be transmitted to the second and later
generations [51]. A disproportionate micromelia (Dmm), an incomplete dominant mouse dwarfism with abnormal cartilage matrix observed
for the first time in these experiments [9] was described in detail
by Brown et al. [52]. The data suggest that a reduced or abnormal
cartilage matrix is the cause of dwarfism. In the experiments of
Selby [10] there was no objective morphological criteria for the determination of the mutation rate. The genetical proof was based on
a subjective evaluation of the presumed mutations.

Considering these differences between the quantification of the
genetic risk based on skeletal and cataract mutations and the sample
size of these experiments the estimations of Table 4 are in good
agreement. The problems of the quantification of the genetic risk
based on these two systems were discussed in detail recently [50].

Also for the induction of dominant skeletal mutations in mouse
spermatogonia we have no information about the effectiveness of exposure with low-intensity. Additionally, no experiments were done
until now with female mice.

Using similar suppositions as in the UNSCEAR-1977 Report [49]
one can calculate the population risk. This calculation is based
on the following assumption: 10.1×10^{-6} dominant skeletal muta-
tions/gamete/R $\times$ 4.6 (multiplication factor for conversion to the
overall mutation rate) $\times$ 0.3 (correction factor for dose rate) $\times$ 1.4
(exposure of female mice) $\times 10^6$ (offspring) equals 20. This estimation of the effect of 0.01 Sv per generation of low dose, low dose
rate, low LET irradiation on a population of one million liveborn
individuals of 20 cases with autosomal dominant diseases is based
on a generalization of the results obtained with the specific locus
method [50]. This figure of 20 cases is identical with the estimation of Ehling (1976) [14] and the UNSCEAR-1977 Report [49].

However, in the UNSCEAR-1977 Report a multiplication for the
conversion to the overall mutation rate of 5 was used. In addition, a multiplication factor of 2 was introduced in an attempt to
allow for the likelihood that many dominant mutations have not yet
been described. Furthermore, the expected overall mutation rate
was divided by two in order to allow for skeletal mutations which
are not of clinical significance. The additional multiplication
(×2) and division (:2) canceled each other out. It is questionable
to use such a rationality. The additional multiplication factor is

only meaningful if the number of loci can be estimated with some accuracy. This is not the case. The estimates of the total number of genes are orders of magnitudes apart [13]. Therefore, the author prefers the determination of a multiplication factor for the overall estimation to be based on the number of well established dominant genes in man [16]. It follows that with increasing knowledge the multiplication factor will become more accurate. In the case of the dominant cataract test it is likely that the multiplication factor will increase. For example, the multiplication factor for the conversion of the cataract data to an overall estimate was determined to be 32.4 based on McKusick's tabulation of 1975 and 36.8 based on the tabulation of 1978 [16].

Likewise, the provision that only one half of all skeletal defects is of clinical importance, is not very helpful. This assumption is an educated guess. Only if we would know the total number of genes could one determine this ratio accurately. However, if one would know the numbers of dominant genes one would directly multiply the induced mutation rate with the number of genes for the overall estimation of the radiation-induced genetic effect of the first generation.

For an improvement of the risk estimation it is necessary to extend the data basis for the induction of dominant mutations in mice, especially information about the mutation frequency in female mice and in the low-intensity range of exposure. In addition, it is necessary to emphasize that a mutation with high penetrance has a better chance to be recovered in these experiments than a mutation with low penetrance. Furthermore, it is very likely that the multiplication factors for the determination of the overall estimates will increase as more genes with dominant inheritance are described. Therefore, these quantifications underestimate the radiation-induced genetic damage of the first generation.

Similar to the quantification of the radiation-induced genetic damage of the first generation the data of Table 2 can be used for the estimation of the expected number of dominant mutations in the first generation after ENU-exposure.

6. CONCLUSIONS

Combining a specific locus experiment with the scoring of dominant cataract mutations increases the number of scorable loci by at least a factor of 4. More than 20 additional loci are easily detectable by the cataract assay.

Comparing the overall frequency of radiation-induced dominant cataract mutations with the frequency of specific locus mutations in spermatogonia, 2.5 times more recessive than dominant mutations were induced. Similarly, in the ENU experiment 3.4 times more

specific locus mutations were induced than dominant mutations. On
a per locus basis these ratios are approximately 1:8 and 1:10, respectively. This difference could be due to the differential mutability of dominant and recessive genes. Another explanation would be
the difference between selected and unselected loci. It is therefore desirable to estimate the induction of unselected dominant mutations directly.

A cataract is defined as an opacity of the lens causing a reduction of visual function. This definition and the similarity of
the organogenesis of the mammalian lens makes an estimation of the
dominant cataract mutations directly "in terms of the amount of human ill health that they are likely to cause" [53] possible.

Until now we have no information about the possible influence
of the genetic background on the frequency of radiation-induced gene
mutations in germ cells of the mouse. This lack of important knowledge is partly due to the fact that the specific locus method is restricted to the investigation of animals with a special genotype.
The cataract test system needs no special strains for the determination of the mutation rate of dominant genes. Therefore, the effect
of the genetic background of exposed animals on the frequency of induced gene mutations in mammals can be studied. In addition, this
method can be used for the determination of the mutation frequency
in different species.

As the number of induced dominant cataract mutations continues
to accrue and to be genetically characterized, the number of mutable
loci with effects on the lens will be determined. It is one aim of
these studies to transform the dominant cataract test system into a
mutation test in which the number of mutable loci is known.

The metabolism of the lens is relatively simple. Therefore,
experiments are in progress to determine the biochemical defects of
the recovered mutations. After isoelectric focusing of soluble lens
proteins on ultrathin polyacrylamide gels, it could be shown that,
in the region of the β- and γ-crystallins, protein bands present in
the wild-type were not visible in the heterozygotes and homozygotes
of one recovered mutant. Activity determinations of various enzymes
of lens extracts revealed that the activities of enolase, lactate
dehydrogenase and hexokinase were enhanced in the mutant genotypes
as compared to the wild-type [54]. The isolated mutations will be
an ideal source to initiate studies in the field of developmental
genetics.

A recessive cataract gene in rabbits (kat-1) in the homozygous
state disturbs the metabolism of the lens and permits the penetra-
tion of water into the lens. This absorption of water, depending
on the state of hydration of the animal during the 2nd month of life,
underlies the evolution to complete opacification. Two groups of

rabbits with sutural cataracts, one receiving a diet poor in water
to cause dehydration of the animals, the other receiving a diet rich
in water were compared. Total cataract developed in 16% of the first
as against 85% in the second group [7]. Similar experiments to investigate the manifestation of cataracts in mice will be performed.
The systematic investigation of all induced cataracts in mice will
be an ideal source to study the genesis of hereditary cataracts in
mice. These studies could be a model for the elucidation of the
manifestation of cataracts in man.

ACKNOWLEDGMENT

Dominant cataract studies were supported by contract No. 305-
81-BIO D, and the specific locus experiments by 136-77-1 ENV D of
the Commission of the European Communities.

REFERENCES

1. J. Francois, Congenital Cataracts, Royal VanGorcum Ltd., Assen,
    Netherlands (1963).
2. E. von Hippel, Die Mißbildungen des Auges, in: Die Morphologie
    der Mißbildungen des Menschen und der Tiere, E. Schwalbe, ed.,
    Teil III, 1. Lieferung, pp. 1-66, Gustav Fischer, Jena (1909).
3. U. Ehling, Vererbung von Augenleiden im Tierreich, 65. Bericht
    Dtsch. Opthalm. Ges. Heidelberg 1963, pp. 228-238.
4. E. Marner, A family with eight generations of hereditary cata-
    ract, Acta Opthalm., 27:537-551 (1949).
5. M. Westhues, Der Schichtstar des Hundes, Arch. Tierheilk., 54:
    32-83 (1926).
6. E. von Hippel, Embryologische Untersuchungen über Vererbung
    angeborener Katarakt, über Schichtstar des Hundes sowie über
    eine besondere Form von Kapselkatarakt, Graefes Arch. für
    Ophthalmologie, 124:300-324 (1930).
7. U. Ehling, Untersuchungen zur kausalen Genese erblicher Kata-
    rakte beim Kaninchen, Z. menschl. Vererb.- und Konstitutions
    lehre, 34:77-104 (1957).
8. S. E. Smith and B. F. Barrentine, Hereditary cataract, A new
    dominant gene in the rat, J. Hered., 34:8-10 (1943).
9. U. H. Ehling, Dominant mutations affecting the skeleton in off-
    spring of x-irradiated male mice, Genetics, 54:1381-1389 (1966).
10. P. S. Selby, Induced skeletal mutations, Genetics, 92:s127-
    s133 (1979).
11. J. Kratochvilova and U. H. Ehling, Dominant cataract mutations
    induced by γ-irradiation of male mice, Mutation Res., 63:221-
    223 (1979).
12. U. H. Ehling, J. Favor, J. Kratochvilova, and A. Neuhäuser-
    Klaus, Dominant cataract mutations and specific-locus muta-
    tions in mice induced by radiation or ethylnitrosourea, Muta-
    tion Res., 92:181-192 (1982).

13. U. H. Ehling, Die Gefährdung der menschlichen Erbanlagen im
     technischen Zeitalter, Fortsch. Röntgenstr., 124:166-171 (1976).
14. U. H. Ehling, Estimation of the frequency of radiation-induced
     dominant mutations, ICRP, CI-TG14, Task Group on Genetically
     Determined Ill-Health (1976).
15. U. H. Ehling, Risk estimations based on germ-cell mutations in
     mice, in: "Environmental Mutagens and Carcinogens," Proceed-
     ings of the 3rd International Conference on Environmental Muta-
     gens, T. Sugimura, S. Kondo, and H. Takebe, eds., pp. 709-719,
     University of Tokyo Press, Tokyo/Alan R. Liss, Inc., New York
     (1982).
16. V. A. McKusick, Mendelian Inheritance in Man, 5th Edition, The
     Johns Hopkins University Press, Baltimore-London (1978).
17. U. H. Ehling, Strahlengenetisches Risiko des Menschen, Umschau,
     80:754-759 (1980).
18. J. Kratochvilova, Dominant cataract mutations detected in off-
     spring of gamma-irradiated male mice, J. Hered., 72:302-307
     (1981).
19. J. Favor, ENU-induced dominant cataract mutations in mice, 13th
     Annual Meeting of the Environmental Mutagen Society (USA),
     Boston, Massachusettes, 25.2-2.3.1982, p. 81.
20. J. Favor, The penetrance value tested of a presumed dominant
     mutation heterozygote in a genetic confirmation test for a
     given number of offspring observed, Mutation Res., 92:192
     (1982).
21. R. A. Fisher, Statistical Methods for Research Workers, Oliver
     and Boyd, Edinburgh (1950).
22. E. L. Crow and R. S. Gardner, Confidence intervals for the ex-
     pectation of a poisson variable, Biometrika, 46:441-453 (1959).
23. M. L. Watson, Blind - a dominant mutation in mice, J. Hered.,
     59:60-64 (1968).
24. R. G. Tissot and C. Cohen, A new congenital cataract in the
     mouse, J. Hered., 63:197-201 (1972).
25. O. E. Paget, Cataracta hereditaria subcapsularis: Ein neues,
     dominantes Allel bei der Hausmaus, Z. Indukt. Abstamm. Verer-
     bungsl., 85:238-244 (1953).
26. A. C. Verrusio and F. C. Fraser, Identity of mutant genes
     "Shrivelled' and cataracta congenita subcapsularis in the
     mouse, Genet. Res., Camb., 8:377-378 (1966).
27. B. V. Konyukhov and N. A. Kolesova, A study of effects of the
     mutant gene CAT^{Fr} single dose in ontogenesis of mice, Onto-
     genesis, Academy of Sciences, USSR, 12:298-305 (1981).
28. M. Hatara, R. Shoji, and R. Semba, Genetic background and ex-
     pressivity of congenital cataract in mice, Japan. J. Genetics,
     53:147-152 (1978).
29. K. Theiler, D. S. Varnum, and L. C. Stevens, Development of
     Dickie's small eye, a mutation in the house mouse, Anat. Em-
     bryol., 155:81-86 (1978).
30. S. Oda, T. Watanabe, and K. Kondo, A new mutation, eye lens
     obsolesence, Elo on chromosome 1 in the mouse, Japan. J. Ge-
     netics, 55:71-75 (1980).

31. J. S. Gower, Research communication, Mouse Newslett., 8:14 (Suppl.) (1953).
32. M. F. Lyon, S. E. Jarvis, I. Sayers, and R. S. Holmes, Lens opacity: a new gene for congenital cataract on chromosome 10 of the mouse, Genet. Res., Camb., 38:337-341 (1981).
33. Mouse News Letter, No. 65, July 1981, A. G. Searle, ed., M. R. C. Laboratories, Carshalton, England.
34. U. H. Ehling, Specific-locus mutations in mice, in: "Chemical Mutagens," A. Hollaender and F. J. de Serres, eds., Vol. 5, pp. 233-256, Plenum Press, New York (1978).
35. L. B. Russell, P. B. Selby, E. von Halle, W. Sheridan, and L. Valcovic, The mouse specific-locus test with agents other than radiations, Interpretation of data and recommendations for future work, Mutation Res., 86:329-354 (1981).
36. U. H. Ehling, In vivo gene mutations in mammals, Proc. of the Symposium "Critical Evaluation of Mutagenicity Tests," Bundes-gesundheitsamt, Berlin, 22.-25.2.1982, Walter de Gruyter, Berlin-New York (in press).
37. H. J. Muller, Radiation damage to the genetic material, Am. Sci., 38:33-59 (1950).
38. M. F. Lyon, D. G. Papworth, and R. J. S. Phillips, Dose-rate and mutation frequency after irradiation of mouse sperma-togonia, Nature, 238:101-104 (1972).
39. W. L. Russell, Studies in mammalian radiation genetics, Nu-cleonics, 23:53-56, 62 (1965).
40. J. Favor, A. Neuhäuser-Klaus, and U. H. Ehling, A comparison of ENU induced dominant cataracts and recessive mutations in spermatogonia of mice, 12th Annual Meeting of the European En-vironmental Mutagen Society, Dipoli, Espoo, Finland, 20.-24.6. 1982, p. 188.
41. U. H. Ehling and A. Neuhäuser, Procarbazine-induced specific-locus mutations in male mice, Mutation Res., 59:245-256 (1979).
42. W. L. Russell, L. B. Russell, and E. F. Oakberg, Radiation genetics of mammals, in: "Radiation Biology and Medicine," W. D. Claus, ed., pp. 189-205, Addison-Wesley, Reading, Mass., USA (1958).
43. U. H. Ehling, Induction of gene mutations in germ cells of the mouse, Arch. Toxicol., 46:123-138 (1980).
44. UNSCEAR-Report (United Nations Scientific Committee on the Effects of Atomic Radiation), Supplement No. 14, United Nations, New York (1966) (Quotation page 8).
45. BEIR-Report (Biological Effects of Ionizing Radiations), The Effects on Populations of Exposure to Low Levels of Ionizing Radiation: 1980, National Academy Press, Washington, D.C. (1980) (Quotation page 82).
46. K. Sankaranarayanan, Genetic Effects of Ionizing Radiation in Multicellular Eukaryotes and the Assessment of Genetic Radia-tion Hazards in Man, Elsevier Biomedical Press, Amsterdam (1982).

47. W. L. Russell, Mutation frequencies in female mice and the es-
     timation of genetic hazards of radiation in women, Proc. Natl.
     Acad. Sci. USA, 74:3523-3527 (1977).
48. S. Abrahamson, M. A. Bender, A. D. Conger, and S. Wolff, Uni-
     formity of radiation-induced mutation rates among different
     species, Nature, 245:460-462 (1973).
49. UNSCEAR-Report (United Nations Scientific Committee on the
     Effects of Atomic Radiation), Sources and Effects of Ionizing
     Radiation, United Nations, New York (1977).
50. W. Jacobi, H. G. Paretzke, and U. H. Ehling, Strahlenexposition
     und Strahlenrisiko der Bevölkerung, GSF-Bericht S-710, Gesell-
     schaft für Strahlen- und Umweltforschung, Neuherberg (1981).
51. U. H. Ehling, Evaluation of presumed dominant skeletal muta-
     tions, in: "Chemical Mutagenesis in Mammals and Man," F. Vogel
     and G. Röhrborn, eds., pp. 162-166, Springer-Verlag, Berlin-
     Heidelberg-New York (1970).
52. K. S. Brown, R. E. Cranley, R. Greene, H. K. Kleinman, and
     J. P. Pennypacker, Disproportionate micromelia (Dmm): an in-
     complete dominant mouse dwarfism with abnormal cartilage matrix,
     J. Embryol. Exp. Morph., 62:165-182 (1981).
53. H. B. Newcombe, Mutation and the amount of human ill health, in:
     "Radiation Research - Biomedical, Chemical, and Physical Per-
     spectives," Proc. 5th Intern. Congr. Rad. Res. 1974, O. F.
     Nygaard, H. I. Adler, and W. R. Sinclair, eds., pp. 937-946,
     Academic Press, New York-San Francisco-London (1975).
54. W. Pretsch, D. J. Charles, and J. Kratochvilova, Untersuchungen
     an Linsenextrakten von Mäusen mit erblicher Katarakt: Ultra-
     dünnschicht-isoelektrische Fokussierung und Enzymanalysen, in:
     "Elektrophorese Formum '80," B. J. Radola, ed., pp. 75-80,
     Technische Universität München (1980).

APPLICATIONS IN GENETIC RISK ESTIMATION OF DATA ON THE

INDUCTION OF DOMINANT SKELETAL MUTATIONS IN MICE*

Paul B. Selby

Biology Division
Oak Ridge National Laboratory
Oak Ridge
Tennessee 37830

1. INTRODUCTION

The study of dominant skeletal mutations has major applications in genetic risk estimation and complements the specific-locus method with respect to the type of information provided. Specific-locus experiments are useful for determining whether heritable gene mutations are induced in mammalian germ cells by an agent. They can also be used quantitatively to make extrapolations to exposure levels encountered by humans, and they are particularly well suited for evaluating the effects of physical and biological variables on the frequency of transmitted mutations. They cannot, however, be used to determine the magnitude of damage that would be encountered in first-generation offspring. Two major reasons for this are (1) uncertainty as to how representative the few specific loci studied are of those genes that can mutate to cause dominant effects, and (2) uncertainty over how many genes can mutate to cause such effects. It is generally recognized that only those newly induced mutations that can cause damage in heterozygotes are of any significant importance for a great many generations to come [1, 2]. Studies on the induction of dominant skeletal mutations [4-6] and of dominant cataract mutations [3] have now provided means of estimating this type of damage. As more comparisons become possible between results obtained for dominant mutations and those for specific-locus muta-

---
*Research sponsored by the Office of Health and Environmental Research, U.S. Department of Energy, under contract W-7405-eng-26 with the Union Carbide Corporation.

tions, a firmer basis will be established for extrapolating from specific-locus mutation-rate data to the magnitude of risk.

      Our first experiment using the dominant skeletal approach [4-6]
made use of breeding tests so that it was possible to determine with
certainty which abnormalities were caused by dominant mutations. The
results of that large experiment, in which 37 dominant skeletal mutations were found in 2646 offspring, have been applied by international and national committees in estimating the genetic hazard to
humans from radiation [1, 2]. To accomplish this with the data then
available, it was necessary to make corrections based on specific-
locus data [7, 8] in order to extrapolate downward from the amount
of damage found following massive exposure to the amount expected
from an exposure of 1 R. The resulting estimate of the induced frequency for 1 R of exposure for dominant skeletal mutations was then
multiplied by 10 (or by the extremes of 5 and 15) to expand from one
body system to all body systems, and by 1/2 (or by the extremes of
1/4 and 3/4) to restrict the estimate to only those mutations
thought to have effects that would, if they occurred in humans, cause
a serious handicap [1, 2]. The breeding-test method of studying the
induction of dominant skeletal mutations has some important merits
but is slow and cumbersome. In an attempt to devise a more rapid
means for studying the induction of dominant skeletal mutations, we
have developed three non-breeding-test methods, which are likely to
have wider application in mutagenicity testing.

2.   NON-BREEDING-TEST METHODS

      The three non-breeding-test methods are (1) the sensitive-in-
dicator method, (2) the multiple-anomaly inferential method, and (3)
the mutational-index method. All three were built on the foundation
of what was learned in the large breeding-test experiment concerning
distinctions between non-genetic variability in the skeleton and
phenotypes resulting from dominant mutations.

      In the non-breeding-test methods, mice are generally killed for
skeletal preparation between 6 and 8 weeks of age. They are coded,
prepared (stained with alizarin and cleared), examined, classified,
and uncoded. The entire skeleton anterior to the pelvic girdle is
prepared and examined for the 12 sensitive-indicator malformations
(Sec. 2.1) and for a few other specific anomalies. However, if any
major abnormalities are observed during the scoring for this list
of malformations, all parts of the skeleton that were prepared are
examined. The list of major effects that could not be overlooked
in scoring for the few specific malformations is very large. Thus,
one actually ends up knowing for each mouse whether or not it had
any one of hundreds of specific malformations. These are easily
kept track of by using a computer.

2.1. The Sensitive-Indicator Method

Any mouse having a sensitive-indicator malformation is a presumed mutant. The 12 sensitive-indicator malformations are as follows:

1. Subdivided interparietal
2. Fusion between a frontal and a parietal
3. Fusion between a squamosal and a parietal
4. Fusion between a triangular and a pisiform
5. Fusion between a lunate navicular and a triangular
6. At least one complete fourteenth rib
7. Twelve or fewer rib pairs
8. At least one rib on the seventh cervical vertebra
9. Vertebra prominens shifted to the third thoracic vertebra
10. Fusion between the first and second ventral ribs
11. Fusion between the first and second dorsal ribs
12. Clavicles only partially formed or absent

All 12 of the sensitive indicators were chosen before any data were
collected using this method. They were selected on the basis of our
earlier work [4-6], which suggested that each one of these malformations hardly ever (and perhaps never) occurs in nonmutants and that
each one can result from an induced mutation at any one of many different genes. None of these malformations occurred in any nonmutants
in our earlier work.

It must be stressed that these traits are sensitive indicators
of mutations in (101 × C3Hf) F_1 mice, but they would certainly not
all be sensitive indicators in all mouse strains. Indeed, some of
these abnormalities are common nonmutant variants in some strains.
Presumably, any strain would have its own array of sensitive indicators.

In earlier publications we applied the term sensitive-indicator method to the combination of what are here called the sensitive-
indicator method and the multiple-anomaly inferential method (Sec.
2.2). In that earlier terminology the mutation frequency was referred to as the probable mutation frequency; now it is termed the
presumed mutation frequency.

2.2. The Multiple-Anomaly Inferential Method

The multiple-anomaly inferential method also permits the identification of presumed mutants. Because the majority of mutants
identified by this method do not have any sensitive-indicator anomalies, this method permits the identification of many presumed mutations that would not be found using the sensitive-indicator method.
In applying this method, a mouse is a presumed mutant if it has two
or more rare (rare = less than 1 in 400 in control) major defects

that do not seem to result from one accident of development. An
alternative set of criteria applied in this method requires bilateral
occurrence of a single extremely rare (extremely rare = less than 1
in 1000 in control) major abnormality to classify a mouse as a presumed mutant.

We thus infer from the multiple anomalies present that the
mouse is the carrier of a dominant skeletal mutation. The two sets
of criteria applied in this method were developed during the course
of our earlier breeding-test experiment, in which all 15 F_1 offspring
meeting the criteria and having offspring were shown by means of
progeny testing to be mutants [6]. Accordingly, only a small fraction of presumed mutants based on these criteria would be expected
to be nonmutant variants.

2.3. The Mutational-Index Method

The mutational-index method provides an almost entirely independent means of determining whether dominant skeletal mutations
have been induced. Rather than yielding a presumed-mutation frequency, it permits the calculation of an index of mutation. Similarly to the sensitive-indicator method, skeletons are examined for
the presence of specific-malformations (termed index abnormalities).
The index abnormalities are as follows:

    B   = any one of the twelve sensitive indicators
    F-50 = absence of vertebra prominens
    R   = partially formed thirteenth rib on the right side
    Z   = partially formed thirteenth ribs, bilaterally
    S   = severe scoliosis or lordosis in the cervical, thoracic,
           or lumbar regions
    T   = various tubercula anteriora abnormalities
    V   = any true rib fusion other than the one that is a sensi-
           tive indicator
    N   = any departure from seven ventral rib pairs connecting
           to the sternum

The symbols shown are computer codes that will be used later in this
paper to symbolically refer to these groups of abnormalities. All
of the index abnormalities were chosen before any of the comparisons
of individual treatments that will be discussed later in this paper
were made.

Index abnormalities are known, or strongly suspected, to be
caused by dominant skeletal mutations. However, except for the
sensitive indicators, which are included with them, they probably
all occur at much too high a frequency as nonmutant variants to be
used by themselves in deciding that a particular mouse is a presumed
mutant. The more index abnormalities present in a given mouse, the
greater the likelihood that this mouse is a mutant. Because of this,

the index of mutation is calculated in such a way that more weight
is given to mice having more index abnormalities.

The first step in the calculation of the index of mutation is
to calculate the adjusted percentage of index anomalies for each experimental group and for the control. This adjusted percentage is
the sum of the index abnormalities (but no more than three for any
one mouse) divided by the number of mice in the group and multiplied
by 100. The index of mutation (or index) for an experimental group
is the experimental adjusted percentage of index abnormalities minus
the adjusted percentage of the control. Because individual mice can
contribute more than one event to the total, the index of mutation
is not a mutation frequency. Most mice with an index abnormality
have only one, however. For statistical evaluation, the index abnormalities are treated as if they are binomial variables; a chi-
square test is performed, provided no expected value in a particular
2×2 contingency table is less than five. In those few cases, the
Fisher exact test is used. There is no reason to expect the index
of mutation to be statistically significantly greater than 0 unless
the frequency of dominant skeletal mutations is higher. Because
sensitive indicators make up only a small fraction of all index abnormalities found, and because the index of mutation is independent
of the criteria applied in the multiple-anomaly inferential method,
the index of mutation is an almost entirely independent measure of
the induction of dominant skeletal mutations.

3. TYPES OF DATA OBTAINED USING NON-BREEDING-TEST METHODS

Each of the three methods could be applied individually in
evaluating the skeletons of the offspring of treated mice for induced genetic damage; however, it is advantageous to apply all three
of them simultaneously. If this is done, the three main types of
data obtained are (1) the frequency of presumed mutations, (2) the
frequency of clinically important presumed mutations, and (3) the
index of mutation.

3.1. The Frequency of Presumed Mutations

If the sensitive-indicator and multiple-anomaly inferential
methods are applied together, the number of presumed mutants is the
sum of those found by each method minus the number of mice classified as presumed mutants by both methods. Frequencies of presumed
mutations found in the experiments that are being used to test and
apply these methods are shown in Table 1. (In earlier publications
the presumed-mutation frequency was termed the probable mutation
frequency.) All experiments reported deal exclusively with offspring
derived from treated spermatogonia. In three of the experimental
groups, presumed-mutation frequencies are statistically significantly
higher than in the control group. The frequencies in these three
experiments are also statistically significantly higher than that in
the 300-R acute X-ray group.

Table 1.   Frequencies of Presumed Mutations after Various Treatments
           of Spermatogonia

| | Number of | | Point estimate of mutation frequency per gamete (%) | |
| | Presumed | F_1 offspring | Observed | Induced |
Treatment	mutants	examined	frequency	frequency
Control	5	1599	0.3	---
600 R acute X rays[a]	10	732	1.4[c]	1.1
100 R + 500 R acute X rays[a]	7	448	1.6[c]	1.2
300 R acute X rays[a]	0	388	0	negative
600 R chronic gamma rays[b]	2	178	1.1	0.8
Ethylnitrosourea at 150 mg/kg	7	331	2.1[c]	1.8

[a] Dose rate of 85-93 R/min.

[b] Dose rate of 0.005 R/min.

[c] Significantly higher than control and 300 R acute X rays, P < 0.02.

     Besides induced mutations, the number of presumed mutations can
include spontaneous mutations and nonmutant variants.   Induced presumed-mutation frequencies are obtained, of course, by subtracting
the control frequency from each experimental frequency.

3.2.   The Frequency of Clinically Important
         Presumed Mutations

     Not all mutations lead to equally serious phenotypes.   In fact,
many mutations at the molecular level are almost certainly unimportant as far as human health is concerned.   Even among skeletal
abnormalities that are easily detectable in heterozygotes, there are
many that have no clinical importance.   It is not yet known whether
the fraction of presumed mutants that is clinically important is
constant regardless of the mutagen.   Any list of presumed mutants
could be evaluated, one by one, by clinical geneticists as to severity.   Once a large number had been thus classified, the same classification could be applied consistently to future experiments.   No

Table 2. Indices of Mutation after Various Treatments of Spermatogonia

Treatment[a]	Number of mice with each anomaly[b] B	F-50	R	Z	S	T	V	N	Sum of anomalies in row	No. of F_1 offspring examined	Index[c] (%)
Control	2	12	12	3	4	10	2	2	47	1599	0
600 R acute X rays	7	7	10	1	4	5	1	1	36	732	2.0[d]
100 R + 500 R acute X rays	4	7	6	6	3	4	4	1	35[e]	448	4.4[d]
300 R acute X rays	0	6	3	1	1	3	0	0	14	388	0.7
600 R chronic gamma rays	0	1	4	1	1	1	1	1	10	178	2.7[f]
600 R chronic gamma rays, corrected	0	1	0	1	1	1	1	1	6	150	1.1
150 mg/kg ENU	4	4	11	2	0	4	4	1	30	331	6.1[d]
250 mg/kg ENU	0	0	0	0	1	1	1	0	3	8	35[d]

[a] Dose rates of radiation exposures are the same as in Table I.

[b] Index anomalies are identified by symbols used in the text.

[c] Index of Mutation = (adjusted percentage of index anomalies in experimental group) - (adjusted percentage of index anomalies in control group).

[d] Statistically significantly higher than 0, P < 0.01.

[e] Because one mouse had 5 index anomalies, 2 must be subtracted from this total when calculating the adjusted percentage of index anomalies (Sec. 2.3).

[f] Uncorrected index; next index down is corrected for likely preexisting spontaneous mutation by omitting 28 offspring of one sire.

attempt is being made here to classify the presumed mutants in Table 1 as to whether their phenotypes are of clinical importance. Such an exercise would be more meaningful when these experiments are farther along. Among the 31 presumed mutations reported (Table 1), there are many whose effects are clearly deleterious and many whose

effects are almost certainly unimportant. If one were attempting to
determine whether an agent would cause serious genetic damage, the
most important end point would be the induced frequency of clinically
important presumed mutations.

3.3.   The Index of Mutation

Table 2 shows the indices of mutation for the experiments described above. Analysis of these data shows that the indices of
mutation are statistically significantly higher in the following
comparisons: 250 mg/kg of ethylnitrosourea (ENU) is higher than all
others (though based on only 8 animals); 150 mg/kg of ENU and 100
R + 500 R of X radiation are both higher than 600 R and 300 R of X
radiation; and the ENU treatments, 100 R + 500 R, and 600 R are
all statistically significantly higher than 0. Because an elevation of the index implies that there is an increase in the mutation
frequency, these results reveal many differences in the induction
of dominant skeletal mutations. It is indeed remarkable that the
non-breeding-test methods of studying the induction of dominant
skeletal mutations have already permitted the demonstration of so
many statistically significant differences on the basis of such small
experiments.

The data for the 600-R chronic gamma-radiation exposure are
especially interesting because, at first glance, they seem to indicate a higher frequency, relative to the 600-R acute result, than
expected from specific-locus data, both for presumed mutations
(Table 1) and for the index of mutation (Table 2). The confidence limits for the presumed mutation frequency are still so wide, however,
that it is premature to worry that the dose-rate effect may be much
less pronounced for dominant mutations. As far as the index of mutation is concerned, perusal of pedigrees reveals that of the 10 index
abnormalities in the chronic experiment, 3 abnormalities of type R
and one of type Z (a very similar malformation) were found in the 28
offspring of a single irradiated male. The combined R + Z frequency
in this sibship is statistically significantly higher than the R + Z
frequency in the remaining 150 offspring in that experiment. It thus
seems likely that the sire of the 28 offspring in question was the
carrier of a dominant mutation having low penetrance for these traits.
Exclusion of his offspring from the analysis thus seems reasonable,
and when this is done, the index is only 1.1%, which is much more
in line with the expectation. (This male had no offspring meeting
presumed-mutation criteria.)

It is interesting to consider how well the relative presumed
mutation frequencies and indices of mutation found for different
treatments correspond to the relationships seen for specific-locus
mutations. Table 3 shows ratios of indices for various treatments,
as well as ratios of induced mutation frequencies for presumed mutations and specific-locus mutations. Only point estimates for the

Table 3. Ratios of Induced Mutation Frequencies and Indices of
          Mutation

Treatments being compared	Ratios for		
	Presumed mutation frequencies	Specific-locus mutation frequencies	Indices of mutation
600 R acute X rays divided by 300 R acute X rays	-3.4	1.6[a]	2.9
100 R + 500 R acute X rays divided by 600 R acute X rays	1.2	1.9[b]	2.2
600 R acute X rays divided by 600 R chronic gamma rays	1.3	3.0[c]	1.8[d]
150 mg/kg ethylnitrosourea divided by 600-R acute X rays	1.7	4.7[e]	3.0

[a] Based on data of W. L. Russell.[12]

[b] Based on data of W. L. Russell.[7,12]

[c] Based on ratios of slopes in paper by W. L. Russell et al..[13]

[d] For the chronic exposure, the corrected frequency (see text) is used.

[e] Based on data of W. L. Russell et al.[14] and W. L. Russell.[12]

ratios are given; their confidence limits are extremely wide, especially for the skeletal results. The ratios for the indices more
closely resemble those for specific-locus mutations than do the
ratios for presumed mutations. In both cases, however, there is a
reasonably good correspondence with specific-locus results. It will
be important to see if the correspondence improves as sample sizes
increase.

4. ASSESSMENT OF THE NON-BREEDING-TEST METHODS

    The non-breeding-test methods yield the following three major
types of information: the frequency of presumed mutations, the index of mutation, and the frequency of presumed mutations thought to
be of clinical importance. The first two of these are important for
determining whether a treatment induces dominant mutations, and, if
so, to what extent. The latter is especially important for risk
estimation.

Table 4.   Uncorrected Frequencies of Presumed Mutations after Pro-
           longed Treatment of Both Parents with Dichlorvos

Treatment	Number of		Point estimate of mutation frequency per gamete (%)
	Presumed mutants	F_1 offspring examined	
Regular control	5	1599	0.3
Dichlorvos[a]	4	517[b]	0.8
Concurrent control	7	505	1.4%

[a] Exposed for 80 days to one-third of a resin strip impregnated with dichlorvos
(Johnson Wax BOLT brand) that was placed on top of pen; mated immediately
afterward.

[b] 26 (or 5 %) of the F_1 offspring came from matings in which only the male had
been exposed; none of these 26 was a presumed mutant.

Experiments need not be huge to yield important results. For
example, the 150 mg/kg experiment with ENU involved less than one
person-month of labor and yet it provided the first demonstration
that a chemical can induce dominant deleterious mutations in spermatogonia. By means of the index of mutation, the 250 mg/kg exposure
of ENU was shown to be the most potent treatment studied so far with
less than one-half day of labor for one person.

5.   A PRACTICAL APPLICATION OF THE NON-BREEDING-TEST METHODS

5.1   Effects of Dichlorvos: A Preliminary Report

The non-breeding-test methods are being used in an experiment
in progress with dichlorvos which has already yielded some interesting results and illustrates a practical application of these
methods. Dichlorvos is widely used to kill insects and intestinal
worms. It has been shown to be a mutagen in a few short-term tests
[9], but it has not been tested for gene-mutation induction in
mammalian germ cells. In early 1980 we used this chemical to eradicate mites from the mice used in our skeletal-mutation experiments.
Applied as described below, dichlorvos is non-toxic. However, before using it routinely in our colony to eradicate mites, we needed
to make sure that it did not induce enough mutations to interfere
with our experiments.

To test this, we exposed C3Hf-strain females and 101-strain
males (the strains used in all experiments described earlier) for

80 days to one-third of a resin strip impregnated with dichlorvos
(Johnson Wax BOLT brand) placed on top of each cage. The total exposure was at least 20 times higher than that sufficient to extermin-
ate all mites. A concurrent control was also set up. Mice were distributed at random from our production stocks to the experimental and
control groups.

Animals were mated immediately after treatment. Thus, the offspring resulted from germ cells exposed chronically at many different stages of development in both sexes. About half of the data reported here come from litters conceived within 7 weeks after the end
of treatment. The treatment had no effect on fertility, so there
is no reason to expect clusters of mutations. The frequency of presumed mutations found is shown in Table 4, with the control described
earlier (and here called the regular control) included for easy comparison. The point estimate of the presumed-mutation frequency in
the exposed group is almost three times that in the regular control;
however, the frequency in the concurrent control is almost five times
that in the regular control. Fortunately, experiments using the non-
breeding-test methods provide additional information to help solve
such a puzzle.

The indices of mutation are shown in Table 5. There is no suggestion that dichlorvos has increased the index of mutation. In
view of past experience (Sec. 3.3), it seems unlikely that there
could be a sizeable increase in the frequency of dominant skeletal
mutations that did not also result in an increase in the index of
mutation. It therefore seems likely that the finding of a higher
presumed-mutation frequency in the dichlorvos group than in the regular control (Table 4) is due to something besides induced mutations.

It is noteworthy that the concurrent control is higher than the
regular control for both the index of mutation and the presumed-
mutation frequency. An examination of the syndromes themselves and
of the pedigrees of the mutants indicates that this is probably due
to two preexisting mutations. Thus, of the 11 presumed mutations
scored in the dichlorvos experiment and its concurrent control, two
(in the concurrent control group) caused the same sensitive indicator and came from the same litter and eight of the remaining nine
presumed mutants had syndromes that seemed quite similar. Each one
had fusions in the thoracic, lumbar, or sacral regions or in some
combination of these. Five had severe scoliosis in either the
cervical, lumbar, or thoracic region. Three had an exceedingly rare
shortening of vertebrae in the lumbar or sacral region, and three
exhibited dorsal rib fusions. All eight of these similar presumed
mutants were females, which may mean that the mutation is more likely
to be expressed in females. Yet another common characteristic of
these eight was that they were all caused by a mutation with low
penetrance. Among the eight different mothers that each produced
an offspring with the syndrome, the numbers of offspring examined

Table 5. Uncorrected Indices of Mutation after Prolonged Treatment
of Both Parents with Dichlorvos

Treatment	Number of mice with each anomaly[a]								Sum of anomalies in row	No. of F₁ offspring examined	Index[b] (%)
	B	F-50	R	Z	S	T	V	N			
Regular Control	2	12	12	3	4	10	2	2	47	1599	0
Dichlorvos[c]	0	5	2	2	2	1	1	1	14	517[d]	-0.2[e]
Concurrent control	5	8	6	2	5	3	7	1	37[f]	505	2.2

[a] Index anomalies are identified by symbols used in the text.

[b] Index of mutation = (adjusted percentage of index anomalies in group of
interest) - (adjusted percentage of index anomalies in regular control group
in this application).

[c] Exposed for 80 days to one-third of a resin strip impregnated with dichlorvos
(Johnson Wax BOLT brand) that was placed on top of pen; mated immediatedly
afterward.

[d] 26 (or 5%) of the F₁ offspring came from matings in which only the male had
been exposed; none of these 26 had any index abnormalities.

[e] If the concurrent control is used instead of the regular control in
calculating the index, the index is -2.4.

[f] Because one mouse had 12 index anomalies and another had 5, 11 must be
subtracted from the total when calculating the adjusted percentage of index
anomalies.

to date range from 14 to 30, with all but two having 21 or more.
Each one of these eight presumed mutants came from a different set
of parents, four from the concurrent control and four from the group
in which both parents were treated. The parental strain in which
the preexisting mutation arose is not yet known with certainty, but
certain indications point to the C3Hf strain. The C3Hf strain will
be assumed to be the strain of origin in the following discussion.
If, in fact, the mutation arose in the 101 strain, the conclusion
reached in this section would be the same.

The explanation for the sudden burst of mutants appearing in
the dichlorvos experiment and its concurrent control is that shortly
before the colony was first treated for mites, we made our first

Table 6. Corrections in the Frequencies of Presumed Mutations

----

| | # of presumed mutations/# of progeny = (%) for | |
Correction made	Concurrent control	Dichlorvos
None	7/505 (1.4%)	4/517[a] (0.8%)
Remove sibs if same syndrome	5/505 (1.0%)	4/517[a] (0.8%)
Remove through first cousins if same syndrome	4/505 (0.8%)	3/517[a] (0.6%)
Remove through first cousins-once-removed if same syndrome	4/505 (0.8%)	1/517[a] (0.2%)
Remove through second cousins if same syndrome	3/505 (0.6%)	1/517[a] (0.2%)
Remove through second cousins-once-removed if same syndrome	1/505 (0.2%	1/517[a] (0.2%)
Remove through third cousins if same syndrome	1/505 (0.2%)	0/517[a] (0%)

[a] As explained in Table IV, 26 of the F_1 offspring came from matings in which only the male was exposed.

cut-back of the C3Hf stock since we had expanded it from three sib pairs derived from Ehling's subline in 1976. By chance, one or both of the two parents to which we cut back must have carried a dominant skeletal mutation with low penetrance (less than 10%) and variable expressivity for serious effects. Because we had cut back to this one pair and then rapidly expanded the stock, we obtained a burst of mutants exhibiting similar syndromes. It should be realized that because this mutation can cause a large number of effects, it has some similarity to many other mutations found earlier in our experiments.

How does one deal with such a burst of spontaneous mutations in analyzing the results of an experiment? Certainly if a strong case can be made that a group of mutants results from a spontaneous mutation present in the stocks before the experiment was initiated, the group should not be treated as induced mutations in the analysis. Section 6 discusses the problem of preexisting mutations and how to handle it in future experiments. For the present experiment, we make use of the fact that the closer the degree of relationship of two mice having what appears to be the same syndrome, the greater the likelihood that their syndromes result from the same preexisting spontaneous mutation rather than from independently induced ones.

Table 7.   Differences in the Indices of Mutation in the Special
           Families[a]

Treatment	Index of mutation in special families	Index of mutation in remaining families[b]
Concurrent control	13	0.3
Dichlorvos	1.8	-0.6[c]

[a]The special families, of which there are 4 in each group, are those known to
be segregating the preexisting spontaneous mutation with low penetrance that
is described in the text.

[b]Regular control group's adjusted percentage of index anomalies was used in
calculating the indices.

[c]26 of the 433 offspring in the experimental group that were used in
calculating this index came from matings in which only the male was exposed.

(It should be noted that this argument could not be used if there
were such a long sterile period that clusters of induced mutations
would be expected.) Table 6 applies this approach to the dichlorvos
experiment. By removing sibs with the same syndrome, we eliminate
both occurrences of one sensitive indicator in the concurrent control. When corrections are made through first cousins-once-removed,
only one presumed mutation remains in the experimental group, and
four in the control. After corrections have been made through third
cousins, only one presumed mutant (having a unique sensitive indicator) remains in the control, and there are no presumed mutants
in the experimental group. In view of these results, it is concluded that there were no induced mutations in the experimental
group.

The explanation for the high index of mutation in the concurrent control (Table 5) appears to be that the preexisting spontaneous mutation with low penetrance described above causes several
of the index anomalies; thus, the presence of these mutations in-
flated the index. Table 7 shows what happens to the indices of
mutation when the four special families (that is, those known to be
segregating the mutation) are separated out of each group. In each
case, the four special families have an elevated index, and the remaining families have an adjusted percentage of index anomalies
similar to that of the regular control.

5.2. Implications of Results Obtained to Human Health

The corrected mutation frequency in the dichlorvos experiment
is 0/491. (These calculations use only the data for treatment of
both sexes.) In calculations below, the frequency of 0/487 will be
used because it would have been difficult to recognize an independent
mutation in the four mutants that resulted from the preexisting mutation. From these data, by using the binomial distribution, we can
rule out a mutation frequency of 0.62% at the 5% significance level.
The frequency of presumed mutations in the historical control is
taken to be 6/2100 or 0.29% (regular control + corrected concurrent
control). Thus, the induced frequency ruled out at the 5% significance level is 0.62-0.29%, or about 0.33%. The standard treatment
for eradication of mites in a mouse colony produces at most one-
twentieth of the exposure that was applied in the experiment. Because there was no effect of the experimental treatment with dichlorvos on fertility (spermatogonial stem-cell survival) and thus
no reason to expect a humped dose-response curve, it will be assumed that a straight-line extrapolation is conservative. The induced frequency ruled out at the 5% level may thus be divided by
20 without presumably underestimating the effect of the standard
treatment used to kill mites. Accordingly, the maximum induced presumed-mutation frequency from a treatment to eradicate mites is
0.016%. This is only a small fraction (about one-twentieth) of the
historical control mutation frequency of 0.29%, and an effect of
this size would not interfere with experiments.

A similar analysis for the index of mutation shows that the
maximum index from a treatment to kill mites is 0.0008%. This is
only an extremely small fraction of the adjusted percentage of index anomalies in the historical control, which is 3.01% with the 4
special families removed. (In this analysis, the special families
were included in the experimental group.) Thus, for both end points,
an exposure to eradicate mites would be expected to have no important
influence on results.

Can these data be used to say anything about genetic risk to
humans from dichlorvos? It is, of course, somewhat unusual to study
mutation induction in a mixture of germ cells from both sexes in a
single experiment. However, this can be done using the non-breeding-
test methods (and also the breeding-test method), and it may be a
rather good approach to use when first testing a chemical in mammals.
A negative result thus says something important about those germ-
cell stages that persist a long time. A positive result can be followed up by more restricted treatments to learn more precisely which
germ cells are at risk. One of the approaches that could be used in
applying the above dichlorvos data in risk estimation is as follows.
It was concluded above that there were no presumed mutations induced
in 487 offspring. The frequency of serious presumed mutations ruled
out at the 5% significance level is 0.62%, and the frequency of

serious presumed mutations in the historical control seems likely to
be 3/2100 or 0.14%. Accordingly, the induced frequency of serious
dominant skeletal mutations ruled out is 0.48%. If we assume, as the
UNSCEAR [1] did, that it is reasonable to multiply by 10 to expand
from the skeleton to all body systems, and if the presumed mutation
frequency obtained using the non-breeding-test methods includes about
three-fourths of all dominant skeletal mutations, then the maximum
risk for humans exposed in the same way as the mice in this experi-
ment would be $(0.0048) \times (10) \times (1.33) \times (1 \text{ million}) = $ about 64,000
serious genetic disorders per million liveborn humans. This is a
maximum risk of 6.4% per child.

Exposure conditions in the experiment can be reconstructed to
permit a determination of the approximate exposure in ppm. Once
this is done, if the exposure level and pharmacokinetic information
suggest that a couple's exposure would be equivalent to, for example, one-tenth of that in our experiment, the maximum risk for one
of their children would be 0.64%. The exposure level in our experiment must have been extremely high compared to that for many people,
so there may well be many couples in the population for which the
extrapolation factor might be as low as 1/1000. For these, the
maximum risk would be 0.0064% for a given child. The normal risk
for a person is about 10.7% of having a serious genetic disorder at
some time during life [2]. As is shown by these calculations, even
our rather small experiment (so far involving about 10 person-weeks
of labor) should permit some meaningful decisions concerning human
genetic risk. The vast amout of effort that has gone into the testing of dichlorvos for possible effects on health [9] did not include any experiments on gene-mutation induction in mammalian germ
cells. It is noteworthy that the first test of this end point now
indicates a relatively small risk.

6. METHODS FOR DEALING WITH SPONTANEOUS MUTATIONS
    PRESENT IN THE STOCKS

Since the beginning of our studies on the induction of dominant
skeletal mutations, we have been concerned about the possibility that
spontaneous mutations present in the stocks (here called preexisting
spontaneous mutations) might seriously affect our conclusions. In
our breeding-test experiment, this possibility can be virtually
eliminated [1, 10].

In experiments using the non-breeding-test methods, the experimental frequencies of presumed mutations consist of induced mutations, new spontaneous mutations, preexisting spontaneous mutations,
and nonmutant variants. To increase our confidence that the latter
two events would balance out well among all groups (experimental and
control), when we have selected the parents to be used in experiments, we have distributed the mice from individual families among
different groups. Because of similarities in the syndromes caused

by a single mutation, there should be no difficulty in recognizing
the occurrence of a preexisting spontaneous mutation if it has reasonably high penetrance. As a further aid in interpretation, parents
of most presumed mutants are preserved for study. Most of these
have not yet been examined.

In circumstances such as were described above (Sec. 5.1.), it
seems permissible to correct mutation frequencies by eliminating
mutations with low penetrance and variable expressivity that cause
similar syndromes if they are in closely related animals. However,
it would seem risky to continue such elimination of mutants back to
times before the cut-back occurred, because, at that time, the family
carrying the preexisting spontaneous mutation would have made up only
a small part of the total number of families from which parents were
selected in experiments. Such an elimination would be extremely
risky when dealing with a mutation that has highly variable expressivity for effects that are known to be caused by many mutations. Thus,
unless one can be relatively confident that a mutation is a preexisting spontaneous mutation, it seems much safer to assume that
preexisting mutations would essentially balance out among the experiments. Otherwise, the chances would be too high of eliminating new
mutations, both spontaneous and induced.

A modification in the maintenance of our 101 and C3Hf strains
should greatly improve the chances of recognizing preexisting spontaneous mutations. The usual procedure in maintaining an inbred
strain is to cut back to one pair of animals about every six generations and then rapidly expand from this pair to the size of colony
that is needed. This forces all animals to be closely related, but,
as the example above shows, it also ensures that there will be a
burst of preexisting spontaneous mutations following the cut-back
if the pair from which the stock is re-expanded includes a carrier.
For skeletal studies, it seems preferable to maintain several independent sublines of each inbred strain, and to identify the subline of origin for each parent used in an experiment. If the incidence of a particular syndrome turned out to be significantly
higher in the offspring originating from one subline than in those
originating from all others, this would indicate that the syndrome
was caused by a preexisting spontaneous mutation, which could be
ignored in analyses.

Even at the present time, the possible confounding factor of
preexisting spontaneous mutations does not seem to be having a big
influence on our ability to demonstrate different effects from different treatments. This is indicated by the approximate correspondence between relative specific-locus mutation frequencies and those
for both presumed mutation frequencies and indices of mutation
(Tables 1-3); it is extremely unlikely that a chance distribution
of preexisting   ntane us mutations could have produced such correspondence.

It is especially important that we have improved our ability
to deal with preexisting spontaneous mutations before we initiate
multigeneration experiments. Interpretation of such experiments
could be greatly complicated by an influx of many preexisting mutations.

7. ADVANTAGES OF USING PRESUMED MUTATION FREQUENCIES
    INSTEAD OF PROVED MUTATION FREQUENCIES
    IN ESTIMATING GENETIC RISK

For obvious reasons, breeding tests are needed to evaluate the
criteria that are used for identifying presumed mutations. Studies
of an untreated control are essential to determine whether these
criteria are revealing an increase in induced mutations. Without
such a firm foundation, presumed-mutation frequencies could easily
lead to erroneous conclusions. The breeding-test method also has
other important applications. However, once ways of identifying
mutations with considerable accuracy are in hand, frequencies of presumed mutations are to be preferred over frequencies of proved mutations when estimating genetic risk. Two arguments supporting this
point of view follow.

7.1. Less Expensive Experiments with Quicker Results

Samples of a given size can be accumulated at least 10 times
faster by using the non-breeding-test methods. The number of animals that must be raised to obtain a conclusive result is greatly
reduced because F_1 offspring need not be provided with mates and
later generations are not produced. There is no delay while waiting
for the F_1 progeny to produce sufficient offspring for an adequate
test of transmissibility. There is no complication if, for any of
a variety of reasons, animals fail to have offspring. Furthermore,
it is much easier to train co-workers to use the non-breeding-test
methods because they do not involve detailed study of the whole
skeleton.

7.2. Ability to Identify Mutations
      Having Low Penetrance

Three kinds of dominant skeletal mutations that would not be
detected in using the breeding-test method are (1) those killing an
F_1 offspring before the age of scoring, (2) those with such low
penetrance that there are too few of the F_1 offspring's progeny
affected to permit a statistical demonstration of transmission, and
(3) those with incomplete penetrance in which the F_1 offspring has
no malformation even though it carries a serious induced mutation
[4, 11]. If one is interested in estimating genetic risk in only
the first generation after exposure, only the first two of these
categories have any importance. Category 2 is especially relevant
to the bulk of human genetic disorders, which are those that are

sometimes described as irregularly inherited disorders. About 9% of liveborn humans are thought to have a serious handicap at some time during life as the result of an irregularly inherited disorder [2]. Indeed, probably the biggest uncertainty in risk estimation concerns this category. An experiment relying entirely on mutations confirmed by breeding tests cannot detect category-2 mutations. The importance of having the capability for identifying this type of mutation is underscored by our finding (Sec. 5.1) of the preexisting spontaneous mutation that has very low penetrance and variable expressivity for clinically important disorders. A well developed capability for detecting category-2 mutations will be of prime importance when multigeneration experiments are started, because one of the most interesting problems to elucidate in such experiments is the extent of accumulation of disorders caused by mutations having incomplete penetrance.

This paper illustrates some of the reasons why considerable progress can be expected concerning our knowledge of the induction of dominant damage. Comparisons between dominant-skeletal-mutation studies and specific-locus studies will be of particular interest. If specific-locus studies predict the relative amount of dominant damage reasonably well for a variety of mutagens, risk estimation from specific-locus data will be enhanced.

ACKNOWLEDGMENTS

The able assistance of S. K. Lee, T. W. McKinley, Jr., and G. D. Raymer was greatly appreciated. We also express our gratitude to D. P. Atkins and R. J. Raridon for helping us set up the computer data-management system for skeletal research that has made some of the advances described possible. We wish to thank Drs. E. F. Oakberg, L. B. Russell, and G. A. Sega for their helpful comments on the manuscript.

REFERENCES

1. UNSCEAR (United Nations Scientific Committee on the Effects of Atomic Radiation), Sources and Effects of Ionizing Radiation, Report to the General Assembly, with Annexes, pp. 425-564. United Nations, New York, Sales No. E.77.IX.1. (1977).

2. BEIR III Committee (Advisory Committee on the Biological Effects of Ionizing Radiation of the United States National Academy of Sciences), The Effects on Populations of Exposure to Low Levels of Ionizing Radiation, pp. 91-180 in typescript ed. and pp. 71-134 in printed ed., Nat. Acad. Press, Washington, D.C. (1980).

3. J. Kratochvilova and U. H. Ehling, Dominant cataract mutations induced by gamma-radiation of male mice, Mutation Res., 63:221-223 (1979).

4. P. B. Selby and P. R. Selby, Gamma-ray-induced dominant muta-
    tions that cause skeletal abnormalities in mice, I. Plan, sum-
    mary of results and discussion, Mutation Res., 43:357-375
    (1977).
5. P. B. Selby and P. R. Selby, Gamma-ray-induced dominant muta-
    tions that cause skeletal abnormalities in mice, II. Descrip-
    tion of proved mutations, Mutation Res., 51:199-236 (1978).
6. P. B. Selby and P. R. Selby, Gamma-ray-induced dominant muta-
    tions that cause skeletal abnormalities in mice, III. De-
    scription of presumed mutations, Mutation Res., 50:341-351
    (1978).
7. W. L. Russell, Effect of radiation dose fractionation on muta-
    tion frequency in mouse spermatogonia, Genetics, 50:282 (1964).
8. W. L. Russell, in: "Peaceful Uses of Atomic Energy," pp. 487-
    500, Intern. Atomic Energy Agency, Vienna (1972).
9. C. Ramel, J. Drake, and T. Sugimura, An evaluation of the ge-
    netic toxicity of dichlorvos, Mutation Res., 76:297-309 (1980).
10. P. B. Selby and P. R. Selby, Response to K. G. Luening and
    A. Eiche, Penetrance and selection, Mutation Res., 44:453-454
    (1977).
11. P. B. Selby, in: "Mutagenicity: New Horizons in Genetic
    Toxicology," J. A. Heddle, ed., pp. 385-406, Academic Press,
    New York (1982).
12. W. L. Russell, Studies in mammalian radiation genetics, Nu-
    cleonics, 23:53-56, 62 (1965).
13. W. L. Russell and E. M. Kelly, Mutation frequencies in male
    mice and the estimation of genetic hazards of radiation in
    men, Proc. Natl. Acad. Sci. USA, 79:542-544 (1982).
14. W. L. Russell, P. R. Hunsicker, G. D. Raymer, M. H. Steele,
    K. F. Stelzner, and H. M. Thompson, Dose-response curve for
    specific-locus mutations induced by ethylnitrosourea in mouse
    spermatogonia, Proc. Natl. Acad. Sci. USA, submitted (1982).

DETECTION OF ENZYME ACTIVITY VARIANTS IN MICE

Daniel A. Casciano, Jack B. Bishop,
Robert R. Delongchamp, and Ritchie J. Feuers

Department of Health and Human Services
Food and Drug Administration
National Center for Toxicological Research
Jefferson, Arkansas 72079

1. INTRODUCTION

The measurement and assessment of mutagenic risk to man partially depends upon further development of test methods which detect a broad spectrum of induced lesions in experimental animals [1, 2]. In addition to the phenotypic specific locus test and test systems which detect transmitted chromosomal damage, several promising biochemical test methods have been proposed [2-4]. However, these latter approaches measure qualitative characteristics of proteins to detect variants having alterations in enzyme electrophoretic mobility. Approximately two-thirds of all possible amino acid substitutions result in no apparent change in net charge but may result in an unstable protein. Additionally, some deletions, chain terminations, etc., may lead to inactive or absent proteins [5]. Modification of electrophoretic methods to detect these types of genetic alteration requires the laborious construction of specific genetic stocks.

As a complementary approach, we are developing and validating procedures which measure both qualitative and quantitative characteristics of proteins to identify enzyme activity variants in mice [6-8]. The goal of this approach is the identification of individuals that are heterozygous at an unspecified locus for a normal and induced mutant allele which alters the activity of one or more of the enzymes analyzed. This paper discusses the basic procedure and the results of a feasibility study of the assay methodology.

2. METHODS DEVELOPMENT

In order to achieve the proposed goal, it was necessary to select an appropriate animal model, develop automated enzyme analysis
techniques and establish baseline values for selected enzymes analyzed in liver and brain homogenates from C57BL/6J mice, establish
statistical criteria to distinguish abnormal from normal animals,
develop a computerized breeding and data analysis procedure, define
F_1 sample size, and define production and analysis of F_1, F_2, and
F_3 animals.

2.1. Species and Strain Consideration

The basic methodology must be a flexible one capable of identifying mutations through quantitative analysis of biochemical parameters. As with any test, species selection is an important early
consideration; however, because of reproductive capabilities and
relative economy, the mouse is generally most appropriate. For initial development, we used the inbred strain C57BL/6J, since, in such
a highly inbred strain, the potential homozygosity at loci which
might affect quantitative enzyme characteristics should result in
reduced population heterogeneity and, consequently, reduced genetic
variability across all generations. Further, numerous electrophoretic
variants are available on this background that could be employed in
backcross studies to characterize new enzyme mutants. However, the
procedure described herein has the potential for use in comparisons
among several strains or species.

2.2. Biochemical Considerations

Since it was necessary to perform mutliple enzyme analysis from
large numbers of individual animals, development of procedures for
use with automated instrumentation was required [6]. Assays were
performed on liver and brain homogenates using the miniature fast
analyzer (MFA) or micro-centrifugal analyzer (MCA). Liver and brain
tissue were homogenized in a solution to provide buffering capacity
and protection to yield highest activities with the lowest coefficient of variant for the largest number of enzymes. Enzyme reactions were conducted using 1, 2, or 5 µl of homogenate, 70 µl of
a specific assay solution in a total reaction volume of 140 µl.
Analysis methods had been developed by optimizing the linearity of
rate with increasing protein concentration or time by altering substrate, cofactor, coenzyme, coupling enzyme, and buffer concentration. Additionally, determinations of Michaelis-Menten kinetic con-
stants, K_m and V_{max}, were performed through specific and appropriate
alternation of an enzyme solution with regard to variation of substrate concentration of interest.

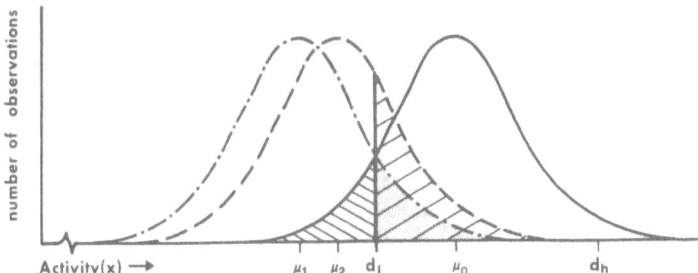

Fig. 1. Distribution of activities for an enzyme and misclassification probabilities for a normal and two mutant populations. (————) Represents the distribution of a population of normal C57BL/6J mice with a mean activity of μ_0
and standard deviation of σ for which $d_1 = [\mu_0(1 + \rho)/2\sigma]$
and $d_h = \mu_0 + 2\sigma$ where $\rho = 0.75$. (— – —) Represents the
distribution of a theoretical mutant population with a mean
activity of $\mu_1 < \mu_0$, where μ_1 happens to equal 0.75 μ_0, and
a standard deviation of σ. (– – –) Represents the distribution of another theoretical mutant population with a mean
activity of $\mu_2 < \mu_0$ but $>\mu_1$, where μ_2 happens to equal 0.80

μ_0, and a standard deviation of σ. ▨ $= \alpha$-error $=$

0.09 as in the average case (Table 1). ▱ $= \beta$-error

for direct analysis of an individual from the mutant
population with mean activity of μ_1. $Z_\beta = [\mu_0(1-\rho)/2\sigma]$.

▢ $= \beta$-error for direct analysis of an individual

from the mutant population with mean activity of μ_2. $Z_\beta = \dfrac{d_1-\mu_2}{\sigma}$.

2.3. Statistical Considerations

In order to determine the normal population variation, enzyme
activity distribution of individuals from a population of 22 litters
(n = 132) of C57BL/6J was determined for 29 brain and liver enzymes
[6]. Each of the 132 samples were analyzed individually for each
enzyme and all values were used in the statistical analysis. The
acceptable range of normal activity was defined using a statistical
model defining upper and lower bounds, d_h and d_1, so that variant
values may be isolated in the experiments with treated mice. If an

animal had normal enzymatic activity, the observation was assumed
to have a Gaussian probability density with mean μ and variance σ^2.
If the animal had abnormally low enzyme activity, the observed activity was assumed to have a Gaussian probability density with a decreased mean, $\rho\mu$ ($0 < \rho < 1$) and variance σ^2.

There are two possible errors associated with the limit d_1. An
α-error results when a normal animal is misclassified abnormal, and
a β-error results when an abnormal animal is misclassified normal.
Ideally, the choice of d_1 yields small α- and β-errors. Unfortunately, in the present case, μ, σ^2 and ρ are considered fixed for a
specific normal or altered enzyme, limiting our freedom to set the
value of α and β through the choice of d_1. As d_1 is increased
toward μ, the α-error increases; as d_1 is decreased toward $\rho\mu$, the
β-error increases. A compromise results in the rule $\alpha = \beta$. When
this rule was imposed on the normal probability distribution, d_1
was placed half-way between $\rho\mu$ and μ (Fig. 1). Then $d_1 = 1/2(\mu +
\rho\mu) = 1/2\mu(1 + \rho)$. In practice, μ is unknown and is determined by
the sample mean from a group of control mice. The α- and β-errors
resulting from this choice can be approximated by determining the
areas under the standard normal distribution curve associated with
Z_α and Z_β (probability functions) as defined by the equation $Z_\alpha =
Z_\beta = [\mu(1-\rho)/2\sigma]$. Little is known quantitatively about the abnormal
conditions which result in high enzyme activities; therefore, the
α-error is unknown and the β-error is approximately 0.025, with the
condition $d_h = \mu + 2\sigma$.

2.4. Other Considerations

The computer support required to perform this experiment was
extensive. A number of programs were developed for dosing and
breeding of animals, and for the collection, collation and analysis
of enzyme activity data. All of these various data compilations,
analysis and information programs were linked so that genetic and
breeding information could be correlated to enzyme activity data for
any F_2 animal. In turn, this information was correlated to data
concerning the F_2's siblings, parent F_1's, and finally the Po from
which the F_2 was originally derived. This feedback information loop
between genetic and biochemical data was an integral portion of the
test system and was essential in all decisions regarding classification of an animal.

3. BASIC PROCEDURE

As stated earlier, the goal of this test method is the identification of individuals that are heterozygous at an unspecified
locus for a normal and an induced mutant allele which alters the activity of one or more of the enzymes analyzed. These individuals
are identified by measuring quantitative biochemical characters using
three levels of screening (Fig. 2). The first screen is the automated

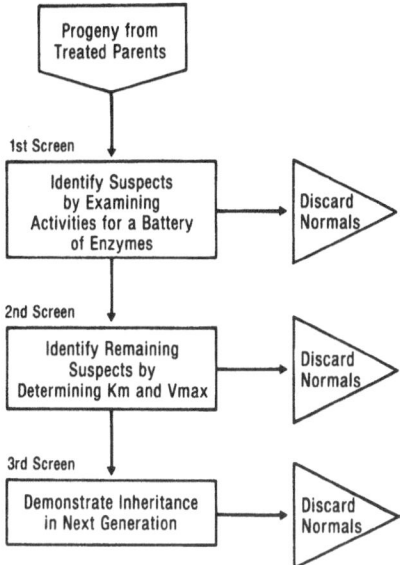

Fig. 2. Basic procedure. Experimental flow diagram of the multi-
          level screen. The flow is of a three level screening pro-
          cedure to determine inheritance of alterations in enzyme
          activity. After the first screen, only apparently abnormal
          enzymes are further studied.

activity analysis of a battery of enzymes from tissues of progeny
of both treated and control animals. Individual analysis values are
compared to a defined "normal" range to identify abnormal values.
The "normal" values of 29 brain and liver enzymes of C57BL/6J mice
and the techniques employed in their determination have been described [6]. Nineteen of these 29 enzymes had desirable characteristics such as low variability, prolonged stability, and metabolic
importance.

As previously stated, classification error may occur with any
quantitative measurement. These errors can be estimated only if
the probability distribution on enzyme activities of both normal and
mutant animals is either assumed or measured. The inability to observe large samples of known induced mutants, and thus to predict
the theoretical change in enzyme activity, limits calculation and
discussion of mutant misclassification to hypothetical situations.
However, to maximize the detection of mutants having even small activity changes, a conservative decision criteria was used to define
"normal" activity. A set of theoretical misclassification errors
for 19 enzymes in normal mice (α-error) calculated according to the
technique described in Section 2.3 are presented in Table 1. A
concurrent control population was used to check and periodically

Table 1.   Enzymatic Activities in Populations of
           C57BL/6J Mice

BRAIN ENZYME	Sex	Mean Activity	C V	α Error
Adenylate kinase	B	4731	0.11	0 12
Pyruvate kinase	B	14185	0 09	0.09
Malic dehydrogenase	B	33246	0 08	0 07
Adenosine triphosphatase	B	455	0.18	0 09
Creatine Phosphokinase	B	1693	0.03	0 01
Phosphoglucose isomerase	F	4237	0.02	0.01
	M	4224	0.04	0 01
LIVER ENZYME				
Glutamic oxaloacetic	F	11216	0.09	0 08
transaminase	M	9355	0.09	0.09
Isocitrate dehydrogenase	F	3342	0.10	0 12
	M	3149	0.12	0.16
Glutathione reductase	F	503	0 07	0.01
	M	550	0.07	0 01
Glutamate dehydrogenase	F	2034	0.16	0 22
	M	1661	0.15	0.20
Cytochrome C reductase	F	1331	0.18	0.08
	M	904	0 13	0.03
Glyoxalic acid reductase	B	17143	0.07	0.04
Alanine amino transferase	F	2527	0.09	0 10
	M	2161	0.09	0.10
Alcohol dehydrogenase	F	198	0 13	0.17
	M	170	0.09	0 08
Fatty acid synthetase	B	268	0 09	0 07
Fructose-1,6-phosphatase	F	175	0.14	0 19
	M	142	0.09	0.10
Lactate dehydrogenase	F	22847	0.13	0 18
	M	24268	0 09	0 10
Fructose-1,6-phosphate	B	376	0.15	0 19
aldolase				
Fructose-1-phosphate	F	933	0.08	0 05
aldolase	M	976	0 08	0 05

Activity = μmoles NAD$^+$ /h/g tissue
C.V. = coefficient of variation
B = Both sexes, F = female and M = male
n = 51 females and 45 males

update the decision criteria used to define "normal range." Upon
completion of the initial screen, several samples will be identified as suspect mutant. However, each will be suspect for only a
few enzymes, and true mutants should be among them.

A more refined classification is achieved by performing a second screen on these suspect progeny. Apparent K_m and V_{max} are determined on aberrant samples using automated Michaelis-Menten procedures [9]. K_m is a concentration term for each combination of
enzyme and substrate which provides information about the physical
state of the active site and about the general structure of the enzyme molecule. A change in K_m is indicative of an alteration in the
molecule which affects its affinity for substrate. V_{max} is a rate
term dependent upon enzyme concentration and reflects the catalytic
efficiency of those molecules present [9-12]. Although K_m and V_{max}
determinations are more time consuming than enzyme activity measurements, they are more definitive and less subject to error because
multiple analysis points provide greater detail about the quality
and quantity of molecules present.

Table 2.   Misclassification Probabil-
           ities for 3 Hypothetical
           Mutants

Aberrant Progeny*	Misclassification Probabilities** when individual β errors are:		
	0.09	0.23	0.40
2/10	0.01	0.03	0.09
3/10	0.05	0.12	0.27

*Minimum progeny with aberrant enzyme activity before parent is classified as mutant

**Calculated with the assumption that individual α-error=0.09

$$P = \sum_{i=0}^{n} \frac{10!}{i!(10-i)!} X^i (1-X)^{10-i} \text{ where } X=0.05 (1-\beta+0.09) \text{ and } n=1 \text{ or } 2$$

In the second screen, kinetic values of samples which contain
an activity aberrancy can be compared to samples of concurrent control animals and/or an internal standard of samples prepared from
pooled tissues of untreated animals. If all values agree, those
suspect animals in the first screen are reclassified as nonmutant.
If there is a lack of agreement between suspect and control animals,
then the suspect animals are subjected to the third screen.

A third screen is performed using progeny of the suspects remaining after the second screen. This screen confirms inheritance
in true mutants while substantially reducing the probability that
any α-error will remain. An induced mutant will be heterozygous,
and the expected Mendelian ratio will be 1:1 for normals and abnormals. A reasonably large population of progeny from suspects
must be sampled to yield a high probability that an altered allele
is represented. The third screen results in mutant classification
of suspects when a defined percentage of the progeny possess aberrant
activity similar to that observed in the first screen. Potential
mutant misclassification can be reduced by carefully choosing and
stipulating the percentage of progeny which must exhibit the aberrancy. For example, the probability that at least 2 of 10 of a true
mutant's progeny will inherit a mutation and that at least 2 heterozygous mutants will be correctly classified is high, provided the
change in activity is beyond the defined "normal" range and the coefficient of variation is sufficiently small. In practice the choice
of these sample sizes and percentages must reflect a compromise between statistical and logistical considerations. Table 2 presents
examples of expected classification errors under specific conditions.

The three screens used in the basic procedure are organized to
efficiently monitor a battery of enzymes with acceptable probabilities of F_1 misclassification. For a normal animal to ultimately
be misclassified as mutant it must be classified incorrectly at every

level. Although the probability of α-error for a specific enzyme may
be high in any one screen, it is unlikely that the same error would
be at each of several screens. The overall chance of misclassification for a normal animal is the product of the error rates at each
level. For example, an error rate as large as 0.25 at each level
will yield an overall rate of 0.016. However, a mutant must be correctly classified at every level if its ultimate classification is
correct. Because mutant misclassification rates are additive, the
decision criteria at each level must emphasize correct mutant classification.

In the first screen, enzymes are monitored using criteria which
keep mutant misclassification small, usually at the expense of an increased misclassification rate among normal animals. The goal in
this screen is to observe as large a portion of the genome as possible in every experimental animal so that subsequent screens can be
focused upon animals which are most likely to have a change in enzyme activity due to a mutation. The second and third screens are
used to confirm the suspected mutant.

4.   FEASIBILITY STUDY USING ETHYL METHANESULFONATE

On the day of treatment, Po males, 10 weeks of age, were randomly allocated to four treatment groups of 30 each and each male
was injected with 100, 175, 200 mg/kg ethyl methanesulfonate (EMS)
or with vehicle only.

4.1. Production of F_1's

Each male was randomly assigned three untreated virgin females,
10 weeks of age, for a 6 day mating interval beginning on day 5 after
treatment to sample treated epididymal and late testicular spermatozoa (day 5-10). At the beginning of days 11 and 22 after treatment, each male in the control and the 100 and 175 mg/kg dosed groups
was reassigned additional sets of three untreated virgin females,
10-12 weeks of age, for 6 day mating intervals to sample treated
mid-to-early spermatids (day 11-16) and primary spermatocytes (day
22-27). Po females, placed in individual cages, were checked twice
weekly for new births beginning 18 days after the first day of a
mating interval. The size of their litters was recorded at birth
and the number of males and females recorded at weaning.

4.2. Production and Analysis of F_2's

At 8-11 weeks of age F_1 males and females within each teatment-
group mating interval were intermated by random assignment of non-
sibs as single pairs. The size of all F_2 litters was recorded at
birth and the number of males and females recorded at weaning. An
attempt was made to obtain a minimum of 10 F_2's for analysis from

each F_1 mating pair. These F_2 animals were killed at 10 weeks of age and their brain and liver tissues removed, stored and homogenized as previously described [6]. Along with the scheduled sacrifice of these F_2 animals, a concomitant group of nontreated C57BL/6J mice were processed in a similar manner except the homogenates of brain or liver tissues were pooled.

For each F_2 animal, a minimum of 3 replicates on each of 19 enzymes, 6 from the brain homogenate and 13 from two fractions of liver homogenate, were performed using the MCA. An analysis for a specific enzyme, as performed in the multi-cuvet rotor of the MCA, included 16 individual F_2 samples, 3 samples from pooled tissue homogenate and a water blank. The relative enzyme activity values for each of the 16 individual F_2 samples, identified by their corresponding animal number, were entered into an IBM 370 computer. Values for the pooled homogenate samples in each analysis were also entered. Activity values from each of the 16 F_2 samples were then multiplied by a "normalization factor" to permit comparison of data across analysis time periods. Computer evaluation of the data involved determination of those normalized enzyme activity values which were aberrant when compared to the "normal range" for that specific enzyme and compilation of the total number of F_2's per F_1 mating pair which had similar aberrancies for the same enzyme(s) through all replicates. F_1 mating pairs having two or more F_2's exhibiting an aberrancy for the same enzyme(s) were classified suspect mutant in this first screen.

Next, additional tissue samples from those F_2 individuals with aberrant activity for the enzyme(s) leading to suspect classification of their F_1's were analyzed for aberrancies in apparent K_m and V_{max} of that enzyme. In a kinetic analysis performed with the MCA the cuvets of a rotor (except for water blanks) each contained a similar sample concentration from a single tissue homogenate but variable substrate concentrations, with 2-4 duplicates of each substrate concentration dependent upon the enzyme analyzed. Daily K_m and V_{max} determinations for a specific enzyme included kinetic analyses of the following tissue sample categories: 1) the F_2 samples having aberrant activity for that enzyme, 2) a sample from one of its siblings having normal activity for that enzyme, 3) a randomly selected sample from an F_2 control group animal with normal activity for that enzyme, and 4) a sample from the pooled homogenate. The K_m and V_{max} values for the F_2 samples having aberrant activity were compared to K_m and V_{max} of the normal samples. If the F_2 samples having aberrant activity did not have K_m and V_{max} values which differed from the normal samples by at least 1.5 standard deviation of their mean, that F_2 sample was considered normal for that enzyme. For an F_1 mating pair to remain classified as suspect mutant through the second screen, they had to have at least two F_2 progeny exhibiting aberrant K_m and/or V_{max} values for the same enzyme.

4.3. Mutant Confirmation

     Those F_1 mating pairs which remained suspect through the second
screen required confirmation as mutants by demonstrating that the
enzyme aberrancy observed in the F_2 generation is transmitted to the
F_3 generation. To generate a population of F_3's, those F_1 mating
pairs which were suspect through the second screen needed to produce
an additional population of F_2 progeny for breeding. Attempts were
made to obtain ten additional F_2 progeny from each of those F_1 mating
pairs which were suspect through the second screen. Thus, all F_1
pair matings were maintained for continuous production of F_2 progeny
until they were classified as nonmutant, became sterile or died.
All available F_2's up to a maximum of 10, for each pair of suspect
F_1's were established as single pair matings with randomly selected control group individuals; or, in some instances, as single
pair sib matings. As with the production of the F_1 and F_2 generations, all F_3 litters were recorded at birth and the number of males
and females recorded at weaning. An attempt was made to produce a
minimum of 10 F_3's per F_2.

     The procedures followed for killing through analysis of the F_3
animals were identical to those previously described for the first
screen. However, as in the second screen, only the enzyme(s) suspected to be aberrant were analyzed. Kinetic analysis was selec-
tively, but not routinely performed on aberrant F_3 samples. Computerized evaluation of the activity analysis data was similar to
that performed in the first screen with F_2's having at least 2 F_3
progeny exhibiting aberrant activity for the same enzyme being
classified as suspect mutant. Evaluation of an F_1 pair in the third
screen involved compilation of both the total number of F_3's per F_2
which had similar aberrancies for the same enzyme(s) and the total
number of F_2's per F_1 mating pair designated as suspect mutant based
upon evaluation of their F_3's. Classification of an F_1 pair as confirmed mutant required that at least two of its F_2's were suspect
mutant by virtue of their having at least two suspect mutant F_3's,
all with a similar aberrancy of the same enzyme(s).

4.4. Results and Discussion

     A total population of 518 F_1 progeny was evaluated according to
the defined methodology. The number of progeny which were derived
from spermatozoa (5-10 day mating interval), spermatids (11-16 day
mating interval), spermatocytes (22-27 day mating interval) of P_0
males in each of the four treatment groups are presented in the
"total" column of Table 3. The numbers in this column actually
represent the total number of males and females from 258 F_1 nonsibling pair matings plus two matings of single F_1 individuals with
control animals. The latter two matings, one from the 22-27 day
mating interval of the control group and one from the 5-10 day mating interval of the 175 mg/kg dosed group, were made because the

Table 3. F_1 Classification Results

EMS DOSE (mg/kg)	Post Treatment Days Sampled	TOTAL	Suspect in 1st Screen(%)	Suspect through 2nd Screen(%)	Confirmed Mutant(%)	Non-Mutant
0	5 - 10	52	5 (10)	0	0	52
	11 - 16	64	8 (13)	0	0	64
	22 - 27	57	11 (20)	1 (2)	-	56
	Total	173	24 (14)	1 (1)	-	172
100	5 - 10	56	13 (23)	3 (5)	2 (4)	52
	11 - 16	44	3 (7)	0	0	44
	22 - 27	54	6 (11)	1 (2)	-	53
	Total	154	22 (14)	4 (3)	2 (1)	149
175	5 - 10	45	10 (22)	4 (9)	1 (3)	40
	11 - 16	20	3 (15)	1 (5)	-	19
	22 - 27	60	10 (17)	1 (2)	1 (2)	59
	Total	125	23 (18)	6 (5)	2 (2)	118
200	5 - 10	66	12 (18)	7 (11)	4 (6)	59

original nonsibling mates died before producing sufficient progeny
for evaluation and the remaining individual was subsequently re-
mated with a control animal. As a result of this manipulation, an
uneven number of total animals is shown for these two treatment-
group mating intervals. For each of the 258 nonsibling pairs, the
probability that both the male and female would carry the same induced mutation contributing to an observed aberrancy in their progeny
was considered extremely small. Therefore, upon evaluation only one
member of each F_1 nonsibling mating pair having at least 2 aberrant
F_2 progeny was designated as suspect or confirmed mutant. The other
member was designated as nonmutant.

The number of F_1 progeny within each of the treatment-group
mating intervals which were suspect mutant through the first or second screens or confirmed mutant through the third screen are also
presented in Table 3. Eighty-one of the total F_1 population of 518
F_1 progeny were suspect mutant in the first screen. Among them, 24
individuals had multiple aberrancies involving 2-4 enzymes and, for
each of 16 out of the 19 enzymes monitored, there was at least one
individual with aberrant activity. No aberrancies were observed involving the enzymes glutathione reductase (GR), phosphoglucose isomerase (PGI) or glutamate pyruvate transaminase (GPT). Glutamate
dehydrogenase (GLDH), fructose diphosphate aldolase (FDP, Ald) and
pyruvate decarboxylase (PDC) were the most frequently observed
aberrant enzymes. No obvious differences were observed between
dosed and control groups; however, the frequency of suspect mutant
F_1's (0.18) in each of the two highest dosed groups were slightly

elevated but not significantly greater than control frequency (0.14).

Eighteen F_1's, 17 in the dosed groups and 1 in the control group, remained suspect mutant after the second screen. Five of those in the dosed group had aberrant K_m and/or V_{max} for more than one enzyme. The one remaining in the control group had been suspect mutant with multiple activity aberrancies but had aberrant K_m for only one enzyme. For each of 11 enzymes, there was at least one individual F_1 with an aberrant K_m and/or V_{max}. After the second screen, the frequencies of suspect mutants from all three dosed groups was significantly higher than the control and increased in a dose dependent fashion. The different suspect mutants exhibited no apparent pattern for specific enzyme aberrancies. For example, of the 7 suspect mutants in the high dosed group, 5 were aberrant for one enzyme only (3 for FDP, Ald, 1 for adenosine triphosphatase (ATPase) and 1 for GLDH) and 2 were aberrant for 2 enzymes (1 for GLDH and PDC and 1 for fatty acid synthetase (FAS) and PDC). Even those individuals which were aberrant for the same enzyme exhibited greatly different K_m and V_{max} values indicating that their specific genetic defects were not the same.

Eight of the 17 suspect mutants from the dosed group were confirmed as mutant through the third screen of progeny in the F_3 generation (Table 3). Four of these, including 2 aberrant for FDP, Ald, 1 for FAS and 1 for ATPase, were from the 200 mg/kg dosed group; two, including 1 aberrant for isocitrate dehydrogenase (ICD) and 1 aberrant for GLDH, were from the 5-10 day and 22-27 day mating intervals, respectively, of the 175 mg/kg dosed group; and two, including 1 aberrant for ICD and 1 aberrant for lactate dehydrogenase (LDH), were from the 5-10 day mating interval of the 100 mg/kg dosed group. There were no confirmed mutants in the control group.

Although our test criteria for classification of F_1 or F_2 breeders as suspect mutant required that they have at least two progeny with the same enzyme aberrancy, the actual number of abnormal and normal progeny observed for most suspect mutant individuals closely approximated the expected Mendelian ratio of 1:1. Also, in most cases, the number of suspect mutant F_2 breeders per mutant F_1 closely approximated the expected 50%. In those instances where sibs were used in establishing F_2 matings, the expected ratio of 3:1 for abnormals to normals among their progeny, assuming both F_2's were carriers, was seldom observed. Individual animal classification error, or sampling error, could account for discrepancies in expected versus observed values. However, for F_2 sib matings, sub-viability of homozygous mutants may have also contributed to such discrepancies.

The aberrant activities observed in F_2 progeny of confirmed mutant F_1's were all lower than normal values with the mean decrease being 25% or greater. The aberrant activities observed in F_3 progeny

of suspect mutant F_2 breeders were similar to those of the F_2 progeny
analyzed. Exceptionally large decreases in the mean activity of F_3
progeny plus the observation of at least one individual having an
extremely aberrant activity suggests that both F_2 breeders were carriers of the mutation. However, it is impossible to provide an un-
equivocal demonstration that homozygous progeny were produced because
F_2 breeders were not subjected to analysis and we do not presently
have sufficient knowledge about the expected variability in activity
values for the types of heterozygous or homozygous mutants which
might be detected in this assay.

The total numbers of animals for each treatment-group mating
interval which were designated nonmutant at any of the three screening levels are shown in the last column of Table 3. Included are
one individual from each of those F_1 mating pairs classified as suspect or confirmed mutant, and both individuals of those F_1 mating
pairs classified nonmutant. Although classification of a mating
pair as suspect or confirmed mutant could be made when 2 or more
progeny had the same aberrancy but less than 10 progeny were evaluated (2 of <10), classification of a mating pair as nonmutant required that at least 8, and preferably 10, progeny be evaluated with
none of them exhibiting the same aberrancy. Thus the total number
of animals in the nonmutant and mutant columns does not always equal
the number in the total column since classification of some F_1 pairs
was incomplete due to insufficient production of F_2 or F_3 progeny.

Twelve of the total 518 F_1 individuals of the test population
could not be designated as confirmed mutant or nonmutant. Nine individuals (1 from the 22-27 day mating interval of the control group;
1 from the 22-27 day mating interval of the 100 mg/kg dosed group;
3 from the 5-10 day and 1 in the 11-15 day mating interval of the
175 mg/kg dosed group; and 3 from the 200 mg/kg dosed group) remained
suspect mutant after the second screen because insufficient F_3 progeny
were available for analysis in the third screen. Three individuals
(2 from the 5-10 day mating interval of the 100 mg/kg dosed group;
and 1 from the 5-10 day mating interval of the 175 mg/kg dosed group)
remained suspect mutant after the first screen because samples lost
due to freezer malfunction were unavailable for kinetic analysis and
the F_1 parents either died or became infertile before additional F_2's
could be produced. It may be that some of these F_1's were truly
mutant with a resultant reduced fertility or viability that contributed to their incomplete classification.

The frequency of confirmed mutants among treated spermatozoa
sampled in both the 200 mg/kg (4/66) and 100 mg/kg (2/56) dosed
groups is significantly greater (P = 0.0035 and 0.0474, respectively)
than the control (0/173). When all incompletely classified suspect
mutants are considered as mutants, the frequency of mutations among
F_1's from treated spermatozoa in all three dosed groups (4/56; 4/45;
7/66) is significantly greater (P = 0.0083; P = 0.0004; and P =

0.0003, respectively) than the control frequency (1/173). In addition, a general trend toward a dose-response is observed across the four treatment levels with the background or spontaneous frequency being less than 1%. Even with the most conservative evaluation of our overall test results where we assume that all suspect mutant individuals in the dosed groups are nonmutant and the one suspect mutant in the control group is mutant, the mutation frequency observed in the highest dosed group (4/66) is significantly greater (P = 0.0160) than the control (1/173).

It is of interest to note that 7 of the 8 mutant lines have been maintained through the fifth generation. The single stock which could not be maintained possessed extremely limited breeding capacity and was eventually lost. The mutant stocks have been maintained and identified using the same analysis, classification, and breeding schemes described for the primary 3 screens.

5.   CONCLUDING REMARKS

Because an enzyme activity represents quantitative expression of a gene product, its measurement may be used to reflect genetic damage altering the amount, stability, or quality of a protein. However, the enzyme activity values resulting from different types and degrees of genetic damage are not easily predicted. A test procedure for identifying an induced enzyme variant must be capable of detecting relatively small changes in activity. The basic procedure described herein, which employs three levels of screening is designed to meet this objective. Our test results demonstrate an ability to detect induced mutants through measurement of enzyme activity employing one specific set of variables out of many possible ones for this type of assay procedure. Our interpretation of these test results is that mutations at loci other than those responsible for the apparent structure of the enzymes monitored or at numerous intragenic sites may be expressed as altered activity. If this interpretation is correct, the test sensitivity may be enhanced exponentially through addition of more enzymes to the battery and/or reduction in classification error through improvement of techniques associated with the enzyme analysis. We are continuing to improve the analytical techniques and evaluation criteria associated with this assay procedure. The absence of qualitative markers to clearly distinguish homozygous and heterozygous mutants from normal animals complicates the use of classical allelism tests in characterization of enzyme activity mutants. However, we hope that studies in progress aimed at further biochemical and genetic characterization of the mutants we have detected will help overcome some of these difficulties. In addition, a further validative experiment is planned where both F_1's and all F_2's will be analyzed subsequent to their production of progeny. With these improvements and validation, this assay procedure should prove to be a valuable tool for use in regulatory assessment of environmental agents.

We perceive that continued use of this technique and/or variations of this technique will provide induced mutants affecting enzyme activity resulting from mutations occurring at various structural and regulatory loci. Isolation and characterization of these
mutants will be beneficial to our understanding of the eukaryotic
gene and of its expression and regulation.

6. SUMMARY

A test procedure is described for detecting germ cell mutations
which alter activity of enzymes in _in vivo_ mammalian tissues. An
experiment with ethyl methanesulfonate (EMS) was conducted to determine the feasibility of detecting chemically induced mutations using
the basic procedure. A test population of 518 F_1 mice from matings
of untreated C57BL/6J females to C57BL/6J males treated with 0, 100,
175, or 200 mg/kg EMS was evaluated according to a set of multi-
level biochemical and genetic screening criteria. The goal of the
assay was identification of F_1 individuals heterozygous at an unspecified locus for an induced mutant allele which altered the activity of one or more of 19 enzymes in brain or liver tissues of
their F_2 and F_3 progeny. Eight individuals among a population of
345 F_1's from the EMS-dosed groups were classified as confirmed mutants having at least two of their F_2 progeny exhibiting the same
enzyme aberrancy and transmitting this aberrancy to the F_3 generation. No confirmed mutants were detected in the control population
of 173 F_1 individuals. The results indicate that this test method
can be used to detect chemically induced mutations expressed as
quantitative alterations in enzyme activity.

ACKNOWLEDGMENTS

We gratefully acknowledge the invaluable technical support for
this project provided by Mr. Olen Domon and Ms. Lynda McGarrity and
the computer programming support provided by Mr. Michael Holland.
We also thank Ms. Carolyn Phifer for her assistance in typing this
manuscript.

REFERENCES

1.   L. R. Valcovic and H. V. Malling, An approach to measuring
      germinal mutations in the mouse, Environ. Health Perspect., 6:
      201-205 (1973).
2.   H. V. Malling and L. R. Valcovic, New approaches to detecting
      gene mutations in mammals, in: "Advances in Modern Toxicology,"
      W. G. Flamm and M. A. Mehlman, eds., Vol. 5, pp. 149-171,
      Hemisphere, Washington, D.C. (1978).
3.   J. Klose, Isoelectric focusing and electrophoresis combined as
      a method for defining new point mutations in the mouse, Ge-
      netics, 92:513-524 (1970).

4. K.R. Narayanana, Detection of biochemical mutants in mice, Arch. Toxicol., 39:61–73 (1979).

5. H. Mohrenweiser, Frequency of enzyme deficiency variants in erythrocytes of newborn infants, Proc. Natl. Acad. Sci. USA, 78:5046–5050 (1981).

6. R. J. Feuers, R. R. Delongchamp, D. A. Casciano, J. G. Burkhart, and H. W. Mohrenweiser, Assay for mouse tissue enzymes: levels of activity and statistical variation for 29 enzymes of liver or brain, Anal. Biochem., 101:895–903 (1980).

7. R. J. Feuers, J. B. Bishop, R. R. Delongchamp, and D. A. Casciano, Development of a new biochemical mutation test in mice based upon measurement of enzyme activities: 1. Theoretical concepts and basic procedure, Mutat. Res., 95:263–271 (1982).

8. J. B. Bishop and R. J. Feuers, Development of a new biochemical mutation test in mice based upon measurement of enzyme activities: 2. Test results with ethyl methanesulfonate (EMS), Mutat. Res., 95:273–285 (1982).

9. T. O. Tiffany, D. D. Chilcote, and C. A. Burtis, Evaluation of kinetic enzyme parameters by use of a small computer interfaced "fast analyzer" – an addition to automated clinical enzymology, Clin. Chem., 19:908–918 (1973).

10. H. N. Kirkman, Enzyme defects, in: "Prog. Med. Genet.," A. F. Steinburg, ed., Vol. 8, pp. 125–168, Grune and Stratton, New York (1972).

11. K. Paigen, Acid hydrolases as models of genetic control, Ann. Rev. Genet., 13:417–466 (1979).

12. K. M. Plowman, Enzyme Kinetics, McGraw-Hill Book Company, Inc., New York (1972).

SOME FACTORS AFFECTING THE MUTAGENIC RESPONSE

OF MOUSE GERM CELLS TO CHEMICALS*

Walderico M. Generoso, Katherine T. Cain,
and April J. Bandy

Biology Division
Oak Ridge National Laboratory
Oak Ridge, Tennessee 37830

ABSTRACT

Accurate evaluation of genetic risk requires a good understanding of the mechanism. This report illustrates this point by demonstrating how factors such as repair, germ cell stage, stock of mice and processes that take place after fertiliztion affect the induction of chromosomal aberrations from both qualitative and quantitative standpoints.

INTRODUCTION

In evaluating the potential of any particular chemical for mutation induction, we often associate it with it's chemical and biological reactivities even though to date no specific reaction with DNA or with other components of the chromosome has been unequivocally associated with point mutational or chromosomal aberration events. So far, molecular studies have, at best, yielded data that are only suggestive of the most likely adducts responsible for mutagenesis. Nevertheless, chemical reactivity and binding with DNA, and perhaps binding with other chromosomal components as well, are almost certainly the initial steps in the complex pathway that culminates in the transmission and expression of mutations.

---

*Research sponsored by the Office of Health and Environmental Research, U.S. Department of Energy under contract W-7405-eng-26 with the Union Carbide Corporation.

This workshop attempts to put together most of the important
mammalian specific locus studies in the context of hazard evaluation
and estimation of genetic risk. However, as we are proceeding in
this exercise it becomes increasingly obvious that, while we strongly
emphasize the importance of understanding fundamental mechanisms, we
are also recognizing the fact that we know very little about them.
Indeed, none of us expects things to be simple. On the contrary, we
all realize that in thinking about mammalian germ cell mutagenesis
we are dealing with an organism at the highest level of biological
complexity. We all know that it is not enough that we are able to
demonstrate interaction between the chemical in question and the
germ cell chromosome because there are intra- and intercellular
factors that act either before or after the chemical chromosome reaction has taken place. Some of these factors are unique in mammalian germ cells.

It may appear odd that in this workshop, where mammalian specific locus endpoints are the subject of our deliberation, the endpoints I will be using belong to the class of chromosomal aberrations. In the context of mechanism, it should not be too unfair to
think that similar level of complexity exists in the induction of
specific locus mutations and chromosomal aberrations and perhaps one
might even go to the extent of assuming that induction of these two
general classes of endpoints is affected by many of the same factors
and share many of the same mechanisms. Accordingly, this report is
aimed at demonstrating how factors such as repair, germ cell stage,
strain of mice and processes that take place after fertilization
affect induction of chromosomal aberrations not only from the quantitative standpoint but also from a qualitative standpoint and how
studies of these factors contributed to our present understanding
of a likely mechanism.

NATURE OF GERM CELLS

It is a well-known fact that the various germ cells of mice may
respond differently, sometimes by orders of magnitude to physical
and chemical mutagens. Perhaps this is not too surprising in view
of the fact that mammalian germ cells may differ greatly in terms
of many biological properties. It is not the intention of this report to elaborate comprehensively on these differences, but one
property for which germ cells may differ and that, for sure, has
some influence in mutagenesis, particularly in the repair of premutational lesions and fixation of mutations, is the length of time
between exposure to mutagen and the next round of DNA synthesis.
For example, all germ cells in the ovary of adult mammalian females
are "nondividing." All oocytes are arrested in practically the same
meiotic state except the very few that are about to be ovulated.
Thus, the length of the interval between the time of binding of
chemical mutagen to oocyte chromosomes and the next chromosome replication can vary, depending upon the stage of follicular development

at the time of exposure, from just a few hours to more than a year in the case of mice or to several decades in the case of humans. What happens to the initial reaction product is a matter of conjecture but the oocyte undergoes a dramatic transformation from the primordial state to its state at ovulation and it does not seem unreasonable to assume that during the period of "quiescence" of the primordial oocyte and the period of follicular growth, processes such as adduct repair, hydrolysis, etc., affect the fate of the premutational lesion.

Similarly, the length of the interval between mutagen exposure and the succeeding chromosome replication can vary greatly from one spermatogenic germ cell stage to another. For example, it has been reported [1] that the various types of differentiating spermatogonia cycle considerably more frequently than the long-cycling spermatogonia stem cells. In the case of meiotic and postmeiotic germ cells the length of this interval can vary from several hours (for fully mature sperm) to several weeks (for early meiotic spermatocytes). How these large time differences from exposure to chromosome replication affects the yield of chemically induced aberrations from both a qualitative and quantitative standpoint is little understood, although in one case there appears to be a satisfactory explanation [2].

The length of time between S phases is only one of the many properties for which various germ cells differ from one another. Other properties that have to be important in mutagenesis are transport phenomena, presence or absence of enzyme-mediated processes or the physical properties of the chromosomes in relation to their susceptibility to attack by chemical mutagens. As various germ cells may differ greatly from one another in these properties, fundamental differences between any specific germ cell stage and somatic cells must also be assumed, particularly in the context of risk and hazard evaluation.

MECHANISM FOR INDUCTION OF DOMINANT-LETHAL MUTATIONS
AND HERITABLE TRANSLOCATIONS

The differences between germ cell stages discussed above strongly imply that the study of mechanisms must take into consideration the germ cell stage. It is reasonable to assume that any particular mutagenesis-related process that operates for one germ cell stage may not operate for another. Accordingly, the present discussion is restricted to mutagen-treated meiotic and postmeiotic male germ cells where the subsequent replication of chromosomes takes place after fertilization.

Dominant-lethal mutations and heritable translocations are endpoints of transmitted chromosome breakage effects in mice that, previously, were believed to be induced together. It was generally as-

sumed that the production of heritable translocations and dominant-
lethal mutations from any two affected chromosomes is simply a function of chromosome-type breakage and exchange processes that randomly
result in symmetrical (balanced) and asymmetrical (dicentric chromosome plus acentric fragment) exchanges. This belief developed as a
result of a series of early data with ionizing radiation and with
certain alkylating chemicals which indicated a positive correlation
in the inducibility of dominant-lethal mutations and heritable translocations in male postmeiotic germ cells [3, 4, 5]. Later, the positive correlation picture was strengthened by the results of comparative dose-response studies with ethyl methanesulfonate (EMS) and triethylenemelamine (TEM) [3, 6, 7]. The highest frequency of heritable
translocations induced by x ray or by these two chemicals appears to
level off at around 30% among viable progeny. This frequency is
reached when the frequency of dominant-lethal mutations is around
60% or more. All these early data created the general impression
that the two endpoints are induced together in relative proportions
that follow the rules of probability and that perhaps the simple
interpretation that the same mechanism, which is capable of producing symmetrical (heritable translocations) and asymmetrical exchanges
and deletions (dominant-lethal mutations), is involved in the induction of these two endpoints is correct. To date, however, we have a
much better picture of the processes that lead to the production of
dominant-lethal mutations and heritable translocations [2]. It now
appears that the chemical induction of these two endpoints does not
necessarily share the same mechanism.

The first indication that this is the case was the observation
that isopropyl methanesulfonate (IMS) [8] and benzo(a)pyrene (BaP)
[9], which are effective inducers of dominant-lethal mutations in
postmeiotic male germ cells, induce very few or no heritable translocations at the same germ cell stages. This observation suggested
that the quality of lesion induced has something to do with the fate
of the affected chromosome. Indeed, this appears to be the case as
we shall see.

All criteria indicate that EMS- and TEM-induced heritable translocations, like those induced by x rays, arose from chromosome-type
exchanges suggesting that the conversion of the corresponding reaction products into exchanges must occur before the first postfertilization chromosomal division [10]. In view of this, we may assume
that in the case of IMS and BaP such exchange failed to occur before
the first postfertilization chromosomal division presumably because
of the stability of the reaction products. As a consequence of the
persistence of the lesion to the time of the first chromosomal division or possibly even to subsequent early cleavage divisions, chromatid-type aberrations are expected to be the primary products as
in the case with somatic cells in short-term culture. What might
be the expected effects of various chromatid-type aberrations in
early cleavage embryos has been discussed in a recent report [2].

In brief, symmetrical chromatid-type exchanges are not likely to contribute significantly to the incidence of heritable transloca-tions because of mosaicism and because of the random contribution of cells during the formation of the inner cell mass. On the other hand, all other chromatid-type aberrations (i.e., asymmetrical ex-changes and deletions) are likely to be expressed as dominant-lethal mutations.

These lines of reasoning guided as in studying what might be the reasons why certain chemicals induce high levels of both domi-nant-lethal mutations and hertiable translocations while others in-duce primarily dominant-lethal mutations. We speculated that the relative rates at which these two endpoints are produced depends largely upon the stability and longevity of the induced chemical lesions or reaction products. What is known at this point is that while many chemicals can induce dominant-lethal mutations in all meiotic and postmeiotic stages including fully mature sperm and sperm that were treated after entry into the egg [11], all heritable translocation data with effective chemicals have been on practically all stages except fully mature sperm. In other words, these data were obtained from germ cells in which the interval between expo-sure to chemicals and the first postfertilization chromosome divi-sion was at least several days long. We thought that by substan-tially shortening the interval between chemical treatment and mating, most of the reaction products would still be present at the time of the first postfertilization chromosome replication and should result primarily in chromatid-type aberrations and consequently in higher proportion of dominant-lethals relative to heritable translocations.

Accordingly, we conducted a study with TEM designed to compare the ratio of heritable translocations to dominant-lethal mutations induced in matings that occurred almost two weeks after treatment (middle spermatids) with that induced in matings that occurred within only two and a half hours after treatment (fully mature sperm) [12]. As expected, there was a marked difference in this ratio be-tween the two germ cell stages. Relative to induced dominant-lethal mutations, the incidence of heritable translocations observed for fully mature sperm was markedly lower than that for spermatids. This observation seems to indicate, as implied earlier, that in the case of the short treatment-to-fertilization interval the bulk of reaction products remained intact and resulted in dominant lethality. Alternative explanations (considered less likely) have already been discussed in detail [12].

It also raised a fundamental question regarding the reaction products in spermatids. Clearly, some kind of transformation must have occurred during the almost two weeks time it took the treated spermatids to reach the ejaculate in order to account for the rela-tively higher incidence of heritable translocations. Either the process of exchange itself gets completed during this time or the

reaction products underwent a transformation into an intermediate
lesion. Two studies appeared to have answered this question. One
study provided a clear demonstration that, unlike x rays, there is
delay in the formation of chemically induced aberrations in pachytene
spermatocytes [10]. Even though clear-cut dominant-lethal mutations
and heritable translocation effects were observed from treated pachytene, no significant chromosomal aberration effects were observed
when these cells were scored cytologically as they progressed to
diakinesis-metaphase I stage. The other study [13] provided evidence that the process of exchange between chromosomes of chemically
treated male postmeiotic germ cells takes place after sperm entry.
It was shown that the yield of EMS-induced heritable translocations
was affected by the stock of untreated females used, even though the
male who received the mutagen treatment was always of the same stock.

Besides the process of chromosome exchange something else may
happen to the chemically treated chromosomes of male germ cells after
fertilization. It was demonstrated that the fertilized egg can carry
out repair of certain chemical lesions that are present in the fer-
tilizing sperm [14]. It now appears that the stable or persistent
reaction products referred to earlier are repairable in the egg or
else they lead to dominant lethality.

The various results described up to this point have contributed
to what appears to be reasonable mechanisms by which dominant-lethal
mutations and heritable translocations are produced from chemical
treatment of male postmeiotic germ cells. The production of heritable translocations requires a transformation of the reaction products into suitable intermediate lesions. These intermediate lesions
are then converted into exchanges after the sperm has entered the
egg but before the first postfertilization chromosome replication -
i.e., heritable translocations arise almost exclusively from chromosome-type exchanges. Dominant-lethal mutations may also result
from chromosome-type aberrations such as asymmetrical exchanges and
deletions. In addition, however, they can also result from chromatid-type aberrations. When the reaction products remain unchanged
and persist up to the time of pronuclear chromosome replication,
chromatid-type aberrations are the primary aberration products. It
should be obvious from this interpretation that for any particular
chemical, the production of chromosome- and chromatid-type aberrations is not necessarily mutually exclusive. On the contrary, both
classes of aberrations may be produced at rates that may differ from
one chemical to another depending upon the array of reaction products
produced.

VARIATION IN MUTAGENIC RESPONSE BETWEEN STOCKS OF MICE

It is not the intention here to review the previously published
studies on the variability of mutagenic response in mammals resulting from difference in genetic background. That such variability

Table 1. Differences between Stock of Male Mice in Response to Dominant-Lethal Effects of EMS

Treatment	Stock of males[a]	Number of mated females[b]	Number of pregnant females	Number of live embryos (ave)	Dead implants (%)	Dominant lethals[b] (%)
EMS[c] (200 mg/kg)	(C3H × C57BL)F$_1$	39	36	5.4	47	50
	(101 × C3H)F$_1$	37	35	2.8	68	74
	(SEC × C57BL)F$_1$	45	43	5.7	44	47
	T-stock	43	39	7.6	30	30
Control[d]	(C3H × C57BL)F$_1$	23	22	10.2	9	
	(101 × C3H)F$_1$	18	17	10.6	7	
	(SEC × C57BL)F$_1$	21	19	10.8	2	
	T-stock	23	19	11.5	4	

[a] FFemales used were from (C3H × C57BL)F$_1$ stock.

[b] Calculated using the formula: $\% \text{ D.L.} = [1 - \frac{\text{average number of living embryos (experimental)}}{\text{average number of living embryos (control)}}] \times 100$. All calculations were based on the pooled control average of 10.8 from all stocks.

[c] Treatment to fertilization interval - 6½ to 9½ days. EMS was administered as a single i.p. injection.

[d] Control females were mated 4½ to 7½ days after injection. They were also used as contemporary control for another study.

Table 2. Stock-Related Difference in Mutagenic Response to EMS

Treatment	Stock of males[a]	Stock of females[b]	Number of mated females	Number of pregnant females	Number of live embryos (ave)	Dead implants (%)	Dominant lethals (%)[c]
EMS[d] (200 mg/kg)	(C3H × 101)F$_1$	T-stock	49	45	2.5	60	63
	T-stock	(C3H × 101)F$_1$	37	35	5.8	19	16
Control	(C3H × 101)F$_1$	T-stock	41	40	6.8	19	
	T-stock	(C3H × 101)F$_1$	50	45	6.9	3	
EMS (150 mg/kg)	(C3H × 101)F$_1$	(SEC × C57BL)F$_1$	60	43	7.1	21	18
	T-stock	(C3H × 101)F$_1$	65	62	6.6	9	3[e]
Control	(C3H × 101)F$_1$	(SEC × C57BL)F$_1$	77	55	8.7	4	
	T-stock	(C3H × 101)F$_1$	52	44	7.1	5	

[a] Males received EMS treatment.

[b] Females were not treated. Treatment to fertilization interval - 6½ to 9½ days.

[c] Calculated using the formula: $\% \text{ D.L.} = [1 - \frac{\text{average number of living embryos (experimental)}}{\text{average number of living embryos (control)}}] \times 100$.

[d] Administered as a single i.p. injection.

[e] No significant increase in dominant-lethal effect was detected.

Table 3. Response of Different Stocks of Male Mice to Dominant-
Lethal Effects of IMS in Spermatids[a]

Treatment	Stock of males	Number of mated females[b]	Number of pregnant females	Number of live embryos (ave)	Dead implants (%)	Dominant lethals[c] (%)
IMS[d] (75 mg/kg)	(C3H × C57BL)F$_1$	44	32	6.5	32	30
	(C3H × 101)F$_1$	46	36	5.8	38	38
	(SEC × C57BL)F$_1$	46	36	5.1	43	45
	T-stock	42	31	7.6	21	18
Control	(C3H × C57BL)F$_1$	23	22	9.3	2	
	(C3H × 101)F$_1$	24	15	9.5	4	
	(SEC × C57BL)F$_1$	23	20	9.6	4	
	T-stock	23	18	8.9	2	
IMS[d] (75 mg/kg)	(C3H × C57BL)F$_1$	47	37	6.6	30	29
	(101 × C3H)F$_1$	36	28	5.8	35	38
	(SEC × C57BL)F$_1$	45	37	6.0	37	35
	T-stock	44	37	7.8	21	16
Control	(C3H × C57BL)F$_1$	45	43	9.8	5	
	(101 × C3H)F$_1$	39	32	10.1	3	
	(SEC × C57BL)F$_1$	40	33	8.2	8	
	T-stock	48	30	8.8	4	

[a]Males were mated 15½ to 18½ days posttreatment. IMS was administered as a single i.p. injection.

[b]All females were from (SEC × C57BL)F$_1$ stock.

[c]Calculated using the formula: % D.L. = [1 − $\frac{\text{average number of living embryos (experimental)}}{\text{average number of living embryos (control)}}$] × 100.

All calculations were based on the pooled control average from all stocks (9.3 for both groups).

[d]These are replicate experiments.

exists is generally known. Rather, published and previously unpublished results from our laboratory are discussed in an attempt to
demonstrate the importance of stock selection (both males and females) in mutagenicity evaluation. Again, as in the previous section, it is essential to identify the germ cell stage involved. Accordingly, discussion of relative mutagenic response is based upon
dominant-lethal mutations induced in male postmeiotic germ cells.
The issue at hand is to what extent does the genetic background of
treated males and of untreated females influence the yield of chemically induced dominant-lethal mutations?

We knew from an earlier study that the yield of dominant-lethal
mutations induced by EMS in male postmeiotic germ cells was affected
by the stock of males treated [15]. In particular, a clear-cut dif-

Table 4. Response of Different Stocks of Male Mice to Dominant-
Lethal Effects of IMS in Spermatozoa[a]

Treatment	Dose (mg/kg)	Stock of males	Number of mated females[b]	Number of pregnant females	Number of live embryos (ave)	Dead implants (%)	Dominant lethals[c] (%)
IMS	75	(C3H × C57BL)F$_1$	41	25	5.2	44	43
		(101 × C3H)F$_1$	43	27	2.9	63	68
		(SEC × C57BL)F$_1$	40	31	2.9	63	68
		T-stock	43	27	4.7	47	48
IMS	65	(C3H × C57BL)F$_1$	46	33	5.5	36	40
		(C3H × 101)F$_1$	37	26	4.6	44	50
		(SEC × C57BL)F$_1$	43	33	5.7	44	37
		T-stock	39	24	6.8	30	25
Control		(C3H × C57BL)F$_1$	23	12	8.3	7	
		(C3H × 101)F$_1$	22	18	9.6	2	
		(SEC × C57BL)F$_1$	24	19	8.6	6	
		T-stock	17	10	9.9	5	

[a]Males were mated ½ to 3½ days posttreatment. IMS was administered as a single i.p. injection.

[b]All females were from (SEC × C57BL)F$_1$ stock.

[c]Calculated using the formula: % D.L. $= [1 - \dfrac{\text{average number of living embryos (experimental)}}{\text{average number of living embryos (control)}}] \times 100$.

All calculations were based on the pooled control average of 9.1 from all stocks.

ference was found between T-stock (low yield) and (101 × C3H)F$_1$ (high yield) males with (SEC × C57BL)F$_1$ males placed in between these two stocks. A repeat of this study, shown in Table 1, again revealed the same order of relative sensitivity and indicate that the yield of dominant lethals for (101 × C3H)F$_1$ was about twice as much as that for T-stock males. We also found more recently that the EMS-induced dominant-lethal mutations were affected by the stock of females used, even though the males receiving the mutagen were always from the same stock [14]. It was observed that T-stock, (SEC × C57BL)F$_1$ and (C3H × C57BL)F$_1$ females yielded similar levels of dominant-lethal mutations which were clearly higher than that for (101 × C3H)F$_1$ females. We performed a study with EMS in which high-yielding stocks of females [T-stock and (SEC × C57BL)F$_1$] were mated with a treated sensitive stock of males [(C3H × 101)F$_1$] and a low-yielding stock of females [(C3H × 101)F$_1$] were mated with a treated resistant stock of males (T-stock). Results of this study shown in Table 2 clearly revealed the marked differences between these two mating groups. At the lower EMS dose of 150 mg/kg a clear dominant-lethal effect was found in the former mating combination compared while no significant dominant-lethal effects were measured in the latter mating combination.

Table 5.   Response of Different Stocks of Male Mice to Dominant-
Lethal Effects of TEM in Spermatozoa[a]

Treatment	Stock of males	Number of mated females[b]	Number of pregnant females	Number of live embryos (ave)	Dead implants (%)	Dominant lethals[c] (%)
TEM[d] (0.2 mg/kg)	(C3H × C57BL)F$_1$	44	37	5.2	47	46
	(101 × C3H)F$_1$	38	37	6.2	36	36
	(SEC × C57BL)F$_1$	43	42	4.4	53	55
	T-stock	43	38	4.4	52	55
Control	(C3H × C57BL)F$_1$	23	20	8.8	8	
	(101 × C3H)F$_1$	16	15	9.7	6	
	(SEC × C57BL)F$_1$	20	19	10.7	2	
	T-stock	20	17	9.7	8	
TEM[d] (0.2 mg/kg)	(C3H × C57BL)F$_1$	45	38	5.6	36	48
	(101 × C3H)F$_1$	32	30	6.3	37	42
	(SEC × C57BL)F$_1$	45	44	5.2	48	52
	T-stock	42	37	5.5	46	49
Control	(C3H × C57BL)F$_1$	44	43	10.6	4	
	(101 × C3H)F$_1$	38	35	11.0	3	
	(SEC × C57BL)F$_1$	47	44	10.9	5	
	T-stock	43	34	10.8	3	

[a]Males were mated 4½ to 7½ days posttreatment. TEM was administered as a single i.p. injection.

[b]All females were from (C3H × C57BL)F$_1$ stock.

[c]Calculated using the formula: % D.L. = [1 − $\frac{\text{average number of living embryos (experimental)}}{\text{average number of living embryos (control)}}$] × 100.

All calculations were based on the pooled control averages in each group from all stocks (9.7 and 10.8).

[d]These are replicate experiments.

An even more dramatic influence of stocks exists in the case
of IMS. There was almost an order of magnitude difference in the
yield of dominant-lethal mutation between two stocks of untreated
females mated to the same stock of treated males [14]. Differences
between stocks of treated males are also clear-cut. These differences are more pronounced in spermatids (Table 3) than in spermatozoa (Table 4). It is noteworthy that T-stock males, which is the
most resistant stock for EMS, is also the most resistant one for
IMS. T-stock males, however, are not always relatively more resistant to mutagenic chemicals. Data in Tables 5 and 6 show that
the T-stock males were no more resistant than the rest of the
stocks studied.

Table 6. Response of Different Stocks of Male Mice to Dominant-Lethal Effects of TEM in Spermatids[a]

Treatment	Stock of males	Number of mated females[b]	Number of pregnant females	Number of live embryos (ave)	Dead implant (%)	Dominant lethals[c] (%)
TEM[d] (0.2 mg/kg)	(C3H × C57BL)F₁	42	39	3.0	62	70
	(101 × C3H)F₁	43	38	3.9	54	61
	(SEC × C57BL)F₁	43	40	2.7	68	73
	T-stock	42	40	3.3	63	67
Control	(C3H × C57BL)F₁	24	23	9.7	4	
	(101 × C3H)F₁	21	19	10.3	1	
	(SEC × C57BL)F₁	22	17	10.5	4	
	T-stock	23	20	9.8	7	
TEM[d] (0.2 mg/kg)	(C3H × C57BL)F₁	46	45	4.0	57	61
	(101 × C3H)F₁	34	31	4.1	57	60
	(SEC × C57BL)F₁	46	43	3.3	63	68
	T-stock	43	37	3.3	61	68
Control	(C3H × C57BL)F₁	40	38	10.2	5	
	(101 × C3H)F₁	34	32	10.9	6	
	(SEC × C57BL)F₁	47	46	9.9	4	
	T-stock	43	40	9.8	5	

[a] Males were mated $11\frac{1}{2}$ – $14\frac{1}{2}$ days posttreatment. TEM was administered as a single i.p. injection.

[b] All females were from (C3H × C57BL)F₁ stock.

[c] Calculated using the formula: % D.L. $= [1 - \frac{\text{average number of living embryos (experimental)}}{\text{average number of living embryos (control)}}] \times 100$.

All calculations were based on the pooled control averages in each group from all stocks (10.0 and 10.2).

[d] These are replicate experiments.

It is important to note that, contrary to T-stock males, (101 × C3H)F₁ and the reverse hybrid (C3H × 101)F₁ have consistently shown relatively high sensitivity to the three compounds discussed in this section. Furthermore, our long experience with these stocks indicate that they have high sensitivity to other chemical mutagens as well. In the case of females, we have observed that random-bred T-stock and inbred SEC females consistently yielded the highest dominant-lethal effects when mated to males treated with chemical mutagens. Thus, the combination of these males and females are expected to provide the best chance of detecting dominant-lethal or heritable translocation effects of a given test chemical.

The demonstrable differences between stocks of females are presumably attributable to differences in the ability of the egg to re-

pair the chemical lesion in the male genome [14]. Because only one
stock of males was used in these studies, the lesions that the sperm
brought into the egg of various stocks of females must be the same
qualitatively and quantitatively. The situation in males is less
tenable. In cases where differences were observed for maturing
sperm, one may rule out repair as the basis. However, in the case
of early spermatids, where certain repair processes are known to
exist [16], repair is a possible basis along with processes that may
affect the rate of binding between the mutagen and the chromosomes.

CONCLUSION

Our present understanding of the processes involved in chemical
induction of chromosome aberrations in male meiotic and postmeiotic
germ cells demonstrates the value of fundamental information in the
evaluation of genetic risk. Obviously, it is important to determine
whether a chemical in question is capable of producing both dominant-lethal mutations and heritable translocations or mainly dominant-lethal mutations. That processes in germ cells are not necessarily the same in somatic cells is clearly illustrated by the
fact that, to date, there is no reliable cytogenetic method that
allows us to predict accurately the relative inducibility of these
two endpoints in mammalian germ cells.

REFERENCES

1.   E. F. Oakberg and C. Huckins, Spermatogonial stem cell renewal
      in the mouse as revealed by 3H-thymidine labeling and irradia-
      tion, in: "Stem Cells of Renewing Populations," A. B. Cairne,
      ed., pp. 287-302, Academic Press, New York (1976).
2.   W. M. Generoso, A possible mechanism for chemical induction of
      chromosome aberrations in male meiotic and postmeiotic germ
      cells of mice, Cytogenet. and Cell Genet. (in press).
3.   W. M. Generoso, K. T. Cain, and S. W. Huff, Inducibility by
      chemical mutagens of heritable translocations in male and fe-
      male germ cells of mice, in: "Advances in Modern Toxicology,"
      W. G. Flamm and M. A. Mehlman, eds., Vol. 5, pp. 109-129,
      Hemisphere Publishing Co., Washington, D.C. (1978).
4.   B. M. Cattanach, Induction of translocations in mice by tri-
      ethylene melamine (TEM), Nature, 180:1346-1365 (1957).
5.   B. M. Cattanach, C. E. Pollard, and J. H. Jackson, Ethyl
      methanesulfonate-induced chromosome breakage in the mouse,
      Mutation Res., 6:297-307 (1968).

6. B. E. Matter and W. M. Generoso, Effects of dose on the induc-
    tion of dominant-lethal mutations with triethylenemelamine in
    male mice, Genetics, 77:753-763 (1974).
7. W. M. Generoso, W. L. Russell, S. W. Huff, S. K. Stout, and
    D. G. Gosslee, Effects of dose on the induction of dominant-
    lethal mutations and heritable translocations with ethyl
    methanesulfonate in male mice, Genetics, 77:741-752 (1974).
8. W. M. Generoso, S. W. Huff, and K. T. Cain, Relative rates at
    which dominant-lethal mutations and heritable translocations
    are induced by alkylating chemicals in postmeiotic male germ
    cells of mice, Genetics, 93:163-171 (1979).
9. W. M. Generoso, K. T. Cain, C. S. Hellwig, and N. L. A. Cacheiro,
    Lack of association between induction of dominant-lethal muta-
    tions and induction of heritable translocations with benzo(a)-
    pyrene in postmeiotic germ cells of male mice, Mutation Res.,
    94:155-163 (1982).
10. W. M. Generoso, M. Krishna, R. E. Sotomayor, and N. L. A.
    Cacheiro, Delayed formation of chromosome aberrations in mouse
    pachytene spermatocytes treated with triethylenemelamine (TEM),
    Genetics, 85:65-72 (1977).
11. K. E. Suter, Chemical induction of presumed dominant-lethal
    mutations in postcopulation germ cells of mice, I. Relative
    sensitivity between pre- and postcopulation germ cells to iso-
    propyl methanesulfonate, Mutation Res., 30:355-364 (1975).
12. W. M. Generoso, K. T. Cain, C. V. Cornett, E. W. Russell, C. S.
    Hellwig, and C. Y. Horton, Difference in the ratio of dominant-
    lethal mutations to heritable translocations produced in mouse
    spermatids and fully mature sperm after treatment with tri-
    ethylenemelamine (TEM), Genetics (in press).
13. W. M. Generoso, K. T. Cain, M. Krishna, E. B. Cunningham, and
    C. S. Hellwig, Evidence that chromosome rearrangements occur
    after fertilization following postmeiotic treatment of male
    germ-cells with EMS, Mutation Res., 91:137-140 (1981).
14. W. M. Generoso, K. T. Cain, M. Krishna, and S. W. Huff, Genetic
    lesions induced by chemicals in spermatozoa and spermatids of
    mice are repaired in the egg, Proc. Natl. Acad. Sci. USA, 76:
    435-437 (1979).
15. W. M. Generoso and W. L. Russell, Strain and sex variations in
    the sensitivity of mice to dominant-lethal induction with ethyl
    methanesulfonate, Mutation Res., 8:589-598 (1969).
16. G. A. Sega, Unscheduled DNA synthesis in the germ cells of male
    mice exposed in vivo to the chemical mutagen ethyl methane-
    sulfonate, Proc. Natl. Acad. Sci. USA, 71:4955-4959 (1974).

QUALITATIVE ANALYSIS OF MOUSE SPECIFIC-LOCUS MUTATIONS:

INFORMATION ON GENETIC ORGANIZATION, GENE EXPRESSION,

AND THE CHROMOSOMAL NATURE OF INDUCED LESIONS*

Liane B. Russell

Biology Division
Oak Ridge National Laboratory
Oak Ridge, Tennessee 37830

1. INTRODUCTION

Mutations scored in specific-locus experiments can serve many purposes. Their primary function, of course, is to detect and measure mutagenicity of an agent [1]. This primary function is fulfilled rapidly and accurately, thanks to the well-defined phenotypes of mutations at the marked loci. However, when resources permit the setting up of stocks and the subsequent genetic analysis of the mutations, the information obtained can make major contributions to several other areas of knowledge. Such information can shed light on the nature of mutations induced by specific agents, and on the manner in which certain types of genetic lesions are expressed on the organismic level. These areas of information are needed, in combination with each other, for an assessment of risk. Thirdly, the findings shed light on the genetic organization in mammalian chromosomes. Finally, they serve to identify tools for further investigations in basic genetics and mutagenesis. This paper summarizes contributions to these areas from findings made on mutations induced in radiation experiments at Oak Ridge.

2. GENETIC MATERIAL ANALYZED

The specific-locus test (SLT) developed by W. L. Russell [2] detects forward mutations at 7 loci. Many hundreds of such mutations were scored in a variety of radiation experiments carried out over almost 3

*Research sponsored by the Office of Health and Environmental Research, U.S. Department of Energy, under contract W-7405-eng-26 with the Union Carbide Corporation.

decades, and a sizable proportion of these have been propagated as stocks. Mutations at 3 of the loci - d (dilute), se (short ear) and c (albino) - have been genetically analyzed. The d and se loci are closely linked (0.2 cM) on Chromosome 9, and the c locus is located on Chromosome 7. The analysis has involved: (1) allelism tests; (2) a variety of phenotypic characterizations of the original mutant animals and their heterozygous and homozygous descendants (classification includes: whole-body vs. fractional; resembling marker allele or different allele; normal or altered viability, fertility, weight); (3) deficiency mapping with nearby markers; (4) fullscale complementation tests of the non-viable subset of mutations, using deaths at various stages of prenatal or postnatal development, body weight, and reduction or absence of various enzymes as phenotypes; (5) cytological analysis of banded chromosomes in a few of the mutants.

Our studies involved 314 independent mutations, distributed as follows: 122 d locus, 43 se locus, 37 d se [3, 4], with over 800 combinations of independent mutants studied; and 112 c locus [5-8], with 469 combinations studied. Numerous complementation groups were delineated in each of the regions, and several new functional units identified. Two of the 112 Oak Ridge c-locus mutations and three derived from experiments at Harwell have also been analyzed by Gluecksohn-Waelsch [9].

The standard 7-locus SLT has detected not only alterations involving the marked loci themselves, but also genetic changes elsewhere on the chromosome acting through a special type of position effect [10-12]. In addition, a specific-locus test in which the hemoglobin loci, Hba and Hbb, were included as markers [13], yielded an aberration affecting one of the standard loci, c [14]. These other products of SLTs will also be included in the subsequent discussion.

3. FINDINGS CONCERNING GENETIC ORGANIZATION

As shown by complementation maps for d and/or se mutations [4, 15] and c-locus mutations (Fig. 1), the procedures used are capable of considerable resolution of mutational types and of the genetic region. The SLT is evidently capable of scoring more than a limited array of genetic alterations. Starting with 2 markers on Chromosome 9, the analysis yielded over 20 complementation groups involving 11-12 functional units, including 8-9 newly identified ones. At the c locus, the analysis yielded 13 complementation groups involving 8 functional units, including 5 newly defined ones. There is evidence (Sec. 4) that some of the complementation groups represent intragenic mutations and that others are deficiencies overlapping the marked locus or loci.

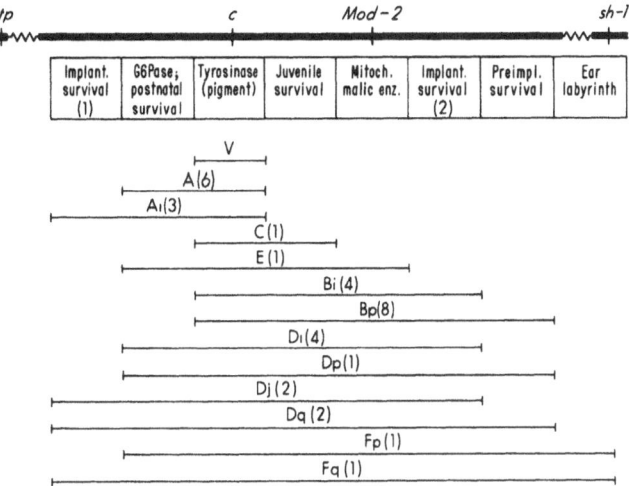

Fig. 1. Complementation map of c -locus mutants developed by Russell et al. [8]. With postulated functional units (shown in boxes below the genetic map), all mutants fit a linear pattern. The number shown in parentheses following each complementation-group designation indicates the number of independent Oak Ridge mutations in the group. In addition, Harwell mutants c^{3H}, c^{6H}, and c^{25H}, can probably be added to groups E, Bi, and Dp (or Dq), respectively. "V" indicates viable albino mutants of which there were 52 among 90 non-mosaic c-locus mutations found in the progeny of irradiated mice.

Large numbers of cistrons must lie within the spans defined by the deficiencies in each of the two regions, and the functional units identified to date provide evidence on only a small fraction of this number. Additional functional units could be defined (a) by future enlargement of the complementation grid through the inclusion of new mutants, and (b) by more extensive investigations of phenotypes in the existing grid. For example, embryological studies of c locus mutants succeeded in subclassifying what was originally one phenotype, prenatal death, according to intrauterine stage at which death occurred. Introduction of these more detailed phenotypes into the complementation studies led to the definition of three functional units - one for preimplantation survival and two for implantation survival [8]. Detailed analysis of physiological, morphological, and developmental characters of the mutants, singly and in all combinations, would undoubtedly produce finer subdivisions than those achieved to date. Although it is not impossible that intracistronic complementation [16] exists in the mouse, it is likely that most functional units that can be separated by complementation represent separate cistrons.

It may be particularly useful to look for new functional units between the c locus and the two units presently mapping adjacent to it – that for neonatal survival, G6Pase, and certain other proteins on one side, and that for juvenile survival, normal size and fertility on the other. Already, over 80 distinct combinations of c-locus deficiencies extending to the left and right of c have produced what appears to be full complementation for all phenotypes studied (except albinism). If more detailed studies also fail to reveal aberrant phenotypes, one may conclude that the sites controlling the neonatal and juvenile survival functions are either immediately adjacent to the c-locus or separated from it by non-coding DNA.

One objective in developing a map of functional units within a given chromosomal segment is to obtain a sample of the mix of vital and non-vital functions in the mouse genome. Because complementation for various types of lethality is relatively easy to study, vital functions are among the first ones localized on the complementation map. Thus in the c-locus region, five new vital "loci" have been mapped (Fig. 1), and, in the d-se region, seven to eight (Russell, unpublished). The study has, however, also succeeded in demonstrating that certain loci are non-vital. Thus, combinations of certain deficiencies that overlap at c (cA or cAi with cBi, cBp, or cC – see Fig. 1) can produce complementation for lethality but not for the visible effect, albinism. The c locus is thus a non-vital locus. A similar demonstration can be brought for the se locus.

There is also indirect evidence that the total absence of the Mod-2 (mitochondrial malic enzyme) locus is fully viable. While combinations of mutants (cE/cBi or Bp) that lack both the MOD-2 and "juvenile-survival" functions die between days 7 and 119 of age and are of reduced size ("Pattern II") [8], the very same syndrome is produced by combinations that lack only the "juvenile-survival" function but carry one dose of Mod-2 (namely, cC/cBi or Bp, $cC/cDi,Dp,Dj$, or Dq, cC/cFp or Fq). Absence of Mod-2 cannot thus be implicated in the Pattern-II syndrome, and a single dose of Mod-2 (as in $+/cE$) is known to be compatible with survival.

The situation is less clear at the d-locus, where it has not yet been possible to achieve complementation for the opisthotonic-lethal function. Thus, combinations of overlapping deficiencies have produced the dilute-opisthotonic phenotype (over 220 such combinations), or prenatal lethality, but not viable dilutes [4, 15]. It is possible that there is an "op" site within the d cistron; or, that dop is the amorph and $\underline{d}$ the hypomorph with regard to a common gene product. Although one case of possible recombinational separation of d from dop was observed, this case could also be explained by a spontaneous reverse mutation, $d \rightarrow +$ [3].

In the c-locus region, it was possible to postulate an align-
ment of functional units by which all analyzed mutations fit a linear
pattern, and there is no compelling argument against the assumption
that all c-lethals are overlapping deficiencies. In the d-se
region, while the majority of dp^l, se^l, and dse mutants fitted a
linear pattern, there were a few which did not. In tests conducted
to determine whether some of these non-conforming types could be the
result of two independent mutations, the results indicated that this
was not the case. However, one mutant provided evidence of conver-
sion-like events in crossover experiments that utilized flanking
markers on both sides [4]. These events have not yet been explained.

The analysis of mutations provides evidence by which the genetic map can be related to the cytological map, since some of the
deficiencies are large enough to produce visible deletions in banded
metaphase chromosomes. Working with a single deficiency known to
contain a given locus, a crude localization can be derived [17]. A
more refined cytological mapping is possible when several deficiences
involving different assortments of loci can be studied in banded
chromosomes. This type of analysis has allowed us to assign c and
Mod-2 to band E1 (proximal portion), sh-1 to band E1 (distal portion)
or band E2, and Hbb distal to E3 [18, 19]. A more extensive analysis
[19], involving, in addition to deficiencies, also some other chromosome aberrations (Sec. 2), provides tentative evidence that relative distances in the genetic map may differ from those in the metaphase map.

4. FINDINGS CONCERNING GENETIC EXPRESSION

Analysis of mutations recovered in SLT experiments has provided
some answers (and will provide more in the future) concerning the
action of several loci and of small chromosomal regions on the organismic or cellular level. The information that can be obtained
from combinations of various overlapping deficiencies extends that
which can be obtained from individual homozygotes and heterozygotes
for studies of the effect of either total or heterozygous absence
of small chromosomal segments.

The available genetic material also permits the study of expression of non-deficiency mutations. Evidence that certain ones
of the mutations are of this type exists for the hypomorphs. Thus,
as discussed (Sec. 3), at $\underline{c}$ and $\underline{d}$, overlapping deficiencies produce
pigment phenotypes indistinguishable from c/c and d/d, respectively,
and deficiencies overlapping at se produce an external-ear phenotype
that resembles se/se. Therefore, any mutation whose phenotypic expression differs from c, d, or se in the above respects may be assumed to be due to intragenic change, rather than deficiency (Table
1). Many such "intermediate alleles" have been found in radiation
experiments.

Table 1.   Relation between Phenotype of Mutant at Specific
           Locus[a] and Probable Chromosomal Nature of
           Mutation

Expression of mutation at marked locus	Change in other phenotype(s)	Probable chromosomal nature of mutation
Hypomorph	No	Intragenic lesion
Null	No	Intragenic lesion; or deficiency of marked locus only[b]
Null	Yes	Multilocus deficiency

[a]Established for c and se loci

[b]or including adjacent non-coding DNA

     While total absence of the c or se loci is known, from the
study of overlapping deficiencies, to produce viable animals indistinguishable from c/c or se/se, respectively, one cannot assume
that the converse is necessarily true; i.e., induced mutations whose
phenotypes in the homozygous state are equivalent to those of c/c
or se/se need not (but may) be deficiencies of the respective cistron.
Such nulls could also be intragenic lesions (Table 1).   Thus, the
homozygous viability of nulls at proved non-vital loci cannot, by
itself, yield any information concerning the nature of the mutational lesion.

4.1.   Total Absence of Small Chromosomal Segments

     The homozygous effects of known deficiencies can vary over the
entire possible viability spectrum, depending on location of the deficiencies.   Thus, total absence of a small chromosomal segment may
be (1) completely viable [e.g., overlapping Df(c)'s, Df(se)'s, or
Df(Mod-2)'s], (2) lethal as late as the young adult (e.g., d^{op}/d^{op},
c^C/c^C, or combinations overlapping at these units), (3) lethal
perinatally (c^A/c^A or appropriate combinations), (4) lethal at or
shortly after implantation (see Fig. 1 for homozygotes and combinations), (5) lethal before implantation (Fig. 1).

     An exceptional finding that is, so far, without explanation
concerns certain non-complementing combinations of c lethals.   Among
221 combinations (13,209 offspring) that kill well before the mid-
point of gestation, there were 9 (614 offspring) in each of which a
single individual escaped from this early death and survived until
birth [8].   Recombinational explanations can be ruled out for most
of these cases.

Table 2.   Selected Findings Concerning Heterozygous
            Deficiencies

A. Maximum lengths recovered as viable heterozygotes

Region	Type	Length (cM)
c	Df (c Mod-2 sh-1)	6 - 11
d-se	Df (d se sv)	2.2 - 11

B. Examples showing that content, more than length, affects viability

Deficiency	Length (cM)	Viability of +/Df
Df(c Mod-2)	2 - 9	Normal[a]
Df(se)[b]	0 - 2.2	Markedly reduced

[a]Not distinguishable from +/+ on mixed background.[5] Not yet analyzed in
co-isogenic stocks.

[b]Mutant se^{207K}. Ref. 34.

For a given region, there appears to be a crude relation between length of the deficiency and time of death of the homozygote;
for example, all deficiencies extensive enough to be readily detectable in banded metaphase chromosomes (c^Fp^1, c^Fq^1, c^Dq^1, and
c^Dp^1 or $Dq]^{25H}$) kill before implantation. However, specific content
of a deficiency is presumably a more important determinant than
length. It is conceivable, e.g., that, on the tp side of c (see
Fig. 1), the nearest functional unit for preimplantation survival
could be at a considerable distance from the last functional unit
mapped; if so, one might recover deficiencies that are longer than
the preimplantation lethal c^Bp group but kill later, namely at or
soon after implantation.

4.2. Heterozygous Deficiencies

The analysis of SLT mutations has shown that surprisingly long
deficiencies are recoverable in heterozygous condition (Table 2).
In the c region, two of the deficiencies are at least 6 cM long (but
neither of these could be greater than 11 cM). In the d-se region,
the longest deficiencies found were at least 2.2 cM, but at most 11
cM, in length. While exact determination of vital effects of
heterozygous deficiences (survival, weight, reproductive capacity,
etc.) must await analysis in coisogenic stocks, it is already clear
that marked viability depression occurs in certain cases. As was

true for homozygous effects, content is probably a more important
determinant of heterozygous viability than is length (Table 2). Thus,
many c locus deficiencies that are longer than 2 cM produce no readily
measurable effects in heterozygotes, while marked viability depression
was found for an se-lethal that is less than 2.2 cM long.

There appears to be some correlation between the presence of
clear heterozygous effect and early time of death of the homozygote.
Thus c-locus-mutant stocks segregating for heterozygous and wild-
type offspring were studied with respect to whether there was a reduction in the expected numbers of the former. In 13 stocks of preimplantation lethals and 21 stocks of lethals that kill at later
stages, heterozygote frequencies were significantly reduced in 31%
and 5% of the stocks, respectively (P = 0.11 for difference between
the two groups of stocks), and obviously (but not necessarily significantly) reduced in 54% and 10% of the stocks, respectively (P =
0.01).

The analysis of specific-locus mutations has provided some clues
about the chromosomal nature and heterozygous effects of the genetic
alterations that may be recovered in recessive-lethal tests in the
mouse [20, 21]. Such recessive lethals could be deficiencies ranging from very small to quite long (at least 6 cM), and some of them
may be intragenic mutations. The heterozygous effects of recessive
lethals could range from not obviously deleterious to markedly so,
and the latter type is expected to be more prevalent among those
lethals that kill before implantation.

4.3.   Associated Phenotypes; Pleiotropy

The large numbers of characterized deficiencies that are available, and the extensive complementation grids that can be constructed with them, permit tentative conclusions as to whether certain
phenotypes that are found associated in some mutants are multiple
end points of a single basic lesion, or the results of the loss of
several separate, though neighboring, cistrons. An example is provided by the association of the lack of glucose-6-phosphatase (G6Pase)
activity and perinatal death that occurs in homozygous c^A-group or c^E-
group mutants. We were able to produce 97 neonatally lethal combinations of independent mutations (involving 9 complementation
groups), and deHamer [22] tested 70 of these for G6Pase activity;
every one was severely deficient. Of 117 combinations that did not
die neonatally, 87 were tested for G6Pase, and in all 87 of these
the activity was in the range of that found in normal littermates.
Thus, the two phenotypes could not be separated in extensive tests
and probably result from the same basic lesion.

A number of additional phenotypes have also been found in certain homozygous c^A- and c^E-group mutants, namely, abnormality of two

other liver-specific enzymes, of serum protein, and of the structure
of subcellular membrane organelles [9]. So far, an association of
these abnormalities with G6Pase deficiency has been demonstrated in
3 combinations of mutants, while 3 other combinations, which had
normal G6Pase levels, were normal in the other respects also [23].
While these numbers are not as extensive as those for the G6Pase-
neonatal-death comparison, they provide some evidence of the same
sort, suggesting that all of the multiple phenotypes are part of a
single basic lesion at what is assumed to be a regulatory locus [9].
By contrast, as discussed in Sec. 3, lesions at the Mod-2 locus presumably do not have associated viability phenotypes.

4.4. Copy Number

The ability to study the effects of certain regions in double,
single, and sometimes zero dose (in certain combinations of overlapping deficiencies) allows the simple determination of whether
product is related to copy number. This is the case for Mod-2 [24],
the structural locus for mitochondrial malic enzyme; on the other
hand, G6Pase activity is the same in heterozygous deficiencies as
in normal genomes, and Gluecksohn-Waelsch [9] has taken this as evidence that a regulatory gene is involved (see also Sec. 4.3). No
specific gene product is known for the other functional units that
control survival at various ages; however, it should be noted that
whenever complementation for lethal effects was found, it was full
rather than partial complementation, indicating that a single copy
of the pertinent region was sufficient.

4.5. Gene Expression at the Cellular Level

Some attempts have been made to study the expression of genes
located at or near the marked loci on levels besides the organismis
one. One major tool for such investigations is provided by X-autosome translocations affecting Chromosomes 7, on which c and p are
located, and Chromosome 4, on which lies the b locus. In such translocations, major portions of the pertinent autosomal region are in-
activated in roughly'half the cells of the body [10-12]. We have
studied the effects of p, c, and b lethals (presumed deficiencies)
carried in the non-translocated autosome opposite reciprocal X-
autosome translocations that carry the pertinent wild-type allele
[25, 26, 15]. As illutrated in Fig. 2, females with such genomes
are functional mosaics of the type $+/Df(m)///0/Df(m)$ [where $Df(m)$
represents a deficiency for the marked locus and probably adjacent
regions, and $+^m$ represents the wild-type allele at the marked locus
and intact adjacent regions]. With respect to dosage of $+^m$, such
females are $1///0$ mosaics, and - if a zero dose of $+^m$ is viable on
the cellular level - there should be phenotypic mosaicism for the
marker (namely, wild type where the cellular genotype is $+^m/0$; and
the color characteristic of the homozygous null allele where the
cellular genotype is $0/0$).

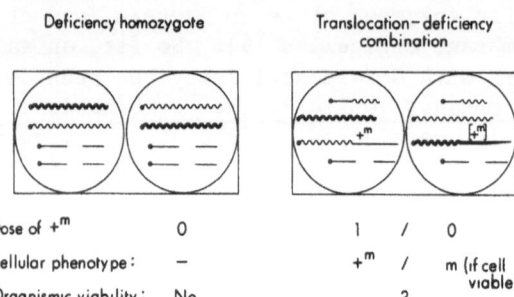

Fig. 2.   Use of X-autosome translocations as tools to explore whether
organismically-lethal deficiencies act also as cell lethals.
Straight lines, autosomes or autosomal segments; wavy lines,
X-chromosomes or X-chromosomal segments; heavy wavy or heavy
straight lines, inactivation due to X allocycly; gap in
autosome, deficiency of marked locus m and, possibly, adjacent loci; [$+^m$], inactive gene. Alternative cellular conditions are shown for a female deficiency homozygote, Df(m)/
Df(m) (left rectangle), and for a female heterozygous for
an X-autosome translocation carrying $+^m$, and for Df(m) in
the intact autosome (right rectangle). See Table 3 for
various observed outcomes.

     The results of these experiments indicate whether the deficien-
cies, in addition to being organismic lethals (a fact already known
from the Df(m)/Df(m) genotypes), also act as cell lethals; if so,
no coat-color mosaicism would be obtained. The findings were different for different deficiencies tested (Table 3): 12 independent
Df(b)'s and a Df(p) were not cell lethal but depressed overall viability in tht T/Df combinations [25], indicating that dosage of some
gene product was important on the organismic level; deficiencies of
the c^A and c^E groups did not markedly affect cell viability or overall viability; and a c^{Dj}-group deficiency prevented survival when in
combination with T(X;7) [15]. Gluecksohn-Waelsch [9] has recently
reported on a similar study in which 5 c lethals were combined with
the TlCt X insertion. Her results for a c^A-group mutant and for two
c^E's parallel ours in terms of showing full, or near-full, overall
viability. Survival was zero for the combination with c^{25H} (probably a c^{Dp-} or c^{Dq}-group mutant [8]), thus paralleling our findings with c^{Dj}. It was clearly depressed with c^{6H} (a c^{Bi}-group mutant), similar to the condition we had reported for Df(b) and D(p)
mutants [25].

     The problem of possible cell-lethal (or semi-lethal) action of
deficiencies arises, in a specialized setting, in the case of the
gametes. It was found some time ago [3] that several of the d se
deficiencies were transmitted with less than normal frequency.
While such reduced transmission ratios could be the result of selec-

Table 3. Phenotypes of Females Carrying X-Autosome
Translocations and Deficiencies Involving Specific Loci

Translocation-deficiency combinations[a,b]		Phenotype	
		Viability	Coat-color mosaicism
$\dfrac{T(X;\underline{7})\ (15)}{c^A\ \text{or}\ c^E}$ $\dfrac{T1Ct\ (9)^c}{c^A\ \text{or}\ c^E}$		near-normal	yes
$\dfrac{T(X;\underline{4})\ (25)^d}{Df(\underline{b})}$, $\dfrac{T(X;\underline{7})\ (25)}{Df(\underline{p})}$, $\dfrac{T1Ct\ (9)}{c^{B1}}$		reduced	yes
$\dfrac{T(X;\underline{7})\ (15)}{c^{Dj}}$ $\dfrac{T1Ct\ (9)}{c^{Dp}\ (or\ Dq)}$		lethal	--

[a]See Fig. 2 for illustration of combination involving reciprocal translocation and deficiency. For extent of deficiencies designated by the c-locus complementation-group symbols, see Fig. 1. Time of death of homozygous deficiencies as follows: c^A/c^A, or c^E/c^E, perinatal; c^{B1}/c^{B1}, c^{Dj}/c^{Dj}, or Df(p)/Df(p), at or shortly after implantation; c^{Dp}/c^{Dp} or c^{Dq}/c^{Dq}, before implantation; the 12 Df(b)/Df(b)'s, not yet determined.

[b]Superscript numbers refer to publication list.

[c]T1Ct, is an insertion of a large segment of Chromosome 7 into the X.

[d]12 independent Df(b)'s were tested.

tion against heterozygous embryos, the evidence for selection in gametes comes from the finding that the depressed ratios also occurred in the progeny of heterozygous males that produced normal litter sizes (see Fig. 6 in Ref. 27). If more stringent evidence confirms this indication of gametic selection, it may be possible to identify gene products expressed in normal gametes which are missing in deficiency-bearing gametes.

5. FINDINGS CONCERNING THE MUTATION PROCESS

5.1. Radiation-Induced Mutations

The characterization of SLT mutations has allowed certain conclusions to be drawn about what types of mutations are (or are not) induced by radiation, and how the biological and physical variables

Table 4. Relation between Circumstances of Exposure and Type of Mutation[a]

Germ-cell stage	Type of radiation	d-se region				c-region			
		No.	Hypomorph %	Null; no other phenotype %[b]	Null; also other pheno. %	No.	Hypomorph %	Null; no other phenotype %	Null; also other pheno. %
Spontaneous	-	28[c]	7.1	89.3	3.6	17[c]	29.4	64.7	5.9
Spermatogonia	Low LET	81	7.4	75.3	17.3	51	17.6	49.0	33.4
Spermatogonia	Neutrons	39	5.1	61.5	33.3	15	0	60.0	40.0
Postgonial stages	Various	25	0	56.0	44.0	8	0	37.5	62.5
Oocytes	Various	29	3.4	24.1	72.4	16	0	37.5	62.5
Postgonial + oocytes	Low LET	45	2.2	40.0	57.8	12	0	50.0	50.0
Postgonial + oocytes	Neutrons	9	0	33.3	66.7	12	0	25.0	75.0

[a]See Table I for probable chromosomal nature of mutations.

[b]Includes d^{op}

[c]Includes fractional (mosaic) mutants from experimental groups (see text).

of the treatment affect the nature of the mutations produced. It could be demonstrated (Sec. 4) that "intermediate alleles," at least at the c, se, and d loci, must be intragenic mutations, and that radiation is thus capable of producing such mutations as well as deficiencies. The majority of radiation-induced mutations at the c and se loci are to the null allele and are homozygous viable (c^{av} or se^{v}). Since these two loci have been shown to be "non-vital" loci, such mutations could be deficiencies no larger (or not much larger) than the locus itself; they might also be intragenic mutations (Table 1). Mutations that produce a null phenotype for the marker locus and affect other functions as well (e.g., viability, though the marked locus is "non-vital") are multi-locus deficiencies.

The type of mutation induced has been shown to be correlated with the germ-cell type irradiated and with the ion density of the radiation applied, but not with dose rate [4, 8]. Irradiation of spermatogonial stem cells produces more restricted lesions (i.e., more hypomorphs, fewer multi-locus deficiencies) than does irradiation of postspermatogonial stages or oocytes. Within the cell-stage groups, low-LET irradiation produces more restricted lesions than does neutron irradiation. These relations are illustrated in Table 4 for the d-se and c regions.

The demontration of the relatively restricted nature of mutations induced even at high dose rates by low-LET irradiation in spermatogonial stem cells made possible the conclusion that single-track events are responsible for most such mutations. The decrease in mutation rate by dose protraction, which had earlier been demonstrated for spermatogonia [28], thus could not be explained in terms of reducing the chance for interaction of two separate hits. The analysis of mutations has therefore strengthened the original hypothesis that the dose-rate effect is accounted for on the basis of single-track mutational lesions, with a swamping of the repair process at high dose rates.

Spermatogonial exposure to ethylnitrosourea produces an even higher frequency of hypomorphs and lower frequency of multi-locus deficiencies (lethals) than does low-LET irradiation of spermatogonia [29]. Thus, the overall relative frequency of nulls to altered-activity mutants depends on the nature of the mutagen and the type of cell exposed. For intragenic mutations alone, keeping mutagen and exposed cell type constant, the relative frequency of nulls and hypomorphs undoubtedly varies with the locus.

5.2. Spontaneous Mutations

The detailed study of SLT mutants has shown that the pathways by which spontaneous mutations arise may differ from those involved in the induction of mutations. The frequency of fractional (mosaic) mutants at all 7 loci is roughly similar in offspring of irradiated

and control mice [25], and the same is true when the study is restricted to the c locus [6]. Radiation thus does not induce fractionals, and fractional mutants in irradiated as well as control
groups are presumably of spontaneous origin. At the c locus, the
large majority of spontaneous mutations were fractionals, and, at d,
almost one-half were. (Fractionals would probably be nondetectable
at se, and more poorly detectable at $\underline{d}$ than at c.) Analysis of segregation ratios derived from 16 c-locus fractionals led to the conclusion that the mutations had occurred in one strand of the gamete
DNA, or in a daughter chromosome derived from pronuclear DNA synthesis of the zygote, or in one of the first two blastomeres prior
to replication.

Another mechanism implicated in the production of certain spontaneous mutations is "double nondisjunction," whereby the offspring
receives 2 copies of the marked chromosome and no copies of the
wild-type chromosome. Several spontaneous $d\ se$ mutants [3] as well
as an $Hbb^d\ c^+$ mutant [13] are apparently of this type. The event is
detectable only where two markers are present on the chromosome involved. When only one marker is present, the result is indistinguishable from a repeat mutation to the marker allele. Notwith-
standing its name, "double nondisjunction" could result from a single
nondisjunctional or recombinational event in the first (or an early)
cleavage. Although such an event produces two genetically distinct
blastomere populations, the embryo proper would often be nonmosaic,
because it arises from only a very small subset of the cleavage
products. There is no evidence that either mosaic or "double non-
disjunction" mutants are induced by radiation or other mutagenic exposure of the germcells. Indeed, one would not expect them to be if
they arise in the zygote or shortly thereafter, as postulated. The
population of spontaneous mutants might therefore have an admixture
of types not present in the population of induced mutants. This
possibility raises doubts concerning the accuracy of the "doubling-
dose" approach in the calculation of genetic risk.

6. FINDINGS THAT PROVIDE TOOLS FOR BASIC INVESTIGATIONS

The analysis of SLT mutations has yielded, in addition to presumed intragenic alterations, a considerable number of chromosomal
aberrations. These include series of deficiencies that overlap to
various degrees and extend to varying lengths in both directions
from the marker [4, 8]; X-autosome translocations that (in about
half the cells of the body) inactivate regions including some of the
marked loci [11, 12]; and a tandem duplication which duplicates the
c-Hbb segment as well as a region on either side [14]. In addition,
it is possible that the se-locus mutations that do not fit the linear complementation map (Sec. 2) may represent small rearrangements, perhaps inversions (Russell, 1971).

Using the array of characterized aberrations, it is possible to construct extensive series of gene dosages ranging (organismically) from 0 to 3 copies in steps of 0.5 (Fig. 4 in Ref. 27). Such dosage series can be used to investigate not only the marked loci, but any cistron included in some of the deficiencies, e.g., *Mod-2*, or its specific regulator, *Mdr-1* [30]. Such dosage series can greatly expand the comparison of 2, 1, and 0 copies, discussed in Sec. 4.4, and may serve to identify additional structural or regulatory loci. In addition to comparisons of gene dosages, the material also permits comparisons of *cis* and *trans* effects by utilizing the *Dp/Df* combination for the former configuration.

The deficiencies can be useful for the mapping of genes whose location is only approximately known (e.g., from cell-hybridization studies), particularly those for which no variants have been identified. For example, it is known that the mouse *Ldh-1* (lactate dehydrogenase α chain) gene is on Chromosome 7. LDH dosage studies in Df(p) mutants might lead to definitive localization of *Ldh-1*. New genes could be identified (and localized) by comparing normal and heterozygously deficient animals with respect to certain enzyme activities. Deficiency mapping can also establish order of genes in cases where only distances are known. Thus, *sv* was localized on the opposite side of *se* from *d* by this method [31].

The characterized mutations can provide material for attempts at genetic "rescue." For example, it might be possible to determine whether the small body size associated with the Chromosome-7 tandem duplication can be eliminated by combining this duplication with certain deficiencies spanning part of the same region. If such attempts were successful, this would lead to an approximate localization of major body-size genes. Genetic rescue has been attempted for the male-sterilizing effects of T(X;7)s - however without success [32].

The use that can be made of T(X;A)/Df combinations in determining whether certain genetic states are cell lethal was discussed in Sec. 4.5.

The various chromosome aberrations identified among SLT mutations may, finally, provide favorable material for studies designed to identify and isolate DNA sequences from genetically defined regions of the mouse genome. The methodology for such studies has been proposed (Russell and Bernstine, 1982), and experiments are in progress to identify DNA restriction fragments which are missing in a large *c*-locus deficiency [33].

7. SUMMARY

Analysis of mouse specific-locus (SL) mutations at three loci has identified over 33 distinct complementation groups - most of

which are probably overlapping deficiencies – and 13 to 14 new functional units. Perhaps due to ease of ascertainment, the complementation maps that have been generated for the d-se and c regions include numerous vital functions; however, some of the genes in these
regions are non-vital, i.e., the mouse can tolerate their total absence as produced by overlapping deficiencies. At such loci, hypo-
morphic mutants (as distinguished from nulls) must represent intragenic alterations, and some viable nulls could conceivably be intragenic lesions also.

Analysis of SL mutations has provided information on genetic
expression. Homozygous deficiencies can be completely viable or can
kill at any one of a range of developmental stages. Heterozygous
deficiencies of up to 6 cM or more in genetic length have been recovered and propagated. The time of death of homozygous deficiencies
and the degree of inviability of heterozygous deficiencies are probably related more to specific content of the missing segment than
to its length. -- Multiple phenotypes that are found associated in
some of the mutants may result from a lesion affecting a single gene,
or from the loss of several neighboring cistrons; and the distinc-
tion between these alternatives has been aided by complementation
analysis. -- Combinations of deficiencies with X-autosome translocations that inactivate the homologous region in a mosaic fashion have
shown that organismic lethals are not necessarily cell lethal.

The spectrum of mutations induced (e.g., the ratio of nulls to
altered-activity mutants) depends on the nature of the mutagen and
the type of germ cell exposed. Radiation of spermatogonia produces
intragenic as well as null mutations. Spontaneous mutations (several of which may arise in the zygote or in early cleavage) have an
admixture of types not present in populations of mutations induced
in germ cells, and this raises doubts concerning the accuracy of
"doubling-dose" calculations in genetic risk estimation.

The analysis of SL mutations has yielded genetic tools for the
construction of detailed gene-dosage series, cis-$trans$ comparisons,
the mapping of known genes and identification of new genes, genetic
rescue of various types, and the identification and isolation of
DNA sequences.

REFERENCES

1.   L. B. Russell, P. B. Selby, E. von Halle, W. Sheridan, and L.
     Valcovic, The mouse specific-locus test with agents other than
     radiation: Interpretation of data and recommendations for
     future work, Mutation Res., 86:329-354 (1981).
2.   W. L. Russell, X-ray-induced mutations in mice, Cold Spring
     Harbor Symposia on Quant. Biol., 16:327-336 (1951).

3. L. B. Russell and W. L. Russell, Genetic analysis of induced deletions and of spontaneous nondisjunction involving chromosome 2 of the mouse, J. Cellular Comp. Physiol., 56, Suppl. 1, 169-188 (1960).

4. L. B. Russell, Definition of functional units in a small chromosomal segment of the mouse and its use in interpreting the nature of radiation-induced mutations, Mutation Res., 11:107-123 (1971).

5. L. B. Russell, W. L. Russell, and E. M. Kelly, Analysis of the albino-locus region of the mouse, I. Origin and viability, Genetics, 91:127-139 (1979).

6. L. B. Russell, Analysis of the albin-locus region of the mouse, II. Fractional mutants, Genetics, 91:141-147 (1979).

7. L. B. Russell and G. D. Raymer, Analysis of the albino-locus region of the mouse, III. Time of death of prenatal lethals, Genetics, 92:205-213 (1979).

8. L. B. Russell, C. S. Montgomery, and G. D. Raymer, Analysis of the albino-locus region of the mouse: IV. Characterization of 34 deficiencies, Genetics, 100:427-453 (1982).

9. S. Gluecksohn-Waelsch, Genetic control of morphogenetic and biochemical differentiation: lethal albino deletions in the mouse, Cell, 16:225-237 (1979).

10. L. B. Russell, Mammalian X-chromosome action: inactivation limited in spread and in region of origin, Science, 140:976-978 (1963).

11. L. B. Russell and C. S. Montgomery, Comparative studies on X-autosome translocations in the mouse, II. Inactivation of autosomal loci, segregation, and mapping of autosomal breakpoints in five T(X;1)'s, Genetics, 64:281-312 (1970).

12. L. B. Russell and N. L. A. Cacheiro, The use of mouse X-autosome translocations in the study of X-inactivation pathways and nonrandomness, in: "Genetic Mosaics and Chimeras in Mammals," Liane B. Russell, ed., pp. 393-416, Plenum Press, New York (1978).

13. L. B. Russell, W. L. Russell, R. A. Popp, C. Vaughan, and K. B. Jacobson, Radiation-induced mutations at the mouse hemoglobin loci, Proc. Natl. Acad. Sci. USA, 73:2843-2846 (1976).

14. L. B. Russell, W. L. Russell, K. B. Jacobson, R. A. Popp, and C. Vaughan, Induced mutations at the mouse hemoglobin loci, Genetics, 83:s65 (1976).

15. L. B. Russell, unpublished.

16. A. Chovnick, V. Finnerty, A. Schalet, and P. Duck, Studies on genetic organization in higher organisms: Analysis of a complex gene in Drosophila melanogaster, Genetics, 62:145-160 (1969).

17. D. A. Miller, V. G. Dev, R. Tantravahi, O. J. Miller, M. B. Schiffman, R. A. Yates, and S. Glueckshon-Waelsch, Cytological detection of the c^{25H} deletion involving the albino (c) locus on chromosome 7 in the mouse, Genetics, 78:905-910 (1974).

18. L. B. Russell and N. L. A. Cacheiro, The *c*-locus region of the mouse: genetic and cytological studies of small and intermediate deficiencies, Genetics, 86:s53–s54 (1977).

19. N. L. A. Cacheiro and L. B. Russell, unpublished.

20. W. Sheridan, The detection of induced recessive lethal mutations in mice, these proceedings (1982).

21. T. H. Roderick, Utilization of inversions to detect and localize recessive lethal mutations, these proceedings (1982).

22. D. L. DeHamer, A biochemical study of lethal mutations at the *c*-locus in the mouse, Ph.D. Dissertation, The University of Tennessee (1975).

23. S. Gluecksohn-Waelsch, M. B. Schiffman, J. Thorndike, and C. F. Cori, Complementation studies of lethal alleles in the mouse causing deficiencies of glucose-6-phosphatase, tyrosine aminotransferase, and serine dehydratase, Proc. Natl. Acad. Sci. USA, 71:825–829 (1974).

24. E. G. Bernstine, L. B. Russell, and C. S. Cain, Effect of gene dosage on expression of mitochondrial malic enzyme activity in the mouse, Nature, 271:748–750 (1978).

25. L. B. Russell, Genetic and functional mosaicism in the mouse, in: "The Role of Chromosomes in Development," Michael Locke, ed.,pp. 153–181, Academic Press, New York (1964).

26. L. B. Russell, J. W. Bangham, and C. S. Montgomery, The use of X-autosome translocations in the mouse for the study of properties of autosomal genes, Genetics, 50:281–282 (1964).

27. L. B. Russell and E. G. Bernstine, Mammalian mutagenesis: future directions, in: "Molecular and Cellular Mechanisms of Mutagenesis," J. F. Lemontt and W. M. Generoso, eds., pp. 345–360, Plenum Press, New York (1982).

28. W. L. Russell, L. B. Russell, and E. M. Kelly, Radiation dose rate and mutation frequency, Science, 128:1546–1550 (1958).

29. W. L. Russell, Factors affecting mutagenicity of ethylnitrosourea in the mouse specific-locus test and their bearing on risk estimation, in: "Environmental Mutagens and Carcinogens," Sugimura, Kondo, and Takebe, eds., pp. 59–70, Tokyo Press, Tokyo, and Alan R. Liss, Inc., New York (1982).

30. E. G. Bernstine, C. Koh, and C. C. Lovelace, Regulation of mitochondrial malic enzyme synthesis in mouse brain, Proc. Natl. Acad. Sci. USA, 76:6539–6541 (1979).

31. L. B. Russell, Order of *d, se, sv*, Mouse News Letter, 38:32 (1968).

32. L. B. Russell, N. L. A. Cacheiro, and C. S. Montgomery, Attempts at genetic "rescue" of male sterility resulting from X-autosome translocations, Genetics, 94:s89–s90 (1980).

33. L. Albritton and E. G. Bernstine, unpublished.

34. L. B. Russell, Heterozygous effects on viability of *d* or *se* mutations that involve a known minimum number of lethal functional units, Oak Ridge National Laboratory Biology Division Annual Prog. Rept., ORNL-4817, pp. 114–116 (1972).

DEVELOPMENTAL GENETICS OF SPECIFIC LOCUS MUTATIONS

Salome Gluecksohn-Waelsch

Department of Genetics
Albert Einstein College of Medicine
Bronx, New York 10461

INTRODUCTION

The remarkable complexities of eukaryotic gene organization
and gene expression have received much emphasis in recent years.
Whereas the organizational characteristics of polycistronic operons
in E. coli are fairly well understood and appear quite straight forward, those of eukaryotes are considerably more complex as exempli-
fied particularly by the analysis of the β-globin gene of the mouse.
The specific substructure of the mammalian gene including coding
sequencies transcribed into mRNA (exons) as well as non-coding intervening sequences (introns), in contrast to the uninterrupted con-
tinuity of coding sequences in bacterial genes allows for various
levels of control and has altered significantly the concept of the
eukaryotic gene, as well as its organization and expression. This
should also have an impact on approaches to the study of specific
locus mutations which may include effects on all elements of the
gene complex, not only on those containing the coding sequences.

LETHAL ALBINO DELETIONS

The specific mutations with which we have been concerned were
identified originally as alleles at the albino locus in the course
of studies reported earlier in this workshop. The developmental
analysis of these mutations has revealed effects additional to those
on pigmentation, and has led to the discovery of a model system for
studies of gene control and regulation in cell differentiation. In
particular, the differentiation of liver cells on biochemical and
ultrastructural levels is affected by these mutations. Our studies
make use of this model system in order to achieve an integrated view
of genetics, developmental, molecular and biochemical phenomena and

259

their interactions in the realization of specific cell differentiation.

Lethality of homozygotes is one of the characteristics of the
radiation induced albino alleles which have served our studies. This
was in fact the attribute responsible for our original interest in
these mutations. Three of them had been produced and identified at
Oak Ridge, and were turned over to us in the hope that we might be
able to elucidate the causes of homozygous lethality. Subsequently,
we received three additional lethal albino mutations from Harwell.
Of the total of six, four were perinatal lethals whereas two were
responsible for early embryonic death, one at the 2-8 cell cleavage
stage, and the other around the time of implantation in the uterus.
The developmental analysis of the latter two was the subject of dissertation research of Susan Lewis who is here with us at this workshop.

In addition to Susan Lewis various other colleagues have made
decisive contributions to the work that I shall discuss, in particular Carl F. Cori, Robert P. Erickson, and more recently in my
own laboratory Phyllis Shaw, Ann Goldfeld, Leslie Pick, Susan
Schiffer, and Craig Robinson.

The present discussion will focus primarily on the perinatal
lethal alleles. Homozygotes were shown to be severely hypoglycaemic
immediately after birth. This led to an investigation of enzymes
involved in glucose metabolism and to the demonstration of practically total absence of glucose-6-phosphatase activity. Many details
of subsequent studies were published in the course of the past 15
years (for review cf. 1), and I shall summarize them here before
proceeding to more recent work.

CELLULAR ABNORMALITIES IN HEPATOCYTES

Figure 1 represents an attempt at a diagrammatic presentation
of the effects of the perinatal lethals on ultrastructure and biochemical differentiation of hepatocytes. In homozygotes, the radiation induced mutation on chromosome 7 results in ultrastructural
membrane abnormalities of microsomal membranes, Colgi apparatus and
the nuclear membrane (gray). The normal arrangement of parallel
cisternae is disrupted in the rought endoplasmic reticulum (RER)
with subsequent vesiculation and dilatation of membranes. There is
widespread focal loss of membrane bound ribosomes and disaggregation
of polysomes. The nuclear membrane is dilated and remains close to
the nucleus in the regions of nuclear pores (black) only. The normally lamellae of the Golgi apparatus are swollen and dilated into
large clear vacuoles (gray). Mitochondria, lysosomes and peroxi-
somes appear normal (black) as does the plasma membrane (black).
These ultrastructural abnormalities characterize liver and kidney
cells of mutatnt homozygotes whereas all other cell types remain

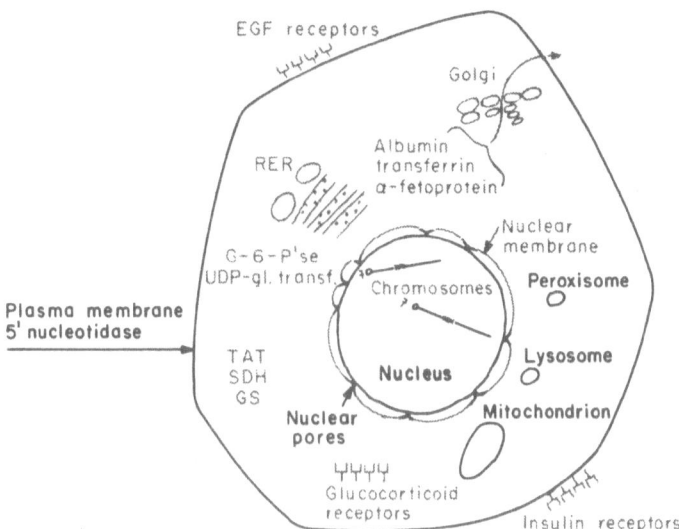

Fig. 1.   Hepatocyte of deletion homozygotes. Diagrammatic presenta-
          tion of hepatocyte of homozygous deletion newborn mouse.
          Gray: abnormal, black: normal. EGF = Epidermal growth
          factor, G-6-P'se = glucose-6-phosphatase, GS = glutamine
          synthetase, RER = rough endoplasmic reticulum, SDH = serine
          dehydratase, TAT = tyrosine aminotransferase, UDP-gl.
          transf. = UDP glucuronosyltransferase.

ultrastructurally normal. The restriction of membranous anomalies
to specific intracellular membranes and to specific cell types reflects the differentiation specific effects of the mutations.

BIOCHEMICAL DEFICIENCES IN HEPATOCYTES

     Biochemically, activities of two microsomal membrane bound en-
zymes were deficient, i.e., glucose-6-phosphatase and UDP-glucuron-
osyl transferase (gray) whereas those of various other enzymes in
the microsomal compartment remained normal. In addition, threee
soluble enzymes were found to be severely deficient in their activities. Furthermore, the quantities of the three major serum proteins
(grays), presumably synthesized by the liver on membrane bound polysomes for export, were decreased in the mutant homozygotes: albumin
by 35%, α-feto-protein by 22% and transferrin by about 40%. Incorporation experiments using ^{14}C-leucine showed a reduction of
plasma protein synthesis to about 20% of normal in newborn mutant
homozygotes even though incorporation into total liver proteins remained nearly normal.

These results strongly indicate specific effects of the radiation induced mutations on particular proteins rather than a general
breakdown of the entire liver protein synthesizing machinery. Finally, more recent work showed several hormone-receptor systems of
hepatocytes to be affected in mutant homozygotes, including a deficiency of insulin receptors (gray), epidermal growth factor receptors (gray) and glucocorticoid hormone-receptor interactions
(gray). The question of the specific hepatic abnormalities ultimately responsible for the perinatal death of homozygotes is still
a matter of speculation.

The multiple effects of the radiation induced albino mutations
raised the suspicion that they might in fact represent overlapping
deletions; this was confirmed by complementation tests and cytogenetic mapping. All perinatal lethals share effects on the enzymes indicated in Fig. 1, deficiencies of plasma proteins, and
ultrastructural abnormalities of subcellular organelles of liver and
kidney cells. The length of the perinatal deletions to be discussed
here is probably in the range of 1.5-2 centimorgans. Cytogenetic
mapping in mitosis and meiosis supported earlier conclusions derived
from complementation tests that the albino mutations represented deletions [2, 3].

Enzyme activities deficient in livers of deletion homozygotes
were normal in deletion heterozygotes. The absence of a dosage effect indicated that the enzyme abnormalities were not due to deletions of the respective structural genes. The deleted genome was therefore suspected of carrying program information for enzyme expression which
was located outside of the structural gene and was concerned with
aspects of gene control and regulation on various levels. In order
to test the hypothesis of intact structural genes, experiments were
designed to study the possible inducibility of the deficient enzyme
tyrosine aminotransferase (TAT) in somatic hybrid cells derived from
fusion of homozygous mutant mouse liver cells to rat hepatoma cells.
The results of this experiment clearly showed inducibility of mouse
TAT in hybrid cells. The missing function(s) of the deleted mouse
gene(s) theregore appear to be complemented by gene products of the
rat cells which interact with the structural gene of the mouse and
induce the expression of mouse TAT activity [4].

Preliminary results obtained by Carl Cori in similar experiments designed to test the inducibility of glucose-6-phosphatase
in hybrid cells derived from the fusion of mutant mouse liver with
rat hepatoma cells, indicate that the structural locus for this enzyme also is intact in liver cells of deletion homozygotes, and can
be activated by gene products of rat hepatoma cells. The structural
genes for these two enzymes appear to be located on different chromosomes since the expression of TAT and that of glucose-6-phosphatase
assort at random in the hybrid cultures tested.

It is noteworthy that in these experiments the proportions of mutant cell hybrids expressing inducible mouse enzyme activity are the same as those of normal cell hybrids. In the latter case re-expression of enzyme activity occurs since this function had been extinguished in primary liver cell cultures. Such reexpression requires an activating stimulus apparently emanating from the rat genome and causing reversion from the dedifferentiated state. The ability of both TAT and glucose-6-phosphatase activities to be induced in mutant liver cell hybrids demonstrates coordinated control of these liver specific enzymes by genes included in the deletion.

Our search for possible mechanisms responsible for the various abnormalities of liver ultrastructure and function in deletion homozygotes was guided strongly by the high degree of specificity of liver cell defects which contrasted with the normal state of many other aspects of liver structure and function. The targets of the deletions' effects appeared to be mechanisms of control of specific aspects of liver cell differentiation. The lack of inducibility of several of the defective enzymes, e.g., glucose-6-phosphatase and tyrosine aminotransferase, in conjunction with the demonstration of intact structural genes for both of these enzymes suggested an investigation of hormone receptor complexes in deletion homozygotes. Their role in the regulation of specific genes has been stressed particularly by Yamamoto and Alberts [5].

The first hormone to be investigated was insulin which, in addition to its role in glucose metabolism and the inducibility of TAT, is thought to be a fetal growth factor and possibly to play a regulatory role in differentiation. All of these attributes seemed to justify a study of insulin receptor interactions in deletion homozygotes. The results of these studies [6] showed a significant decrease of insulin binding activity in newborn mice homozygous for the deletion c^{14CoS} in comparison to their littermates. All results of experiments with intact hepatocytes as well as partially purified liver membranes fit the interpretation that the decrease in binding activity is due to a decrease in numbers of insulin receptors with apparently normal affinity. It is noteworthy that the decrease in plasma membrane bound insulin receptors is the first indication of any abnormality in this particular membrane which previously was shown to be normal ultrastructurally as well as in the activity of 5' nucleotidase, a plasma membrane marker.

Another receptor bound to the plasma membrane is that for the epidermal growth factor (EGF). The investigation of EGF hormone receptor complexes is in progress in our laboratory and is being carried out by Dr. Phyllis Shaw. Preliminary results point to a reduction in number of EGF receptors in deletion homozygotes analogous to that described for the insulin receptor.

Finally, glucocorticoid hormone receptor complexes were investigated in the search for mechanisms responsible for the lack of inducibility of glucose-6-phosphatase, tyrosine aminotransferase and serine dehydratase activities. In contrast to the two plasma membrane bound receptor elements just discussed, glucocorticoid receptors are compartmentalized in the cytosol of the cell.

CONCLUSIONS

The analysis of the effects of the radiation induced albino deletions points to the conclusion that the genes included in the corresponding normal region of the chromosome, i.e., the counterdeletions, represent a series of regulatory genes concerned specifically with the differentiation of certain attributes of liver cell structure and function. Attempts to arrive at an approximate estimate of the numbers of genes included in a deletion such as that in the perinatal lethals amounting to 1.5 to 5 centi Morgans give results ranging from 15 to 50 actually transcribing genes in addition to non-transcribing sequences. These are surprisingly finite numbers rendering the size of the counterdeletions manageable and the transcribing genes' functions perhaps amenable to analysis. Estimates of gene numbers concerned with assembly and function of subcellular organelles in other systems and organisms are within similar orders of magnitude. The genetic complexity underlying liver cell differentiation with structural and regulatory genes dispersed throughout the genome is not unique. Another example is the genetic control of tubulin and microtubule assembly which depends on a considerable number of widely dispersed genes [7]. In the system discussed here, genes dispersed over a number of chromosomes include structural genes for the various cell specific enzymes and serum proteins as well as those concerned with their compartmentalization into subcellular organelles, their integration, activation, and secretion. At least 5 and probably more of the 20 chromosomes of the mouse appear to be involved at this time, and the region of chromosome 7 identified by the radiation induced mutations represents one of the gene clusters concerned with the assembly and functions of one of the most important organ systems of a mammalian organisms, i.e., the liver.

ACKNOWLEDGMENTS

Work carried out in the author's laboratory was supported by grants from the National Institutes of Health (GM-27250 and GM-19100) and the American Cancer Society (CF-38).

REFERENCES

1.   Salome Glucksohn-Waelsch, Genetic Control of Morphogenetic and Biochemical Differentiation: Lethal Albino Deletions in the Mouse, Cell, 16:225-237 (1979).

2.   D. A. Miller, V. G. Dev, R. Tantravahi, O. J. Miller, M. B.
     Schiffman, R. A. Yates, and S. Gluecksohn-Waelsch, Cytological
     Detection of the c^{25H} Deletion Involving the Albino (c) Locus
     on Chromosome 7 in the Mouse, Genetics, 78:905-910 (1974).
3.   G. Jagiello, J. S. Fang, H. A. Turchin, S. E. Lewis, and S.
     Gluecksohn-Waelsch, Cytological Observations of Deletions in
     Pachytene Stages of Oogenesis and Spermatogenesis in the Mouse,
     Chromosoma, 58:377-386 (1976).
4.   Carl F. Cori, S. Gluecksohn-Waelsch, H. P. Klinger, L. Pick,
     S. L. Schlagman, L. S. Teicher, and H. F. Wang-Chang, Comple-
     mentation of Gene Deletions by Cell Hybridization, P oc. Natl.
     Acad. Sci. USA, 78:479-483 (1981).
5.   K. R. Yamamoto and B. M. Alberts, Steroid Receptors: Elements
     of Modulation of Eukaryotic Transcription, Ann. Rev. Biochem.,
     45:721-746 (1976).
6.   Anne E. Goldfeld, C. S. Rubin, T. W. Siegel, P. A. Shaw, S. G.
     Schiffer, and S. Gluecksohn-Waelsch, Genetic Control of Insulin
     Receptors, Proc. Natl. Acad. Sci., 78:6359-6361 (1981).
7.   Don W. Cleveland, M. A. Lopata, R. J. MacDonald, N. J. Cowan,
     W. J. Rutter, and M. W. Kirschner, Number and Evolutionary Con-
     serviation of α- and β-Tubulin and Cytoplasmic β- and γ-Actin
     Genes Using Specific Cloned cDNA Probes, Cell, 20:95-105
     (1980).

DOMINANT AND RECESSIVE EFFECTS OF ELECTROPHORETICALLY

DETECTED SPECIFIC LOCUS MUTATIONS

Susan E. Lewis* and Franklin M. Johnson†

*Life Sciencies and Toxicology Division
Chemistry and Life Sciences Group
Research Triangle Institute
Research Triangle Park, North Carolina 27709

†Laboratory of Genetics
National Institute of Environmental Health Sciences
Research Triangle Park, North Carolina 27709

ABSTRACT

The heterozygous and homozygous effects of mutations recovered from a biochemical screen of mouse samples are described. Of the total of 28 electrophoretically detected mutations only one, a Pgm-2 mobility mutation appears to have caused a change to a previously known allelic form. All but one electrophoretically detected allele have been shown to be homozygous viable. Two mutations identified by other means are homozygous lethal; the first is a mutant detected because of its reduced PK-3 activity in heterozygotes, and the second is a morphologically detected mutation at the d locus. Three dominant visible mutations are also described briefly.

INTRODUCTION

We have recently discovered a number of mutations in the course of a mouse mutagenesis and biochemical screening program. Although the primary emphasis was on the induction and detection of electrophoretically expressed mutations [1, 2], externally visible mutants as well as an enzyme activity mutation [3] were identified. Experiments using procarbazine [2, 4], ethylnitrosourea (ENU) [5, 6], triethylenemelamine TEM) [7, 8] as mutagens were the source of most of these mutations. Experiments employing methylmethanesulfonate (MMS) generated two of the three dominant morphologically expressed mutations.

Table 1. Mutations Detected during Screening of (C57BL/6J × DBA/2J)F$_1$
          Hybrids

Lab Exp#	Locus	Detected[a]	Origin	Strain of Origin	Comments
A	Mod-1	E	Spontaneous (Pre-existing)	C57	null
W9	Mod-1	E	Induced (ENU)	DBA	mobility-fast
B2	Mod-1	E	Induced (ENU)	DBA	null
L2	Mod-1	E	Induced (ENU)	DBA	slow, reduced activity
N2	Mod-1	E	Induced (ENU)	DBA	shadow band
P9	Pep-3	E	Induced (Procarbazine)	DBA	null
Q9	Pep-3	E	Induced (ENU)	C57	null
T9	Pep-3	E	Induced (ENU)	C57	null
Q2	Pep-3	E	Induced (ENU)	C57	null
V2	Pep-3	E	Induced (ENU)	DBA	null
EE-FF	Idh-1	E	Spontaneous (Pre-existing)	C57	missing homodimer
S9	Idh-1	E	Induced (ENU)	C57	mobility - slow
T2	Idh-1	E	Induced (ENU)	DBA	mobility - fast
MM	Pgm-1	E	Spontaneous (Pre-existing)	C57	null
J2-K2	Pgm-1	E	Induced (ENU)	DBA	cluster of two, null
--	Pgm-1	E	Induced (TEM)	DBA	null[b]
C2	Es-1	E	Induced (ENU)	C57	null
--	Es-1	E	Spontaneous (Pre-existing)	DBA	null[b]
R9	Pep-2	E	Induced (ENU)	C57	mobility - slow
Z9	Pep-2	E	Induced (ENU)	C57	fast and light

Lab Exp#	Locus	Detected*	Origin	Strain of Origin	Comments
G2	*d*	V	Induced (ENU)	C57	cluster of two, homozygous lethal
--	*Gpi-1*	E	Induced (TEM)	DBA	null, homozygous lethal[b]
Y9	*Hba*	E	Induced (ENU)	DBA	mobility - slow
E	*Hbb*	E	Induced (Procarbazine)	DBA	mobility - apparently not structural
--	*Ldh-1*	E	Spontaneous (Pre-existing)	DBA	mobility[b]
D2	*Mi?*	V	Induced (ENU)	DBA	coat color and eyes (homozygous deleterious)
P2	*Pep-7?*	E	Induced (ENU)	DBA	null
A2	*Pgm-2*	E	Induced (ENU)	DBA	mobility - slow
M	*6 Pgd*	E	Induced (Procarbazine)	DBA	fast and light (lost mutation)
WI-XI	*Pk-3*	A	Spontaneous (Pre-existing)	C57	decreased activity, null homozygous lethal
--	*Va?*	V	Induced (MMS)	DBA	runted, varigated coat postnatal dominant lethal
--	*?*	V	Induced (MMS)	DBA	dominant, semi-lethal, belly spot and tail

[a]E = Electrophoretically, A = Enzyme Activity Analysis, V = External Visible Examination.
[b]From Soares (9).

Both preexisting spontaneous and induced mutations were found. A list of all the mutations is presented in Table 1. The loci are listed in descending order according to the frequency of mutations at each locus.

LOCUS DISTRIBUTION OF ELECTROPHORETIC MUTANTS

Differential susceptibility among loci to both spontaneous and induced mutations has been reported by W. Russell in these proceedings and by others elsewhere [9, 10]. In the electrophoretic system, 21 loci are currently examined. Although a sample of sufficient size to establish significant differences in the sensitivity of these loci to mutation induction has not yet been collected, possible patterns are beginning to emerge.

So far, mutations have been detected at 13 of the loci examined in the electrophoretic system in our experiments and in those of Soares (Table 1). There remain eight loci at which no mutations have been found. Of the loci which differ between the strains and at which mutations are easily detected, no mutations have been found at Es-3, Gpd-1, and Pgm-3. At the eleven loci which do not differ, mutations have not been identified at the Pep-1, Es-2, Ldh-2, and Gpd-x loci as well as one additional esterase specified by an unknown locus.

The mutations that have been found represent a variety of origins. Preexisting spontaneous mutations were discovered at six different loci. Included are one null mutation each at the Mod-1 [11], the Es-1 [8], and Pgm-1 loci [12], a missing homodimer Idh-1 mutation [13], and a Pk-3 activity mutant [3]. A preexisting mutation at Ldh-1 locus causing altered mobility of the enzyme was also found [7]. Three mutations induced by procarbazine were identified at three different loci; Hbb, Pgd, and Pep-3 [2, 4]. Two mutations, one at the Pgm-1 locus and one at the Gpi-1 locus were induced by TEM [7, 8].

Most of the mutations detected by electrophoresis were induced by ENU [5, 6]. Eight of the seventeen mutations induced by ENU were equally distributed between the Mod-1 and Pep-3 loci. There were two each at the Pep-2 and Idh-1 loci: the remainder occurred singly at several other loci (see Table 1). The distribution of mutations among loci is not significantly different from the Poisson distribution.

When electrophoretically expressed mutations of all origins are taken into account, ten of the 27 are equally distributed between the Pep-3 and Mod-1 loci. Some electrophoretically expressed loci may eventually prove to be more subject that others to mutational alteration than others by various agents, and there may be others which prove to be especially resistant.

NATURE OF ELECTROPHORETICALLY DETECTED MUTANTS

There is considerable polymorphism at biochemically expressed loci in the mouse demonstrated by surveys of a large number of inbred strains and wild mice for variation at different loci [14-16]. However, most of the mutant alleles we detected during electrophoretic screening are not mutations to previously known alleles. Null alleles were discovered at eleven loci (Table 1) but at only one of these loci, Es-1, has a null allele been previously reported [16]. Without molecular studies, it is impossible to tell whether the Es-1 null allele induced by ENU [6] and the spontaneous preexisting null allele described by Soares [7, 8] are identical to the spontaneous null alleles reported by Selander [16].

Of the electrophoretically detected mutations that change the mobility of the proteins specified by the altered locus, only one, a mobility mutation at the Pgm-2 locus could have possibly caused a change of the gene to a known allele. A comparison of the electrophoretic behavior of the PGM-2 specified by our ENU-induced allele with that specified by the Pgm-2^b allele by SM/J mice [17] showed them to be indistinguishable on the starch gel system used in our laboratory [2]. Although sequencing is needed to establish their identity absolutely, these results are compatible with a mutation of a parental Pgm-2^a allele found in both C57BL/6J and DBA/2J [17] to the Pgm-2^b allele.

Neither the spontaneous missing-band Idh-1 [13] mutant nor either of the two Idh-1 mobility mutants we induced with ENU are likely to be a mutation to other known alleles. The slow form specified by the Idh-1^{s9} mutation has been demonstrated to be different from the Idh-1^c allele carried in the M. molossinus strain (J. Womack personal communication). The fast form specified by the mutant Idh-1^{t2} allele is also likely to be yet another new electrophoretic form. Similarly, none of the three Mod-1 alleles with altered mobility induced with ENU appear to be the same as any Mod-1 allele as yet reported.

The screening system has uncovered mutations at three loci at which electrophoretic variation has not been previously described (Hba, Pep-2, and Pep-7). One is an electrophoretically expressed mutation at the Hba locus. Considerable interstrain variation has been found at Hba, but this polymorphism is detectable by solubility differences [18, 19] or by differences in isoelectric focussing [20] and not by electrophoretic differences. Electrophoretically expressed variation at the Hba locus has not been encountered before among a number of inbred and wild strains surveyed [21]. Poor viability of carriers of Hba alleles with altered mobility such as the mutation we found can not be the reason for failure to find such variation. Homozygous and heterozygous carriers of our electrophoretically altered Hba allele are fully viable and fertile. De-

tails of the biochemical and genetic analysis of the mutant hemoglobin will be published elsewhere.

In addition, ENU induced mutations were found at two loci at
which there have been no previously reported variants of any kind;
Pep-2 and Pep-7. Two electrophoretically expressed mutations were
discovered at the Pep-2 locus. One of these, Pep-2^{Z9} also has an
apparent reduction in enzyme activity because of the light appearance of the Pep-2^{Z9} band. A new null mutation was induced by ENU
at the Pep-7 locus as well. Syntenic analysis has already permitted assignment to certain chromosomes of these and other genes
which have no known genetic variation [22]. The discovery of new
alternate alleles such as those we have discovered during screening
will permit more precise mapping.

The nature of Hbb mutant induced by procarbazine [2] remains
an enigma. Although the starch gel pattern is consistently abnormal
with a slightly fast diffuse minor band, the isoelectric focussing
pattern of the mutant protein is indistinguishable from that specified by the diffuse allele carried by the DBA/2J strain. Protein
sequencing revealed no structural abnormality (Raymond Popp, personal communication). Whatever causes the subtle but distinct abnormality is most likely outside the structural gene itself.

Although, some of the last mutations to be discovered, the
Idh-1, Mod-1, and Pep-3 mutants in experiments T2, N2, and V2 respectively have yet to be tested in the homozygous or hemizygous
state, the rest of the electrophoretically detected alleles mutants
in our study have so far been found to be homozygous or hemizygous
viable.

HOMOZYGOUS EFFECTS OF NULL ALLELES

A total of 10 homozygous viable null alleles have been recovered
in our experiments at a five loci: Mod-1, Pep-3, Pgm-1, Es-1, and
Pep-7. Soares [7, 8] also reported a viable null allele at two of
these loci; Es-1 and Pgm-1. Mice homozygous for these alleles totally
lack the enzyme activity in question but are all viable and fertile.
An additional null allele identified by Soares at the Gpi-1 locus is
lethal in the homozygous state.

The ability of homozygous null animals to survive the absence
of peptidase or esterase isozymes is perhaps not surprising when one
considers that mammals in general have several loci specificying
such enzymes which overlap in substrate specificity as well as
tissues distribution [23, 24] and thus potentially in their in vivo
functions. Thus, the gene product from one locus could compensate
for the absence of the gene product from another locus.

In contrast, there are only two loci which specify malic enzyme; the Mod-1 locus and the Mod-2 locus [25]. Differential tissue expression of these enzymes exists with MOD-1, the soluble enzyme, predominating in liver and kidney and MOD-2, the mitochondrial enzyme, the primary isoenzyme in heart [26]. In spite of the fact that homozygous Mod-1 null animals have, at the most trace amounts of malic enzyme activity in liver or kidney [11] we have found, during maintenance of the Mod-1 null strain, that homozygous null animals are fully viable and fertile under laboratory conditions. Furthermore, examination of the specific activities of several other kidney and liver enzymes in Mod-1 null mice reveals them to be normal as well [11]. Thus, there has been no reported effect of absence of the soluble malic enzyme on the physiology or viability of these mice.

Various overlapping deletions at the c locus identified in visible specific locus tests have been shown to include the closely linked Mod-2 locus as well [27]. Homozygous Mod-2 null mice are runted and sterile, lacking appreciable body fat [28, 29]. Examination of enzymes specified by several other loci in heart, liver, and kidney revealed that levels of soluble malic enzyme are increased by approximately 50% and that liver glucose-6-phosphate dehydrogenase and 6-phosphogluconate dehydrogenase are elevated significantly over normal values as well [30]. No increase in soluble malic enzyme activity was detected in heart or kidney in which mitochondrial malic enzyme accounts for significantly more of the total malic enzyme activity than it does in liver [30].

Three null alleles of independent origin were found at the Pgm-1 locus. One was due to a preexisting spontaneous mutation [2, 12]; another was induced by ENU [6]; the third was induced with TEM by Soares [7, 8]. Homozygotes for two of the three Pgm-1 null mutations have been recovered and are fully viable and fertile under laboratory conditions [12].

There are two additional loci specifying enzymes with PGM specificity; Pgm-3 [12] and Pgm-2 [17]. These, like the Mod-1 alleles have distinctive tissue specificities. PGM-1 is the predominant form only in ovary and erythrocytes but no malfunction of either of these tissues is apparent in Pgm-1 null homozygous animals [12]. However, all of these mutants have been studied only under laboratory conditions, but there is evidence from studies of wild mice that Pgm-1 genotypes may affect liver glycogen utilization and survival of the animals in nature [31]. The alternative alleles present in the wild are not nulls but are expressed as differences in electrophoretic mobility. A homozygous null allele could potentially have even greater effects on glycogen utilization and survival than do changes in electrophoretic mobility. It is thus possible that genotypes which appear to be neutral in the laboratory, could have some effect on metabolism and viability under more stringent environmental conditions.

Table 2. Mod-1 Genotypes of Progeny of Dilute
          Cluster Mutation

| | Genotypes of Progeny | | | |
# Litters	$Mod\text{-}1^b/Mod\text{-}1^b$	$Mod\text{-}1^a/Mod\text{-}1^b$	$Mod\text{-}1^a/Mod^a$	Total
8	2	24	10	36

RECESSIVE LETHAL EFFECTS

We have detected some mutations in our screening efforts that
do have profound effects on the viability of the organism. The
Gpi-1 null allele induced by TEM [7] is apparently a homozygous
lethal although the details about the homozygous lethal phenotype
have not been published (A. Peterson, personal communication).

Another mutation has produced a new allele at the Pk-3 locus
which causes reduced pyruvate kinase activity in kidneys of heterozygous carriers [3]. Prenatal lethality of the homozygote was suspected because matings of heterozygotes inter se failed to produce
liveborn mutant homozygotes. Preliminary prenatal studies of litters
from such heterozygous matings have revealed that the homozygotes
die before birth, around the time of implantation. As the nature of
mutational lesion responsible is not known, it is not yet clear
whether the homozygous lethality of this mutant is due to deficiency
of pyruvate kinase or to an effect on a closely linked chromosomal
segment.

Another lethal mutation at the dilute locus was induced by ENU
[9]. This mutation arose as a cluster consisting of one male and
one female progeny. The original mutant animals were mated to each
other and the Mod-1 genotypes of their progeny were determined (Table
2). There was a significant deficiency of $Mod\text{-}1^b$ homozygotes among
the F_2 progeny indicating that the closely linked dilute mutation is
a homozygous lethal. Prenatal death of homozygotes is suggested as
there was no evidence of any abnormality or increased mortality
among the liveborn progeny. The majority of induced and spontaneous
mutations at the dilute locus cause pre- or post-natal death of the
homozygotes [32].

Homozygous lethal mutations such as these that cause death during the early prenatal period may have their primary impact on human genetic risk from subtle effects on heterozygous carriers [33,
34] rather than from homozygous lethality. An embryo dying during
early prenatal development may not even be detected as a pregnancy.
Thus, the existence of homozygous lethal progeny may not be perceived as a health risk per se. However, subtle effects of prenatal

lethal genes on the viability and reproductive competence of the heterozygous carrier have been reported [33, 34] and could pose an important health hazard especially if the genes became widespread in the population.

DOMINANT VISIBLE MUTATIONS

Dominant visible mutations that cause detectable changes in the external appearance of the heterozygous carrier, tend to have such low spontaneous and induced frequencies that they are not particularly useful as mutation indicators [9, 35, 36]. However, this type of mutation may have certain relevance to problems of genetic risk because of the demonstrated deleterious effects on heterozygous and/or homozygous carriers.

Three dominant visible mutations were detected during our screening program. The first two, which were induced during the post-meiotic period by MMS, were clearly deleterious even in the heterozygous state. One of these was a runted female with a tricolor fur pattern which resembled that of Va/+ mice [37] although she displayed none of the neutrological abnormalities associated with heterozygous Va mice. She died at 18 weeks of age without producing any offspring. Skeletal studies revealed a very stunted and abnormally shaped pelvis although all other bones were macroscopically normal.

The other visible mutant induced by MMS was a severely runted male that had a shortened kinky tail and a small but distinct white belly spot. He produced only one litter when backcrossed to C57BL/6J females: a phenotypically normal son and a daughter with a similar but even more extreme phenotype than her father's. She had a much larger belly spot and spastic paralysis of the hind limbs.

The third dominant mutation was induced by ENU and caused a generally lightened coat color and a belly spot in the original F_1 mutant animal. Although this was originally thought to be a mutation at the W locus [5] the trait is not linked to Pgm-1, and homozygous are runted red-eyed blind white animals. It is thus most likely to be a mutation at the Mi locus [38].

ACKNOWLEDGMENTS

The authors would like to thank Chris Worthy and Lois Barnett for their expert technical assistance with the mouse breeding experiments and electrophoretic analysis. We appreciate the efforts of Teresa Erexson in typing the manuscript. The critical comments on the manuscript of Frank H. Deal are also appreciated.

This work was supported in part by Contract No. N01-ES-0-002 from the National Institute of Environmental Health Sciences and Contract No. 68-02-3626 from the Environmental Protection Agency.

REFERENCES

1. H. V. Malling and L. R. Valcovic, A biochemical specific locus mutation system in mice, Arch. Toxicol., 38:45-51 (1977).

2. F. M. Johnson, G. T. Roberts, R. K. Sharma, F. Chasalow, R. Zweidinger, A. Morgan, R. W. Hendren, and S. E. Lewis, The detection of mutants in mice by electrophoresis: Results of a model induction experiment with procarbazine, Genetics, 97: 113-124 (1981).

3. F. J. Johnson, F. Chasalow, G. Anderson, P. MacDougal, R. W. Hendren, and Susan E. Lewis, A variation in mouse kidney pyruate kinase activity determined by a mutant gene on chromosome 9, Genetical Research, 37:123-131 (1981).

4. F. M. Johnson and S. E. Lewis, Mouse spermatogonia exposed to a high, multiply fractionated dose of a cancer chemotherapeutic drug: Mutation analysis by electrophoresis, Mutation Res., 81:197-202 (1981).

5. F. M. Johnson and S. E. Lewis, Electrophoretically detected germinal mutations induced in the mouse by electrophoresis, Proc. Natl. Acad. Sci. USA, 78:3138-3141 (1981).

6. F. M. Johnson and S. E. Lewis, The human genetic risk of airborn genotoxics: An approach based on electrophoretica techniques applied to mice, Brookhaven Symposium (in press).

7. E. R. Soares, TEM-Induced gene mutations at enzyme loci in the mouse, Environmental Mutagenesis, 1:19-25 (1979).

8. E. R. Soares, Identification of a new allele of Es-1 segregating in an inbred strain of mice, Biochemical Genetics, 17:577-583 (1979).

9. G. Schlager and M. M. Dickie, Natural mutation rates in the house mouse. Estimates for five specific loci and dominant mutations, Mutation Res., 11:89-96 (1967).

10. A. G. Searle, Mutation induction in mice, Adv. Radiation Biol., 4:131-207 (1974).

11. C.-Y. Lee, Shwu-Maan Lee, Susan E. Lewis, and Frank M. Johnson, Identification and biochemical analysis of mouse mutants deficient in cytoplasmic malic enzyme, Biochemistry, 19:5098-5103 (1980).

12. F. M. Johnson, F. Chasalow, R. W. Hendren, L. B. Barnett, and S. E. Lewis, A null mutation at the mouse Phosphoglucomutase-1 locus and a new locus, Pgm-3, Biochemical Genetics, 19:599-615.

13. S. Lewis, G. Anderson, L. Barnett, P. MacDougal, and F. M. Johnson, A new electrophoretically expressed allele at the Idh-1 locus in the mouse (in internal review process).

14. J. Womack, Single gene differences controlling enzyme proper-
    ties in the mouse, Genetics 92 Suppl., Proceedings of the Work-
    shop: Methods in Mammalian Mutagenesis S5-S12 (1979).
15. T. H. Roderick, F. H. Ruddle, V. M. Chapman, and T. B. Shows,
    Biochemical polymorphisms in feral in inbred mice (Mus musculus),
    Biochem. Genet., 5:457-466 (1971).
16. R. K. Selander, S. Y. Yang, and W. Craig Hunt, Polymorphism in
    esterase and hemoglobin in wild populations of the house mouse,
    Studies in Genetics V, 271-338.
17. T. B. Shows, F. H. Ruddle, and T. H. Roderick, Phosphogluco-
    mutase electrophoretic variants in the mouse, Biochem. Genet.,
    3:25-35 (1969).
18. R. A. Popp, Studies on the mouse hemoglobin loci: Heterogeneity
    of electrophoretically indistinguishable single-type hemoglobins,
    J. Hered., 53:75-77 (1962).
19. R. A. Popp, Studies on the mouse hemoglobin loci IV, indepen-
    dent segregation of Hb and Sal: Effect of the loci on the
    electrophoretic and solubility properties of hemoglobins, J.
    Hered., 53:77-80 (1962).
20. J. B. Whitney, III, G. T. Copland, L. C. Skow, and E. S. Russell,
    Resolution of products of the duplicated hemoglobin α-chain
    loci by isoelectric focusing, Proc. Natl. Acad. Sci. USA, 76(
    867 (1979).
21. J. B. Whitney, III, Mouse hemoglobinopathies: Detection and
    characterization of thalassemias and globin-structure muta-
    tions. Animal Models of Inherited Metabolic Diseases, E. Liss,
    Inc., in press (1982).
22. P. A. Lalley, J. D. Minna, and V. Francke, Conservation of
    autosomal gene synteny groups in mouse and man, Nature, 274:
    160162 (1978).
23. James E. Womack, Esterase-6 (Es-6) in laboratory mice: Hor-
    mone-influenced expression and linkage relationship to oligo-
    syndactylism (Os), Esterase-1 (Es-1), and esterase-2 (Es-2)
    in chromosome 8, Biochem. Genet., 13:311-322 (1975).
24. S. Rapley, W. H. P. Lewis, and H. P. Harris, Tissue distribu-
    tions, substrate specifications and molecular sizes of human
    peptidases determined by separate gene loci, Am. Hum. Genet.,
    34:307-320 (1971).
25. T. B. Shows, V. M. Chapman, and F. H. Ruddle, Mitochondrial
    malate dehydrogenase and malic enzyme: Mendelian inherited
    electrophoretic variants in the mouse, Biochem. Genet., 4:
    707-714 (1970).
26. D. Brdiczka and D. Pette, Intra- and Extramitochondrial iso-
    zymes of (NADP) malate dehydrogenase, Eur. J. Biochem., 19:
    546-551 (1971).
27. R. P. Erickson, E. M. Eicher, and S. Gluecksohn-Waelsch,
    Demonstration in mouse of X-ray induced deletions for a known
    enzyme structural locus, Nature, 248:416-418 (1974).

28. S. Gluecksohn-Waelsch, M. B. Schiffman, J. Thorndike, and C. F. Cori, Complementation studies of lethal alleles in the mouse causing deficiencies of glucose-6-phosphate, tyrosine amino-transferase and serine dehydratase, Proc. Natl. Acad. Sci. USA, 71:825-829 (1974).

29. S. E. Lewis, A. Turchin, and T. H. Wojtowicz, Fertility studies of complementing genotypes at the albino locus of the mouse, J. Reprod. Fertil., 53:197-202 (1978).

30. H. W. Mohrenweiser and R. P. Erickson, Enzyme changes associated with mitochondrial malic enzyme deficiency in mice, Biochem. Biophys. Acta, 587:313-323 (1979).

31. A. J. L. Brown, Physiological correlates of an enzyme polymorphism, Nature, 269:803-804 (1977).

32. L. B. Russell, Definition of functional units in a small chromosomal segment of the mouse and its use in interpreting the nature of radiation-induced mutations, Mutation Research, 11:107-123.

33. W. Sheridan, The dominant effects of a recessive lethal in the mouse, Mutat. Res., 5:323-328 (1968).

34. K. G. Luning and Sheridan, Dominant effects of recessive lethals in mice, Hereditas, 59:289-297 (1968).

35. W. L. Russell, X-ray induced mutations in mice, Cold Spring Harbor Symp. Quant. Biol., 16:327-336 (1951).

36. M. F. Lyon and T. Morris, Gene and chromosome mutation after large fractionated or unfractionated radiation doses to mouse spermatogonia, Mutation Res., 8:191-198 (1969).

37. A. M. Cloudman and L. E. Bunker, The varitint-waddler mouse: A dominant mutation in Mus musculus, J. Hered., 36:254-263 (1945).

38. A. B. Grobman and D. R. Charles, Mutant white mice, J. Hered., 38:381-384 (1947).

SOME IDEAS ON FUTURE TEST SYSTEMS

Anthony G. Searle

Medical Research Council
Radiobiology Unit
Harwell, Didcot
Oxon OX11 ORD
United Kingdom

1. INTRODUCTION

The specific locus method for determining mutagenicity in vivo
has been used very successfully for over 30 years and doubtless will
continue to be used for many years to come. It has the great advantage of giving clear-cut results in the F_1 generation. Another
advantage in my opinion is that it can be treated as a kind of
"basic kit" to which all sorts of other useful tests and analyses
of the F_1 generation can be added on. These include searches for
various kinds of dominant mutation, or for specific biochemical defects, as well as fertility tests to look for the presence of translocations or other chromosome anomalies. Thus, by building on to
this one test, some idea of the nature and magnitude of the extra
genetic load associated with particular mutagenic exposure can be
obtained.

1.1. Limitations of the Specific Locus Test

Let us consider the limitations of the original specific locus
test system and see to what extent these can be remedied, if this
is thought necessary. Of course the variety of loci under test is
limited, though perhaps less so than one would think from the state-
ment that all but one of the loci are concerned with coat color.
Thus the c locus can be regarded as an enzyme one, since it contains
the structural gene for tyrosinase [1]. Alleles at the dilute (d)
and pink-eyed loci (p) may show very different kinds of behavior
defect as well as effects on pigmentation. One allele at the p
locus (p^{un}) shows a high frequency of reversion to wild type [2]

while the original dilute (d) "mutation" seems to be the result of
integration of an ecotropic murine leukaemia virus genome at this
site [3]. However, "repeat mutations" to dilute are found in specific locus experiments [4], although rare.

Apparent "repeat mutations" can also be caused by the complementation of gametes from the multiple recessive tester stock
which are disomic for one of the chromosomes carrying a tester stock
recessive with nullosomic gametes involving the same chromosome.
Such events can be detected when linked tester stock loci are involved, as with d and se and with c^ch and p, and have indeed been
found with the former pair [5]. There are difficulties with the
complementation process when Chr 2, on which the a locus is situ-
ated, is involved [6], while the high proportion of homozygous
lethal mutations found at the s locus in radiation experiments [7]
shows that complementation involving Chr 14 is very rare, if it happens at all. On the other hand, about half the mutations induced
by radiation at the b locus are homozygous viable and most of these
are apparent b "repeats." It seem quite possible that some of these
are not true mutations at all but the result of complementation of
a tester stock gamete disomic for Chr 4 with a wild type gamete
nullosomic for that chromosome. One way to eliminate this uncertainty in the future would be to incorporate an eighth recessive
into the tester stock, namely misty m. This is linked to b on Chr
4 [8], so results of complementation of the type described would be
brown misty, lighter in color than brown alone.

This is a minor limitation of the test; another is the fact that
it measures forward mutation only, since the treated stock is wild
type at all the loci concerned, except at the a locus, where it is
A^W/+. This last fact means that mutations from A^W to + cannot be
detected. Perhaps a more serious limitation is that, as normally
carried out, it only detects specific locus mutations which have
successfully survived to weaning age and preferably have bred. This
may lead to a considerable underestimate of the total number of mutations, depending on the severity of action of the mutagen. Evidence for this has come from analyses of the results of specific
locus experiments after protracted exposures of males to γ-rays or
fission neutrons [9], in which litters were examined at birth and
on other occasions before weaning. Out of 132 confirmed and presumptive specific locus mutations, 28 or 26% became extinct before
weaning age. This was double the overall pre-weaning mortality.
Forty-nine or 37% became extinct before completion of the allelism
tests. Seventy percent of these were presumptive s mutants, double
the proportion in the confirmed mutations. This finding, together
with the fact that clusters of s mutants are very rare [7] suggests
that dominant deleterious mutations at this particular locus, already reported to produce such mutations by Russell and Russell
[10], are the main reason for the extra pre-weaning mortality of
specific locus mutations. Such mutations are likely to form a sub-

Table 1. Closely Linked Gene Pairs Suitable for Detection
of Deficiences in the Mouse

Chromosome	Symbols	Names	Map units apart
2	a...bp	Non-agouti, brachypodism	0.4
7	p...ru-2	Pink-eyed dilution, ruby-2	2.5
9	d...se	Dilute, short ear	0.2
19	ru..ep	Ruby, pale ear	1.5

For further information, see reference 11

stantial component of the human genetic load and we need to know
more about their prevalence. Lüning [11] has shown that it is not
a general property of induced recessive lethals to have dominant
effects, but the actual proportion of loci behaving thus remains
unknown. Perhaps this is more a property of a particular chromosome region than of a specific locus, which might be connected with
the induction of small deficiencies. Heterozygotes for deficiencies at the d and se loci frequently show adverse effects also [4].

2. SOME OTHER SPECIFIC GERMINAL TESTS

2.1. Detection of Deficiencies

It seems important to find out more about the properties of
chromosomal deficiencies in mammals. In the standard specific locus
test these can be detected visually if they involve the d and se
loci, but otherwise only by complementation tests. As Table 1 shows,
a number of other closely linked pairs of loci seem suitable for
linkage detection; in fact the a-bp pair have been used already in
the Harwell specific locus stock [12]. We have built up a de-
ficiency-testing stock which combines a bp with d se and ru ep, as
well as having p ru-2 as a separate stock. We intend to use these
in various ways to try and generate a range of deficiencies for
study, especially after irradiation of oocytes and post-meiotic
stages in the male, which generate the highest frequency of d-se
deficiencies [4].

2.2. Forward v Reverse Mutations

In mammals, incidences of spontaneous reversions in crosses between and within inbred strains have been calculated by Schlager
and Dickie [13], with respect to mutation at the a, b, c, d, and ln
loci. However, no studies seem to have been made on the induction
of reversions. Clearly it would be desirable to compare rates of
induction of forward and reverse mutations in the same loci at the

Table 2. A System for the Detection
of Both Forward and Reverse
Mutation in the Mouse

Alleles used	Names
a^t	black-and-tan
b^c	cordovan
c^{ch}	chinchilla

$$a^t a^t\ b^c b^c\ c^{ch} c^{ch}\ \times\ a^e a^e\ bm/bm\ c^{ch} c^{ch}$$

$$F_1\ a^t a^e\ b^c +/bm\ c^{ch} c^{ch}$$

black and tan   cordovan   chinchilla

Loci	Types of mutation	Phenotypic change
a	Reverse	Agouti or yellow
	Forward	Grey belly
b	Reverse	Blacker
	Forward	Browner
c	Reverse	Yellow pinna-hairs
	Forward	Very light

same time, by the use of intermediate alleles. Table 2 shows one way in which this can be achieved in ths mouse, and the necessary stocks have now been constructed. The tester stock includes extreme non-agouti a^e since this is recessive to non-agouti a, also misty m as well as b, in order that double non-disjunctional events involving Chr 4 (as discussed earlier) could be detected, since brown misty mice are much lighter than cordovan ones.

2.3. Dominant Deleterious Effects

It seems important to find out to what extent specific locus mutations induced by chemical mutagens have dominant deleterious effects like those already described as occurring after irradiation. In order to discover this it is necessary to pay particular attention to offspring between birth and weaning. Pigmentation develops in the mouse epidermis very shortly after birth, so that if litters from specific locus experiments were systematically examined at birth and about 4 days later then most new mutations at all seven loci except short-ear (se) should be recognized. Their fate could then be followed in comparison with that of sibs and controls.

In many ways it would be desirable to have a wider range of
loci under surveillance. One way of achieving this would be to use
as tester stock the F_1 between the original tester (T) stock and the
Harwell (HT) one. Such mice would be black, with all gametes carrying a and 50% carrying each of the eleven other specific locus recessives. Of this total of twelve loci, the effects of most recessive mutations at the b, bp, c, d, fz, ln, p, pa, and s loci should
be discernible at an early age. From several points of view the use
of a wider spectrum of loci would seem useful and it seems likely
that the F_1 between two rather inbred stocks of very different origin
would show hybrid vigor, thus tending to offset any decrease in the
incidence of mutations in the hterozygous stock. Use of such an F_1
would allow a direct comparison in progeny from the same parents of
the relative sensitivity to these loci to chemical mutagens, to see
if the greater sensitivity of the original set of 7 loci holds for
chemicals as well as for radiation.

A somewhat different approach to the problems of identifying
dominant deleterious mutations would be to use the fact that quite
a high proportion of human mutations of this type (if we exclude
those of late onset) will lead to juvenile (i) growth retardation
or (ii) difficulties in locomotion of one sort or another. A simple
selection system, which could probably be partly automated and
should, in my opinion, include a test of swimming ability, could be
used at weaning age to separate off mice of these categories among
the offspring of control and mutagenically exposed parents. Those
able to breed could then be tested genetically. However, it might
be very difficult, if not impossible, to find discriminating criteria which would pick up mutants with a high enough efficiency to
prevent excess swamping of results with background non-genetic
"noise," especially in view of such complicating factors as differences in litter-size, affecting weight. Nevertheless, such a system might be worth trying out with a powerful mutagen like ENU, as
positive results would be particularly relevant for experimental
studies on the contribution of chemical mutagens to the human genetic load.

2.4. Dominant Mutations with Visible Effects

There are various categories of dominant effect which can be
detected experimentally in different stages in life. Among these
are dominant lethals in utero, mainly in the form of early embryonic
deaths and adding little to the genetic load. Then there are congenital malformations induced in the F_1 generation and detectable
around birth, as has been shown recently by Nomura [14, 15] and by
Kirk and Lyon [16]. Of the various kinds of dominant mutation which
can be detected at weaning age or later, most attention in recent
years has focussed on those involving the skeleton [17, 18] and
the eye [19], from which estimates have been made of the total load
of harmful dominant mutations which would be expected to arise from

Table 3.   Ratios of Various Categories of Dominants and Specific
           Locus Recessives in Radiation Experiments on the Mouse

Category of dominant	Category of recessive	Type of irradiation	Number of		No. dom. no. rec.	Reference
			dominants	recessives		
Visible	7-locus	chronic	30	83	0.36	9
Visible	7-locus	chronic	7	27	0.26	21
Visible	7-locus	acute	6	16	0.38	22
Visible	6-locus	acute	9	7	1.29	22
Visible	6-locus	acute	18	14	1.29	22
Cataract	7-locus	acute	15	38	0.39	19
Skeletal	7-locus	acute	37	4*	9.25	18

*expected, on basis of other data

a given radiation dose.   Both of these types of dominant mutation
demand special techniques for their detection, so it is worth re-
membering that certain categories of dominant mutation (given the
blanket term of "dominant visibles") can be detected by careful external examination of mice at weaning age.   These include those
affecting:

    (i)   coat color (especially white-spotting) and texture, also
         coat loss;

    (ii)   head, limbs, tail, e.g., shortened or twisted limbs,
         kinked tail;

    (iii) digits, to give poly-, oligo- or syndactyly;

    (iv)   eyes and ears, to give marked reduction in size, or
         closed eyes.

All of these, apart from tail anomalies, have parallels in human heredity.   They comprise a valuable extra source of easily available information on harmful dominant conditions which, in my opinion, has received less attention than it should have.   These dominant visible mutations behave in essentially the same way as specific locus mutations in radiation experiments.   Thus they show a
dose-rate effect of about the same magnitude after spermatogonial
treatment, as well as a very similar fission neutron: γ RBE after
chronic exposure [7].   As Table 3 shows, the ratios of their frequencies to those of specific locus mutations are very similar in
different experiments, except for the expected rise to a higher
level when the less sensitive 6-locus stock was used.   The dominant-

recessive ratio is similar to that found for cataracts but much
lower than that estimated for skeletal mutations, but the time and
skill involved is also much lower. If a regular methodical routine
is used for looking at the various parts of the body which may be
affected then any subjective element in the detection can be much
reduced. Lyon [20] has shown (this symposium) that it is possible
to equate these categories of visible dominants with human hereditary
conditions in such a way as to derive an expected overall dominant
load in surviving F_1, as has been done with respect to cataracts
and skeletal mutations.

3. SOMATIC MUTATIONS In vivo

In recent years there has been a great expansion in the number
of test systems for detecting specific locus mutations in human
cells, arising both in vitro and in vivo. Experimental work on the
relationships between mutagenesis and carcinogenesis requires a similar array of different methods for studying the induction of somatic
mutations in vivo in the mouse and other experimental mammals. At
present the well-known "spot" test of Russell and Major [23], which
involves the detection of small mutant patches in the coat, has few
rivals, but clearly it would be desirable to have a method which
worked at the cellular or sub-cellular level and which was amenable
to automated techniques like that described by Bigbee [24] at this
meeting.

3.1. An Attack at the Cellular Level

We have recently published a method [25] which goes part of the
way towards fulfilling these criteria, in that it does work at the
cellular level although not yet amenable to automation. It is based
on the profound changes in melanocyte morphology which take place
with mutaion at the dilute (d) or leaden (ln) loci, so that the normally highly dendritic cells become rounded up and almost without
processes [26], leading to a considerable increase in the intensity
of their melanin pigmentation. This drastic cellular change from a
normal nucleofugal to a mutant nucleopetal morphology can be seen
clearly in some follicular pigment cells of 3-day mice heterozygous
for d and ln, shortly after the onset of melanogenesis and usually
in the form of small clones of mutant cells. About half a million
follicular cells can be scored per mouse from mounts of the dorsal
skin at this age. For both x-rays and procarbazine treatment dose-
response relationships showed good fits to both a quadratic and a
power-law curve [25].

One difficulty in the use of automated scanning for such a system is that follicles of different hair-types (guard-hairs, zig-
zags, etc.) lie at different levels in the skin, so that much focussing up and down is needed. Certain hair mutants, such as sleek
Dl^{slk}, may help to overcome this. However, another line of inquiry

which we are now investigating concerns a possible descent from the
cellular to the subcellular or granular level. A number of coat
color mutations are know known in the mouse which lead to a marked
alteration in the size and shape of the pigment granules synthesized
by the mutant melanocytes [26]. Thus pp melanosomes are small and
shred-like, bb are spherical, bgbg are large and irregular, and so
on. If ways can be found of detecting the few mutant melanosomes
in the coat of hair of a heterozygote for the recessive mutation
concerned among the many million normal ones, then the resolving
power of the resultant test system should be very great indeed.

4. CONCLUSIONS

The specific loci I have been discussing are concerned with
visible morphological effects and thus form only one category out of
many which go to make up the mammalian genome. The need to investigate other functional categories, in order to gain an overall picture of mutagenic effects, is obvious. Studies on histocompatibility loci, which show no signs of induced mutation with radiation
[27] but do with certain chemicals [28, 29] show that no uniform
response to particular mutagens can be expected. As we have seen,
studies on various types of biochemical loci are in progress, but
mutability studies on gene regulation systems, which may be particularly associated with repeated DNA sequences [30], would be of
particular interest, as would others on sites of integration of
viral genomes or on loci connected with disease resistance. However, I hope I have shown that there are still a number of important
problems which can be tackled by use of loci with more easily visible
effects.

5. SUMMARY

The classical specific locus test has some limitations, in that
mutations with deleterious effects in the heterozygote may be eliminated before classification and verification, while some presumptive "repeat mutations" may really be the result of non-disjunction.
Ways to overcome these drawbacks are discussed. An important extra
need, especially with chemical mutagens, is the simultaneous detection of both forward and reverse mutations. The existence of intermediate alleles at the a, b, and c loci has allowed such a stock to
be constructed. Four closely linked gene pairs are suitable for the
construction of special stocks for deficiency detection and three
have been combined for this purpose. In specific locus tests, more
attention should be paid to dominant mutations with externally visible effects, since observational bias seems small when young are
screened in a methodical manner. Methods are being developed for
the detection of somatic mutations in vivo at the cellular level.
These employ some of the same loci used in germinal specific locus
tests and are based on the morphological changes to melanocytes and
their products which are caused by recessive mutations at the loci

concerned. The need is stressed for specific locus tests encompassing entirely different types of loci from those studied at present.

REFERENCES

1. M. Foster, Mammalian pigment genetics, Adv. Genetics, 13:311-
    339 (1965).
2. R. W. Melvold, Spontaneous reversion in mice: effects of pa-
    rental genotype on stability at the p-locus, Mutation Res.,
    12:171-174 (1971).
3. N. A. Jenkins, N. G. Copeland, B. A. Taylor, and B. K. Lee,
    Dilute (d) coat color mutation of DBA/2J mice is associated
    with the site of integration of an ecotropic MuLV genome,
    Nature, 293:370-374 (1981).
4. L. B. Russell, Definition of functional units in a small chro-
    mosome segment of the mouse and its use in interpreting the
    nature of radiation-induced mutations, Mutation Res., 11:107-
    123 (1971).
5. L. B. Russell and W. L. Russell, Genetic analysis of induced
    deletions and of spontaneous non-disjunction involving chro-
    mosome 2 of the mouse, J. Cell. Comp. Physiol., 56, Suppl. 1,
    169-188 (1960).
6. A. G. Searle and C. V. Beechey, Complementation studies with
    mouse translocations, Cytogenet. Cell Genet., 20:282-303 (1978).
7. A. G. Searle, Mutation induction in mice, Adv. Radiation Biol.,
    4:131-207 (1974).
8. M. C. Green, Genetic Variants and Strains of the Laboratory
    Mouse, Gustav Fischer Verlag, Stuttgart and New York (1981).
9. A. L. Batchelor, R. J. S. Phillips, and A. G. Searle, A com-
    parison of the mutagenic effectiveness of chronic neutron- and
    γ-irradiation of mouse spermatogonia, Mutation Res., 3:218-229
    (1966).
10. W. L. Russell and L. B. Russell, The genetic and phenotypic
    characteristics of radiation-induced mutations in mice, Radia-
    tion Res., Suppl. 1, 296-305 (1959).
11. K. G. Luning, Do recessive lethals have dominant deleterious
    effects in mice? Mutation Res., 3:340-345 (1966).
12. M. F. Lyon and T. Morris, Mutation rates at a new set of spe-
    cific loci in the mouse, Genet. Res., Camb., 7:12-17 (1966).
13. G. Schlager and M. M. Dickie, Natural mutation rates in the
    house mouse: estimates for five specific loci and dominant
    mutations, Mutation Res., 11:89-96 (1971).
14. T. Nomura, Changed urethan and radiation response of the mouse
    germ cell to tumor induction, in: "Tumors of Early Life in
    Man and Animals," L. Severi, ed., Perugia Quadrennial Int.
    Cong. on Cancer, Perugia, Italy, pp. 873-891 (1979).
15. T. Nomura, Induction of heritable tumors and anomalies in mice
    by parental exposure to x-rays and chemicals, Nature, 296:575-
    577 (1982).

16. M. Kirk and M. F. Lyon, Induction of congenital anomalies in offspring of female mice exposed to varying doses of x-rays, Mutation Res., 106:73-83 (1982).

17. U. H. Ehling, Dominant mutations affecting the skeleton in offspring of x-irradiated male mice, Genetics, 54:1381-1389 (1966).

18. P. B. Selby and P. R. Selby, Gamma-ray induced dominant mutations that cause skeletal abnormalities in mice, I. Plan, summary of results and discussion, Mutation Res., 43:357-375 (1977).

19. U. H. Ehling, J. Favor, J. Kratochvilova, and A. Neuhäuser-Klaus, Dominant cataract mutations and specific-locus mutations in mice induced by radiation or ethylnitrosourea, Mutation Res., 92:181-192 (1982).

20. M. F. Lyon, Problems of extrapolation of animal data to humans, this symposium.

21. R. B. Flavell, in: "Chromosomes Today," M. D. Bennett, M. Bobrow, and G. M. Hewitt, eds., Vol. 7, pp. 42-54, Allen and Unwin, London (1981).

22. M. F. Lyon and T. Morris, Gene and chromosome mutation after large fractionated or unfractionated radiation doses to mouse spermatogonia, Mutation Res., 8:191-198 (1969).

23. L. B. Russell and M. H. Major, Radiation-induced presumed somatic mutations in the house mouse, Genetics, 42:161-175 (1957).

24. W. L. Bigbee, W. E. Branscomb, and R. H. Jensen, Counting of RBC variants using rapid flow techniques, this symposium.

25. A. G. Searle and D. Stephenson, An in vivo method for the detection of somatic mutations at the cellular level in mice, Mutation Res., 92:205-215 (1982).

26. W. K. Silvers, The Coat Colors of Mice, Springer-Verlag, New York (1979).

27. H. I. Kohn, X-ray mutagenesis: results with the H-test compared with others and the importance of selection and repair, Genetics, 92:863-866 (1979).

28. I. K. Egorov and Z. K. Blandova, Histocompatibility mutations in mice, Chemical induction and linkage with the H-2 locus, Genet. Res., 19:133-143 (1972).

29. H. I. Kohn, H-gene (histocompatibility) mutations induced by triethylenemelamine in the mouse, Mutation Res., 20:235-242 (1973).

30. R. B. Flavell, in: "Chromosomes Today," M. D. Bennett, M. Bobrow, and G. M. Hewitt, eds., Vol. 7, pp. 42-54, Allen and Unwin, London (1981).

PROBLEMS IN EXTRAPOLATION OF ANIMAL DATA TO HUMANS

Mary F. Lyon

Medical Research Council
Radiobiology Unit
Harwell, Didcot
Oxon OX11 ORD
England

1. INTRODUCTION

In extrapolating animal data to humans, in order to attempt to estimate genetic hazards, there are several separate types of extrapolation which have to be made (1) from experimental to environmental dose levels, (2) from one cell type to another, e.g., from somatic to germ cells, or from male to female germ cells, (3) from one species to another, and (4) from estimated mutation rates to cases of genetic disease in man.

2. TYPES OF EXTRAPOLATION

2.1. Extrapolation of Dose-Response

Dose can be expressed or measured in various forms [1].

(a) Exposure dose

The amount emitted or otherwise present in the environment.

(b) Pharmacological dose

The amount entering the body.

(c) Tissue dose

The amount which reaches the tissues after transport, metabolism, activation, detoxication, excretion and so on.

(d) Target dose

    The amount present in the cell nucleus in the neighborhood of
the DNA or other relevant target.

(e) Molecular dose

    The amount which actually reacts with the DNA.

    After this there is the response in term of production of
mutations.

    At each of the transitions from one measure of dose to the next
there is the possibility of non-linear relationships. For instance,
at the tissue level the induction of enzymes for activation or the
saturation of enzymes for detoxication can affect the results. Thus,
it is very likely that the exposure dose or the pharmacological dose,
which may be the ones most easily measured, will be non-linearly related to the effect on the DNA or the response [1]. For this reason, for the future one badly needs some means of measuring the dose
at the tissue, or even better the molecular level. Alkylation of
hemoglobin has been suggested as a possible measure of tissue dose
[2, 3]. Alkylation of DNA [4, 5] or determination of DNA adducts
[6] are possible measures at the molecular level [1].

    Ehrenberg et al. [1] point out that at low doses or concentra-
tions the reaction kinetics approach linearity. Thus, in extrapolation of dose, the principle should be to use data from doses nearest
to those encountered environmentally. If the form of the dose-response relationship for mutagenic effects in germ cells is known
this information should be used. Otherwise linear extrapolation
should be made.

2.2. Extrapolation among Cell Types

    Concerning extrapolation from one cell type to another it is
frustrating that information from somatic cells, whether in vivo or
in vitro, is so much easier to obtain than that from germ cells, but
at present there is no reliable way to extrapolate quantitatively
from somatic cells to germ cells. It remains a hope for the future
that some suitable system for extrapolation from somatic cells will
be found.

    In the case of germ cells, the problem lies in extrapolating
from one cell stage to another, and from one sex to another. Chemicals vary widely in their effects on various germ-cell stages, and
at present the basis of these variations is not understood [7, 8].
Thus, only data from germ cells, rather than somatic cells, should
be used for estimating genetic hazards, and wherever possible these
data should be from the germ cell stage most relevant to the problem being considered.

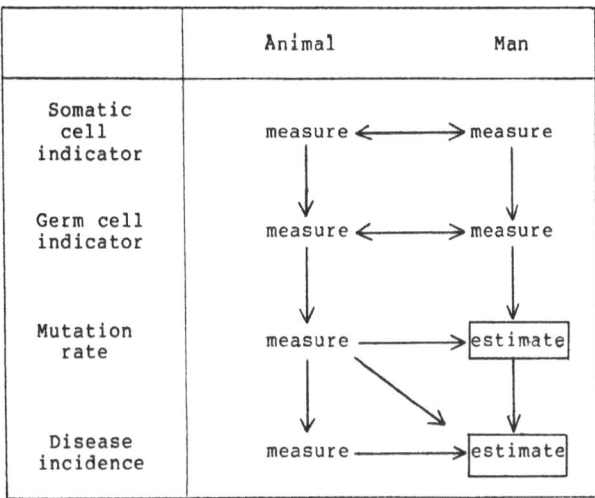

Fig. 1.   Diagrammatic representation of 'parallelogram' method for
          extrapolation from animals to man in estimation of mutation
          rates and resultant increase in genetic disease.

2.3.   Extrapolation from One Species to Another

In extrapolation from species to species there is the obvious
problem of species differences in metabolism, leading to different
tissue doses from the same pharmacological dose. In addition, there
is the possibility that, even if the molecular dose reaching the DNA
is comparable in two species, the amount of resultant genetic damage
may differ, through differences in repair or other factors. This is
seen with ionizing radiation. Although the target dose can be determined with some accuracy and equalized, one sees differences among
species not only between orders, such as between rodents and primates,
but also within the rodents [9] or within the primates [10, 11, 12].
Thus, data from primates other than man may not necessarily be of
any better predictive value than those from the mouse. Because of
these problems of extrapolation either from somatic cells or from
other species, for the future methods of measuring germ cell damage
in man would be very valuable. For the present, extrapolation should
be made from data on the species closest to man.

2.4.   Parallelogram Method of Extrapolation

Various systems have been proposed for overcoming these problems, by the so-called 'parallelogram' method [2, 13, 14, 15]. The
idea is to find some suitable indicator of genetic damage, either in
somatic cells or germ cells, which can be measured in experimental
animals and in man (Fig. 1). In animals one will be able to relate
the indicator to changes in mutation rate in germ cells, as measured

Table 1. Some Estimates of Genetic Hazards of Radiation in Terms of Cases of Genetic Disease per Million Livebirths

| | UNSCEAR 1977(22) | | | | ICRP(21) | | BEIR 1980(20) | |
| | Direct | | Doubling dose | | | | | |
	Gen 1	Total	Gen 1	Total	Gen 1 + 2	Total	Gen 1	Total
Gene mutation								
Autosomal dominant X-linked	20	64	20	100	36	100		40-200
Recessive			slight				5-65	very slow increase
Irregularly inherited			5		30	160		20-900
Chromosome changes								
Numerical	Low	Low			30	30		Increase only slightly
Structural	2-10	51-261	38	40	29	30	<10	
Total	22-30	115-325	63	140	125	320	15-75	60-1180

Unit dose of radiation - 1 rad at low dose rate.

by the specific locus test or some similar suitable test. Then, if
one assumes the same ratio of indicator to mutation rate, one can
predict the increase in mutation rate in human germ cells. It is to
be hoped that one or more of the systems at present being worked on
will prove successful. It will be particularly valuable if it proves
possible to measure mutation rates directly in human sperm, using
the method of Malling [16, 17, 18] or Rudak et al. [19].

For the present, however, it is necessary to extrapolate from
animal systems in order to estimate mutation rates in man. Furthermore, it is also necessary to do so in order to estimate the effect
in terms of human disease of a given increase in the mutation rate.

2.5. Extrapolation from Mutation Rate
to Disease Incidence

Considering this last type of extrapolation, extensive work
has been carried out over the years on estimating the increase in
genetic disease likely to result from an increased mutation rate due
to radiation [20, 21, 22] (Table 1) and one may draw upon this in
attempting a similar estimate for chemicals. Indeed, some workers
suggest that one should express the mutagenic effect of chemicals
in terms of radiation equivalents, equating a certain dose of chemical to so many units of radiation [2, 3]. This, however, seems
inappropriate since the whole response of animals to chemicals may
differ qualitatively from that to radiation, both in the spectrum
of genetic changes produced, and in the relative sensitivity of
various germ cell stages [23].

3. MUTATION INDEX METHOD

It therefore seems appropriate to attempt to develop a new system for dealing with chemicals, while making use of the experience
gained with radiation.

The basis of the method is to use the principles employed in
estimating genetic radiation hazards to prepare a table of cases of
genetic disease arising from a known proportional increase in mutation rate, and then to express the expected rise in mutation rate
from an environmental dose of chemical in such a way that the table
can be consulted.

Basically, there are two methods of predicting increases in
disease incidence resulting from a rise in mutation rate [20, 21,
22].

(a) The proportional method in which it is postulated that a given
     rise in mutation rate will produce a proportional rise in the
     incidence of those conditions in man which depend on recurrent
     mutation for their frequency.

Table 2. Incidence of Genetic Disease Expressed
          as Cases per Million Livebirths and
          Persistence in Generations

	Incidence (bpm)	Persistence (gens)
Gene mutations		
Autosomal dominant or X-linked	10,000	5
Recessive	2,500 [a]	many
Complex or uncertain		
Congenital malformations	24,000 [b]	
Constitutional and degenerative diseases	64,600 [c]	10
Childhood malignant neoplasms	1,400 [b]	
Chromosomal changes		
Structural		
Balanced	1,900	5 [d]
Unbalanced	500	3 [e]
Numerical		
Sex chromosomes	2,000	
		1
Autosomes	1,200	

bpm = livebirths per million; gen = generation

[a] Carter [30]

[b] Leck [31]

[c] Arrived at by substraction of Leck's figures
    from total of 90,000 used by UNSCEAR and others

[d] Evans [32]

[e] Unbalanced types themselves have a persistence
    of 1, but $2/3$ of unbalanced cases arise not by
    mutation but by segregation from balanced
    types [32]

Table 3.   Cases of Dominant Genetic Disease Induced by
            Radiation Estimated from Dominant Mutations
            in Mouse

Type of mutation	Mutation rate $\times 10^6$	Multiplier	Estimated cases of disease per 10^6
Skeletal	3.4	5-10	17-34
Cataract	0.17	16-37	3- 6
Dominant visible	0.2	5-10	1- 2

Mutation rate per gamete per rad of low dose-rate

radiation. Data for skeletal and cataract mutations from

BEIR[20] and Ehling[24,25] and for dominant visibles from

Searle[33]

Multiplier is ratio of total dominant conditions in

McKusick's[26] catalogue to number involving studied

category. Lower multiplier obtained from 328 clinically

significant conditions and higher from 736 total

conditions.

(b) The direct method in which it is argued that an absolute rise
     in mutation rate to a certain class of genetic change, such as
     translocations, in the experimental animal will correspond to
     a similar mutation rate in man, with due allowance for species
     differences and so on.

     In practice, a combination of both methods is typically used to
arrive at a suitable estimate. With both methods there are considerable uncertainties.

3.1. Proportional Method

     For the proportional method one needs to know the current in-
cidence of the various types of genetic abnormality, usually expressed as cases per million live births (Table 2). One also needs
to know their persistence in the population. Because many mutant
genes persist in the population for several generations before dying
out, the current incidence is the product of the mutation rate times
the persistence in generations. Thus, the full effect of an increase
in the mutation rate is spread over several generations. For dominant gene mutation the persistence is usually considered to average

about 5 generations. A 1% rise in mutation rate therefore gives $^{1}/_{3}$ of 1% rise in incidence in the first generation and a total over all generations of 1%. There is a problem with irregularly inherited diseases in that it is not known what proportion are maintained by mutation and it is thus uncertain what the proportional rise may be. BEIR [20] suggested the mutationally maintained component might lie between 5 and 50%.

3.2. Direct Method

Using the direct method and considering gene mutation, one can argue from dominant skeletal mutations [20], cataract mutations [24, 25] or dominant visibles (Table 3). Although it has been said that skeletal and cataract mutations give quite similar estimates for the total number of dominants induced, dominant visibles give a much lower figure, and the range of estimates for all three systems is about 30-fold. The majority of the conditions detected in the dominant visible method would be of clinical importance in man, and correspond to about 70 conditions in McKusick's [26] catalogue. Many of them, including such things as polydactyly, syndactyly and brachydactyly, are very easy to detect and have relatively little effect on viability and fertility. This means that they are easily diagnosed as genetic, and any slight differences in effect in different human families are readily noted, resulting in listing as a separate condition. Probably then these conditions are over represented in McKusick's catalogue. This serves to illustrate the uncertainties in the use of this method of argument.

For the irregularly inherited conditions, one can also use the direct method, by taking the results of a test for the induction of congenital malformations [27]. Male or female mice were treated with various doses of radiation, mated at suitable intervals later, to normal mates, and the offspring were scored for retarded growth and congenital malformations at 18 days gestation. In females, a clear dose-related response, and an effect of interval to mating were obtained (Table 4). The interval effect is consistent with expectation if these were indeed genetic effects being scored. In males, there was again an effect, in that the incidence of malformations was significantly higher in the treated than in the control series (Table 5), but further data are needed in order to give good dose-response data. Thus it seems that conditions such as congenital malformations, generally classified as of irregular inheritance, are induced by radiation and probably by other mutagens. Again there is the problem of finding a suitable multiplier for extrapolation to man. Some of the mouse malformations were retardations. Possibly some would not have been permanently retarded, and thus not really malformed. On the other hand, it is probable that owing to the small size of the animal many malformations, including for instance many heart malformations which are a numerous category in man, would have been undetected in the mouse. Possibly one should

Table 4. Incidence of Abnormal Fetuses from Irradiated Females (expressed as abnormal fetuses/live fetuses %)

Interval (day)	Absorbed dose (rad)			
	108	216	360	504
1- 7	1.8 ± 0.9	0.0	1.8 ± 0.8	2.8 ± 1.3
8-14	0.8 ± 0.7	3.6 ± 1.3	5.6 ± 2.0	5.4 ± 1.9
15-21	1.5 ± 1.0	4.3 ± 2.5	8.4 ± 5.5	12.5 ± 3.2
22-28	1.2 ± 1.1	4.6 ± 5.8	3.3 ± 1.9	11.8 ± 4.0

Control 1.1 ± 0.4

From Kirk and Lyon[27]

Table 5. Induction of Congenital Malformations in Offspring of Male Mice by Exposure of Spermatogonia to X-Rays and Extrapolation to Man

X-rays dose (rads)	Malformed embryos (%)	Induced %	Ind. per rad $\times 10^{-5}$
0	9/1217 (0.68)	-	-
216	8/399 (2.00)	1.32	6.1
504	5/263 (1.90)	1.28	2.5

Mean induced per rad $= 43 \times 10^{-6}$

Reduction factor
  for low dose rate ($^1/3$) $= 14 \times 10^{-6}$

Correction, for
  underdetection (x 2) $= 28 \times 10^{-6}$

Multiplier to total
  conditions $= 90/24 \times 28 \times 10^{-6}$

$= 100 \times 10^{-6}$

Table 6.   Summary of Cases of Genetic Disease per Million Live
Births Resulting from a 1% Increase in Mutation Rate

	Gen 1 (bpm)	Gen 2 (bpm)	Gen 1 + 2 (bpm)	All gens. (bpm)
Gene mutation only	200	150	350	1000
Gene mutation and chromosome strucutal changes	205	154	360	1024
Chromosome numerical changes	33	-	33	33
All types of change	240	154	400	1060

(Figures have in some cases been rounded to avoid undue impression
of precision)

multiply by two to get the likely induction of malformations in man,
but this is uncertain.

There is then the question whether one can extrapolate from congenital malformations to irregularly inherited conditions in general.
One may say that as one class of these conditions is increased by a
rise in mutation rate then others are likely to be so also, but it
is not clear by what proportion.

One could postulate that the mutation dependent fraction of all
the various irregularly inherited diseases is on the average similar,
and thus that the increase would be in proportion to the incidence.
From this the expected number of cases in the first generation for
all irregularly inherited diseases would be in the ratio of the incidences, i.e., $^{90}/_{24}$ of the effect on congenital malformations or
approximately 100 cases per million live births. If, as has been
suggested, the genes persist on the average for about 10 generations
then the total over all generations would be 10 × 100 or 1000. However, there are of course wide limits of uncertainty. The BEIR Committee [20] gives a >10 fold range for total cases caused by unit
dose of radiation.

For the future, one needs better methods of extrapolating from
animals to man in terms of cases of disease resulting from a given
increase in mutation rate. One advantage of the congenital malformation method is that, since it uses an endpoint that can also be
studied epidemiologically in man, it may at some future date be
possible to get direct comparisons.

For the present, taking somewhat arbitrary intermediate figures
(Table 6), the table gives the estimated effect of 1 rad of low dose-

Table 7. Relative Mutagenic Effect (RME) of Some Cytotoxic Drugs for Sperma-togonia and Post-Meiotic Stages of Male Mice

Unit dose = 1 mg/kg

Drug	Spermatogonia			Post meiotic		References
	Specific locus	Sperm abs.	Trans.	Specific locus	Trans.	
Adriamycin	-	-	4.11†	-	1.07	34
Cyclophosphamide	-	.008	0.0	.039	0.557	35,36,37
Mitomycin C	0.81	0.77	0.0	0.0	2.32	35,37,38
Myleran	0.284*	.069	-	-	-	35,36
Natulan	.009	.008	-	0.0075	0.1275	35,36,39,40
TEM	3.09	-	0.0	51.5	1200	36,37
Thio TEPA	-	0.5	12.3	-	42.2	35,41

Trans = Translocations

* Upper fiducial limit

† Based on aberrations in spermatocytes

rate radiation, and this is thought to increase the mutation rate by about 1%. Thus, this can be taken as a table of effects of a 1% mutation rate increase. Let us now consider ways of using this to find effects of known doses of chemicals.

3.3. Calculation of Mutation Index

First, using the specific locus or other suitable method, one finds the induced mutation rate per unit dose. Following the practice of the BEIR Committee one expresses this as a proportion of the spontaneous rate, and this is called the relative mutagenic effect, or RME.

This seems a valuable concept for work with chemicals, since it can be sufficiently flexible to deal with problems such as non-linearity. If the dose-response is non-linear and widely different doses are being received in different cases, one can calculate separate RME's for each dose level. To deal with the problem of varying mutational spectra one can calculate RME's for particular types of genetic change separately. In cases of mixed exposure one can sum RME's from the various chemicals.

Thus, if the dose to which humans are exposed is known, one can calculate a mutation index or MI, for the dose or exposure actually received. The MI is defined as

$$MI = RME \times D \times S \times 100$$

where D is dose, and S any relevant sensitivity factor or factors. Since the table of cases of disease relates to a 1% increase in mutation rate it is convenient to designate unit MI as corresponding to a 1% mutation rate rise and this explains the 100 × multiplier.

3.4. Cytotoxic Drugs Used as Example of Method

As an example of the use of the method one may consider some cytotoxic drugs. The reason for taking these as examples is that data from mouse germ cells are available.

First, one finds the RME per unit dose (Table 7). Information is available for gene mutation and chromosome structural changes, and for spermatogonia and post-meiotic germ-cell stages. Next one needs to know the doses received and in this example the doses chosen are those which might have been received in a 3 month course of treatment. From the product of dose and RME one gets the MI (Table 8). It is to be noted that there is an enormous range of variation among the drugs. Some act mainly on spermatogonial stages; and others on post-meiotic stages. In post-meiotic stages there may be a strong induction of chromosomal structural changes, but in spermatogonia in general not.

Table 8. Mutation Index of Some Cytotoxic Drugs at Doses
Equal to about 3 Months Human Treatment

Drug	Dose assumed (mg/kg)	Gonia Gene mut.	Trans.	Post-meiotic Gene mut.	Trans.
Adriamycin	9.6	200*	3950	50*	1030
Cyclophosphamide	300	240	0.0	1170	16,700
Mitomycin C	1.0	80	0.0	12*	230
Myleran	4.55	31	-	-	-
Natulan	250	255	-	187	3190
TEM	1.0	309	0.0	5150	120,000
Thio TEPA	2.0	100	2450	422*	8440

Data on doses taken from Wade[42] and from Ehling and

Neuhauser[39] for Natulan. Figures have been rounded to

avoid undue impression of precision.

*Assumed values on the basis that MI of gene mutation is

$1/20$ of translocations

The MI may then be translated into numbers of cases of disease
by multiplying the MI by the table for unit MI. Owing to the wide
range of uncertainty all the results are rounded to 2 significant
figures (Table 9). As well as listing the cases per million live
births, one can also give the relative risk by comparing the expected incidence with the current incidence. The effects produced
in post-meiotic stages will relate to conception occurring in the
first three months, and the spermatogonial effects to all later conceptions. Of course, very few conceptions would be expected to occur within three months after treatment with cytotoxic drugs, and
even later there will not be many [28, 29], although the number of
survivors of treatment with these agents in early life is rising.
It should be emphasized that the reason for choosing cytotoxic drugs
as an example was the availability of animal germ-cell data, and not
that these drugs pose any great genetic hazard.

4. CONCLUSIONS

The method described here should be appropriate for other
classes of mutagenic agent, such as those encountered in occupational exposure. A point to notice is that the method considers

Table 9. Cases of Genetic Disease Expected to be Induced per 10^6
Births by about 3 Months Treatment with Some Cytotoxic
Drugs, and Relative Risk

First three months (post-meiotic stages)

Drug	Generation 1 Cases	R.R.	Generation 2 Cases	R.R.	Total all gens. Cases	R.R.
Adriamycin	14,000	1.1	12,000	1.1	74,000	1.7
CP	300,000	3.9	240,000	3.3	1,600,000	16.0
MMC	3,500	1.03	3,000	1.03	17,000	1.2
Myleran	-	-	-	-	-	-
Natulan	53,000	1.5	44,000	1.4	260,000	3.5
TEM	1,600,000	16.0	1,400,000	14.0	8,000,000	77.0
Thio TEPA	130,000	2.2	97,000	1.0	620,000	6.9

After three months interval (spermatogonial stages)

Drug	Generation 1 Cases	R.R.	Generation 2 Cases	R.R.	Total all gens. Cases	R.R.
Adriamycin	60,000	1.6	46,000	1.4	300,000	3.8
CP	48,000	1.5	36,000	1.3	240,000	3.3
MMC	16,000	1.1	12,000	1.1	80,000	1.8
Myleran	7,400	1.07	4,800	1.05	33,000	1.3
Natulan	54,000	1.5	35,000	1.3	240,000	3.3
TEM	62,000	1.6	46,000	1.4	310,000	4.0
Thio TEPA	32,000	1.3	25,000	1.2	160,000	2.5

R.R. = relative risk, assuming normal incidence of 105,000 cases per
10^6 of genetic disease

CP = cyclophosphamide; MC = mitomycin C

the risk to descendants of a given individual rather than a population risk as with radiation. This is because individual exposures
to chemicals may in some cases be quite high and one may get better
protection by considering the individual's risk and giving appropriate genetic counselling. For environmental pollutants, where individual exposures are low, one can of course still consider the
population risk. Major difficulties at present lie in finding suitable means of extrapolation, at all stages from dose-response to
estimation of disease incidence. If reliable extrapolation methods

could be obtained then estimation of genetic hazards would become a much more manageable problem.

REFERENCES

1. L. Ehrenberg, E. Moustacchi, and S. Osterman-Golkar, Dosimetry of genotoxic agents and dose response relationships of their effects, Mutation Res. (in press) (1983).
2. L. Ehrenberg, in: "Banbury Report. I. Assessing Chemical Mutagens: the Risk to Humans," V. K. McElheny and S. Abrahamson, eds., Cold Spring Harbor Laboratory, pp. 152-190 (1979).
3. L. Ehrenberg and S. Osterman-Golkar, Alkylation of macromolecules for detecting mutagenic agents, Teratogenesis, Carcinogenesis, and Mutagenesis, 1:105-127 (1980).
4. W. R. Lee, in: "Chemical Mutagens," A. Hollaender and F. de Serres, eds., Vol. 5, Chap. 8, Dosimetry of chemical mutagens in eukaryotic germ cells, pp. 177-202 (1978).
5. W. R. Lee, in: "Banbury Report. I. Assessing Chemical Mutagens: the Risk to Humans," V. K. McElheny and S. Abrahamson, eds., Cold Spring Harbor Laboratory, pp. 191-200 (1979).
6. F. P. Perera and I. B. Weinstein, Molecular epidemiology and carcinogen-DNA adduct detection: new approaches to studies of human cancer causation, J. Chronic Diseases (1981).
7. M. F. Lyon, Sensitivity of various germ cell stages to environmental mutagens, Mutation Res., 87:323-345 (1981).
8. W. L. Russell, This symposium (1982).
9. M. F. Lyon and B. D. Cox, The induction by X-rays of chromosome aberrations in male guinea-pigs, rabbits, and golden hamsters. III. Dose-response relationships after single doses of X-rays to spermatogonia, Mutation Res., 29:407-422 (1975).
10. J. G. Brewen, J. R. Preston, and N. Gengozian, Analysis of X-ray-induced chromosomal translocations in human and marmoset spermatogonial stem cells, Nature, 253:468-470 (1975).
11. M. F. Lyon, B. D. Cox, and J. H. Marston, Dose-response data for X-ray induced translocations in spermatogonia of rhesus monkeys, Mutation Res., 35:429-436 (1976).
12. P. P. W. van Buul, Dose-response relationship for X-ray induced reciprocal translocations in stem cell spermatogonia of the Rhesus monkey (Macaca mulatta), Mutation Res., 73:363-375 (1980).
13. B. A. Bridges, An approach to the assessment of the risk to man from DNA damaging agents, Arch. Toxicol. Suppl. 3, 271-281 (1980).
14. F. H. Sobels, in: "Progress in Mutation Research," Elsevier/North Holland, Amsterdam, The parallelogram: an indirect apprach for the assessment of genetic risks from chemical mutagens, Vol. 3, pp. 323-327 (1982).
15. G. Streisinger, Extrapolation from species to species and from various cell types in assessing risks from chemical mutagens, Mutation Res., 114:93-105 (1982).

16. H. V. Malling, This symposium (1982).

17. A. A. Ansari, M. A. Baig, J. G. Burkhart, and H. V. Malling, Detection of presumptive point mutation in sperm from mice treated with procarbazine and mitomycin C, Genetics, 94:s3 (1980a).

18. A. A. Ansari, M. A. Baig, and H. V. Malling, In vivo germinal mutation detection with "monospecific" antibody against lactate dehydrogenase-X, Proc. Natl. Acad. Sci., 77:7352-7356 (1980b).

19. E. Rudak, P. A. Jacobs, and R. Yanagimachi, Direct analysis of the chromosome constitution of human spermatozoa, Nature, 274:911-913 (1978).

20. BEIR, The Effects on Populations of Exposures to Low Levels of Ionizing Radiation, National Academy of Sciences, Washington, D.C. (1980).

21. P. Oftedal and A. G. Searle, An overall genetic risk assessment for radiological protection purposes, J. Med. Genet., 17:15-20 (1980).

22. UNSCEAR, Sources and Effects of Ionizing Radiation, United Nations, New York (1977).

23. W. L. Russell, Comments on mutagenesis risk estimation, Genetics, 92:s187-s194 (1979).

24. U. H. Ehling, Strahlengenetisches Risiko des Menschen, UMSCHAU, 80:754-759 (1980).

25. U. H. Ehling, J. Favor, J. Kratochvilova, and A. Neuhauser Klaus, Dominant cataract mutations and specific-locus mutations in mice induced by radiation or ethylnitrosourea, Mutation Res., 92:181-192 (1982).

26. V. A. McKusick, Mendelian Inheritance in Man, John Hopkins Press, Baltimore, 5th edition (1978).

27. M. Kirk and M. F. Lyon, Induction of congenital anomalies in offspring of female mice exposed to varying doses of X-rays, Mutation Res., 106:73-83 (1982).

28. F. Vogel, Approaches to an evaluation of the genetic load due to mutagenic agents in the human population, Mutation Res., 29:263-269 (1975).

29. F. Vogel and P. Jager, The genetic load of a human population due to cytostatic agents, Humangenetik, 7:287-304 (1969).

30. C. O. Carter, Monogenic disorders, J. Med. Genet., 14:316-320 (1977).

31. I. Leck, Congenital malformations and childhood neoplasms, J. Med. Genet., 14:321-326 (1977).

32. H. J. Evans, Chromosome anomalies among livebirths, J. Med. Genet., 14:309-312 (1977).

33. A. G. Searle, Mutation induction in mice, Adv. Radiation Biol., 4:131-207 (1974).

34. W. W. Au and T. C. Hsu, The genotoxic effects of adriamycin in somatic and germinal cells of the mouse, Mutation Res., 79: 351-361 (1980).

35. A. J. Wyrobek and W. R. Bruce, in: "Chemical Mutagens. Principles and Methods for Their Detection," A. Hollaender and F. J. de Serres, eds., Vol. 5, Chaper 11. The induction of sperm-shape abnormalities in mice and humans, pp. 257-285, Plenum Press, New York (1978).

36. U. H. Ehling, in: "Chemical Mutagens," A. Hollaender, ed., Vol. 5, pp. 233-256, Specific locus mutations in mice, Plenum Press, New York (1978).

37. W. M. Generoso, J. B. Bishop, D. G. Gosslee, G. W. Newell, C. J. Sheu, and E. von Halle, Heritable translocation test in mice, Mutation Res., 76:191-215 (1980).

38. U. H. Ehling, Induction of gene mutations in germ cells of the mouse, Arch. Toxicol., 46:123-138 (1980).

39. U. H. Ehling and A. Neuhauser, Procarbazine-induced specific locus mutation in male mice, Mutation Res., 59:245-256 (1979).

40. R. K. Sharma, G. T. Roberts, F. M. Johnson, and H. V. Malling, Translocation and sperm abnormality assays in mouse spermatogonia treated with procarbazine, Mutation Res., 67:385-388 (1979).

41. A. M. Malashenko and N. I. Surkova, The mutagenic effect of thio-TEPA in laboratory mice. I. Chromosome aberrations in somatic and germ cells of male mice, Soviet Genetics, 10:51-58 (1974).

42. A. Wade, Martindale. The Extra Pharmacopoeia, 27th edition, Pharmaceutical Press, London (1977).

SPECIFIC LOCUS MUTATIONS IN THE MONITORING OF HUMAN

POPULATIONS FOR GENETIC DAMAGE

Arthur D. Bloom

Professor of Human Genetics and Development
and of Pediatrics
College of Physicians & Surgeons
Columbia University
630 W. 168 Street
New York, N. Y. 10032

INTRODUCTION

My primary argument in this paper is that in terms of the genetic monitoring of human populations, induced somatic mutation must
be our major future concern. This is not to say that germ cell
effects in humans are medically trivial. It is simply to say that
the best evidence we now have is that somatic mutation may be a major cause of disease, particularly in genetically at-risk persons.
We have no such evidence for the effects of induced germinal mutations in man; and I would, therefore, submit that our focus in future research efforts should be the identification of these genetically at-risk individuals.

THE TWO-HIT CANCER MODEL

The original two-hit cancer hypothesis of Knudson and colleagues [1] was derived mainly from extensive studies of retinoblastoma patients. In these patients, bilateral, early onset tumors were characteristic of the genetic form of the disease, as opposed to the unilateral, later onset more characteristic of the
sporadic form. From this and related studies of childhood neoplasms,
the outline of this cancer hypothesis was sketched, to whit, that
some individuals are born with a genetic predisposition to cancer,
that this predisposition is the result of an inherited single gene
mutation in many of the childhood tumors, and that superimposed on
this mutation is a somatic mutation, which, in fact, triggers the
process of malignant transformation.

We have come, since the mid-'70's, to fill in, at least in
broad strokes, the hypothesis. We recognize, for example, that the
so-called "inherited predisposition" may be the result: 1) of single
autosomal dominant genes, which predispose to such as retinoblastoma,
pheochromocytoma, neuroblastoma, and Wilm's tumor; 2) of homozygosity
for certain autosomal recessive genes, which result in the chromosome breakage syndromes, which, in turn, result in an increased risk
of leukemia, lymphoma and a variety of solid tumors; and 3) of chromosomal abnormality, such as trisomy, deletion and translocation.
In this latter regard, even a seemingly balanced translocation may
predispose to tumor formation, though few such instances are known
in humans. The report in the New England Journal of Medicine [2]
of a family in which a translocation between a number 3 and number
8 chromosome (t 3:8) was segregating revealed an 87% risk of renal
carcinoma by age 59 in translocation carriers. Thus, even in a seemingly balanced rearrangement, the rearrangement per se may increase
risks. And I am certain we shall see many more such associations between subtle chromosomal alterations and neoplastic disease.

IDENTIFICATION OF CARRIERS OF CANCER GENES

Clearly, one of the major tasks facing investigators now and in
future years is the identification of those individuals who are genetically so at-risk. In the presence of a well-defined syndrome,
the risks of leukemia, lymphoma et alia may be estimated with some
ease. But many of the disorders discussed above are variable in
their expressivity and may go either undiagnosed or be misdiagnosed
clinically. Furthermore, a major question yet to be resolved is the
risk for the heterozygous carrier of certain recessive genes, as well
as the risk for carriers of cytogenetic abnormalities. Some of these
recessive genes seem to result in deficiencies of DNA repair and in
enhanced mutability in vitro. The McKusick Catalogue of Mendelian
Disorders in Man [3] lists approximately 200 clinical genetic disorders which are associated with neoplasms. Of these, about one-
half are dominantly inherited, one-third are recessively inherited
and the remainder are largely X-linked. Further, it is estimated
[4] that perhaps 5% of individuals in the population are carriers
for these so-called "cancer genes."

How, then, to approach identification of this subpopulation?
This is an issue of major significance in terms of occupational exposures, but more broadly, even, in terms of medical care and other
environmental exposures of the individuals. An important beginning
in this process has been made by Auerbach et al. [5, 6]. In a
series of papers on Fanconi's anemia (FA), Auerbach and colleagues
have developed an effective kind of chemical stress test. By using
diepoxybutane (DEB), a difunctional epoxide that crosslinks to DNA
at two adjacent guanines and is a known carcinogen in several species, Auerbach et al., first of all demonstrated that the frequency
of chromosomal aberrations in cells from FA homozygote affected is

increased several fold above the frequency seen in cells of puta-
tively normal individuals. Furthermore, heterozygotes for the gene
show an intermediate response to DEB in their peripheral blood lymphocytes. More recently, their group has demonstrated that in am-
niotic fluid cells cultured at different doses of DEB, it is feasible
to distinguish the heterozygote fetus from both normal and affected
fetuses.

We have now initiated experiments on neurofibromatosis cells.
Neurofibromatosis is a dominant disease characterized by cafe au
lait spots, neurofibromas and a variety of other tumors. It has
been shown that 2-29% of patients with neurofibromas have maligant
degeneration of the lesions [7, 8], and it has also been suggested
that there may well be an increased incidence of childhood leukemia
in neurofibromatosis patients [9, 10]. In addition, independent
malignant tumors may be increased in adults with neurofibromatosis
[7, 8]. The problem is that the disease is highly variable in expressivity, and identification of carriers is important, both in
utero and beyond. Our preliminary evidence is that NF cells may
well show small increases in complex aberrations when exposed to
DEB. We are now extending the clastogenesis work, and are also
studying sister chromatid exchange (SCE) induction as another measure of response to this mutagen. The point of all of this is that
one major approach to identification of carriers of potentially deleterious mutations is the in vitro chemical stress test. There are
others, of course, including HLA associations, linkage to other
genes, etc.

SOMATIC MUTATION In vivo

We have heard, earlier in this Conference, of some approaches
to the detection of in vivo mutation. These have involved lymphocytes mutant at the hpt locus (Albertini), red blood cells mutant
at the hemoglobin locus (Stamatoyannopoulos and Mendelsohn) and
serum protein variants (Mohrenweiser). The continued pursuit of
such in vivo measures of somatic cell, specific locus mutation is
clearly in order, for the development of a variety of in vivo assays,
just as we have >35 in vitro mutagenesis assays at our disposal.
There are, however, only a small number of selective systems available for isolation of human cells mutant at specific loci with the
hpt and ouabain-resistance loci being most frequently used, and many
questions will have to be answered about the mutability of these
loci vis-à-vis other loci in the genome, even after questions of
efficiency of cell selection and isolation of mutant proteins are
resolved.

On an even more basic, molecular level, it is clear that we re-
quire direct assays of DNA damage to assess somatic cell mutagenesis.
There are numerous assays under development, including assays for
DNA adducts in body fluids, hemoglobin alkylation, etc.; but the

kind of work being done by Haseltine and colleagues [11, 12], at
Harvard, is to me among the most promising. The principle here involves isolation of DNA from exposed cells, or the use of naked DNA
itself, and the sequencing of the bases of DNA restriction fragments
to determine what alterations have taken place. Haseltine has described, among other phenoman, "hot spots" of mutation in specific
sequences of DNA, and this approach is particularly informative for
study of the mechanisms of mutation in mammalian cells. Whether this
will ultimately prove useful for human population monitoring is not
known, but certainly the sampling of cellular DNA from exposed individuals for molecular changes in base sequences is highly attrac-
tive. We should indicate that the Haseltine experiments do not as
yet involve genes coding for specific known products - they essentially involve sequences of undetermined function.

As an aside, we should mention here that the recent revival of
interest in the oncogenes, based largely on the work of Weinberg and
Cooper [13], is altering for many the basic view of the mechanism of
carcinogenesis, particularly in relation to environmental agents.
In their transfection experiments, the DNA of transformed mouse
cells was first introduced into normal mouse fibroblasts, with transformation of the normal cells resulting. Next, DNA isolated from a
variety of human tumor cell lines was used to transform the mouse
cells, with evidence accumulating that different tumors seem to have
a single tissue-specific transforming gene, specific at least for
the tissue of origin of the tumor. In the presence of, or in response to, appropriate environmental mutagens/carcinogens, it appears that there may be a loss of control of gene expression, perhaps via genetic rearrangement, with resultant neoplastic transformation which is oncogene determined.

So there are now numerous new approaches to the monitoring of
human populations for somatic cell damage, and our views of the consequences of this damage are changing. Our expectation is that newer
cellular and molecular approaches to this monitoring will be developed, to provide evermore sensitive measures of genetic damage.

For the remainder of this paper, I shall discuss other approaches which have been used and are still being refined for human
studies. First, let us consider cytogenetic monitoring in relation
to single gene mutation; then, clinical monitoring, as evidence of
specific locus mutation.

CYTOGENETIC MONITORING

The two major end points in cytogenetic monitoring are, of
course, chromosomal aberrations and SCE's. The technology is well
established for in vivo studies of clastogenesis, less so in vivo
for SCE's. Chromosomal aberrations constitute one type of mutational event, and their induction in somatic cells in vivo implies,

but does not prove, a germ line genetic effect as well. Many clastogens are also carcinogens, as best we can tell in animal studies.
Furthermore, determination of aberration frequencies has become
widely used for dosimetry purposes in radiation-exposed populations.
In terms, however, of exposure to chemical clastogens, in which the
effects are often chromatid rather than chromosome-type aberrations,
we have little in vivo human data. And, as one reviews the literature, there is a paucity of solid baseline, or control, data on the
frequencies of various types of aberrations. Thus, as recommended
in the "Guidelines" [14] of the 1981 March of Dimes Conference, we
need more data on chromatid aberrations in humans, before proceeding
with human population monitoring for exposure to putative chemical
clastogens.

It goes without saying that both chromatid and chromosome-type
aberrations we have no clear human health effects. It is surprising that after all this time we are unable to draw definitive conclusions about the biological effects of induced aberrations. A
study needs to be done - perhaps among the Hiroshima and Nagasaki
survivors - correlating disease with aberration frequency. Though
I suspect there will be a negative correlation, this needs to be
done if only to allay public fears about the significance of the
presence of induced aberrations, which aberrations may be no more
or less meaningful clinically than the specific locus mutations which
at present go undetected.

SCE's correlate better as indirect measures of mutations induced by chemicals than do chromosomal aberrations. In fact, for
many agents, SCE induction is detectable at levels of exposure well
below those required to produce detectable increases in chromosomal
aberrations. One of the major problems in the monitoring of human
populations for SCE's is the paucity of data from controls. The
range of intra-individual and inter-individual varaition appears to
be considerable, and the extent of that variation needs to be studied
more fully, before we can conclude that monitoring of exposed populations for SCE induction is intrinsically worth the doing. Certainly, the in vitro evidence suggests that it may well be; we simply
need more in vivo data. And also, we must remember that SCE's will
be far less a reliable measure of exposure to ionizing radiation exposure than chromosomal aberrations.

We should mention that the issue of the mutagenicity of nonion-
izing radiation exposure is becoming increasingly more pressing as
ultrasonography becomes more widespread in its use, particularly for
fetal monitoring. The problem of monitoring human populations for
the effects of physical agents is really typified by this issue.
There is considerable evidence that high frequency ultrasound produces heat in the target area, as one mechanism of its action.
There are others, as cavitation and mechanical effects. Liebeskind
and colleagues [15] at Albert Einstein have demonstrated numerous

in vitro effects of diagnostic frequencies of ultrasound on mammalian
and human cells. These effects include SCE induction, cell killing,
loss of contact inhibition, and alterations in cell motility. Further,
Kaufman et al. [16] at Argonne have demonstrated an increase in 6-
thioguanine resistant V79 hamster cells after in vitro ultrasound exposure. And yet, even if these in vitro effects, several of which
are clearly genetic, are verified the question remains: how does one
study the potential clinical effects of this mutagen? This brings me
to the final section of this paper, namely, a discussion of the relationship between demonstrated mutagenicity in vitro or even in
vivo, and the evaluation of possible clinical effects.

CLINICAL MONITORING

     In terms of the clinical effects of mutagens, we are, of course,
especially concerned about teratogenicity, carcinogenicity, and reproductive effects (other than teratogenic ones), as well as the
effects of mutagens on a variety of specific systems (the immune
system, in particular). High dose radiation studies have amply
demonstrated teratogenic effects in man and other mammals, but low
dose studies in humans are singularly lacking. When we consider that
2 to 4% of all livebirths are associated with a congenital malformation, and begin to extrapolate up, as we would do in an epidemiological study, we recognize that if we are monitoring, for example, 1000
first-trimester pregnancies (the period of usual maximal sensitivity
to teratogens) we would "normally" expect 20 to 40 anomalies.
Clearly, given this range of variation, a 10 to 20% increase might
easily be undetectable. And yet, extrapolating from animal experiments with radiation, it is clear that teratogenicity is, in fact,
one of the major somatic effects we might expect to see with mutagens. The most likely such agents to be detectable are those with
effects on specific target organs or developing systems.

     Similarly, carcinogenicity is a major clinical effect of some
mutagens. Here, too, however, the ability to detect mutagen-induced
cancers is low, because of the high frequency of cancers in the adult
population. Again, as for congenital malformations, detection of
this effect is only likely when there is a specific, discernible target organ, manifested in statistically significant increases of specific types of cancer.

     Among the major reproductive effects of mutagens in humans are
spontaneous abortion, infertility, induced abnormalities in sperm
counts and morphology, and other such parameters of damage to the
genetic material. None of these effects are at present likely to be
clear indicators of exposures to genotoxic compounds, though with
proper epidemiological study design, they may so become.

In summary: we have a wide range of clinical and laboratory parameters for chemically-induced genetic damage. There are, however, major problems in extrapolating from in vitro to in vivo systems, and especially to man. A major need exists for continued development of new laboratory techniques for in vivo monitoring of humans to mutagenic hazards. The available laboratory techniques for human monitoring are still basically cytogenetic, as of 1982. And, finally, a major need exists for rigorous standards of epidemiological monitoring of the biological and clinical effects of mutagens. We require both short-term and long-term follow-up after exposures to known doses of specific agents, be they chemical or physical, like ultrasound. Before in vivo human studies will ever be definitive, we need extensive baseline population data on many of the parameters discussed above.

REFERENCES

1. A. G. Knudson, H. W. Hethcote, and B. W. Brown, Mutation and childhood cancer: a probabilistic model for the incidence of retinoblastoma, Proc. Natl. Acad. Sci. USA, 72:5116-5120 (1975).
2. A. J. Cohen, F. P. Li, S. Berg, D. J. Marchetto, S. Tsai, S. C. Jacobs, and R. S. Brown, Hereditary renal-cell carcinoma associated with a chromosomal translocation, N. Engl. J. Med., 31:592-595 (1979).
3. V. A. McKusick, Mendelian Inheritance in Man, The Johns Hopkins University Press, Baltimore (1979).
4. A. G. Knudson, Persons at high risk of cancer, N. Engl. J. Med., 301:606-607 (1979).
5. A. D. Auerbach and S. R. Wolman, Susceptibility of Fanconi anemia fibroblasts to chromosomal damage by carcinogens, Nature, 261:494-496 (1976).
6. A. D. Auerbach, B. Adler, and R. S. R. Chaganti, Prenatal and postnatal diagnosis and carrier detection of Fanconi anemia by a cytogenetic method, Ped., 67(1):128-135 (1981).
7. D. A. Barone, Neurofibromatosis: A clinical overview, Postgrad. Med., 66(2):73-82 (1979).
8. W. A. Knight, W. K. Murphy, and J. A. Gottlieb, Neurofibromatosis associated with malignant neurofibromas, Arch. Dermatol., 107:747-750 (1973).
9. J. L. Bader and R. W. Miller, Neurofibromatosis and childhood leukemia, J. Ped., 92(6):925-929 (1978).
10. D. S. Ginsburg, E. H rnandez, and J. W. Johnson, Sarcoma complicating von Recklinghausen disease in pregnancy, Obstet. Gynecol., 58(3):385-387 (1981).
11. S. M. Grunberg and W. A. Haseltine, Use of an indicator sequence of human DNA to study DNA damage by methylbis (2-chloroethyl)amine, Proc. Natl. Acad. Sci. USA, 77(11):6546-6550 (1980).
12. R. F. Martin and W. A. Haseltine, Range of radiochemical damage to DNA with decay of iodine-125, Science, 213:896-898 (1981).

13. G. M. Cooper and P. E. Neiman, Two distinct candidate trans-
     forming genes of lymphoid leukosis virus-induced neoplasms,
     Nature, 292:857-858 (1981).
14. Guidelines for Studies of Human Populations Exposed to Muta-
     genic and Reproductive Hazards, A. D. Bloom, ed., pp. 1-35,
     March of Dimes Birth Defects Foundation, White Plains, New York
     (1981).
15. D. Liebeskind, R. Bases, F. Mendez, F. Elequin, and M. Koenigs-
     berg, Sister chromatid exchanges in human lymphocytes after ex-
     posure to diagnostic ultrasound, Science, 205:1273-1275 (1979).
16. G. E. Kaufman, Ultrasound is a weak mutagen in mammalian cells,
     Abstracts, Annual Meeting of the Radiation Research Society
     (1982).

SOME RESEARCH NEEDS TO SUPPORT MUTAGENICITY RISK

ASSESSMENTS FROM WHOLE MAMMAL STUDIES

Ernest R. Jackson, John R. Fowle III,
and Peter E. Voytek

Reproductive Effects Assessment Group
Office of Health and Environmental Assessment (RD-689)
U.S. Environmental Protection Agency
401 M Street, SW
Washington, D.C. 20460

INTRODUCTION/BACKGROUND

The primary goal of regulatory agencies concerned with human health is to ensure that exposure to chemical substances does not present unreasonable risks. This involves assessing the potential of chemical substances to cause toxic effects and weighing these effects against cost and benefit considerations. If the toxicity and exposure data allow, attempts are made to estimate health damage to the human population. For most environmental chemicals, human data are unavailable. On those occasions when chemical exposure can be associated with a toxicological response in humans, many people have already been exposed and irreparable damage may have occurred. Examples of such events include the kepone incident, neurological disorders; the dibromochloropropane incident, sterility; and the vinyl chloride incident, liver angiosarcomas. Therefore, it is imperative that animal toxicity studies be conducted to assess the likely human health risk before such effects are observed.

The U.S. Environmental Protection Agency (EPA) is faced with the task of making health risk assessment judgments on the thousands of chemicals presently in the environment as well as on new industrial chemicals that are entering the environment in increasing numbers (Table 1). Unfortunately, for many of those chemicals little if any animal toxicological data exist. For example, animal and/or in vitro data were available for only 208 out of 400 organic chemicals identified at Love Canal [1]. Furthermore, in most cases such data are inadequate to make a judgment as to the likely toxicological effects in humans.

315

Table 1. Status of Chemicals in the Environment

60,000 CHEMICALS PRESENTLY IN COMMERCIAL PRODUCTION

365 NEW CHEMICALS INTRODUCED IN 1980*

659 NEW CHEMICALS INTRODUCED IN 1981*

164 NEW CHEMICALS INTRODUCED IN JANUARY AND FEBRUARY OF 1982*

*WHILE THESE CHEMICALS WERE PROPOSED FOR MANUFACTURING,
EXPERIENCE INDICATES THAT LESS THAN HALF ARE ACTUALLY
INTRODUCED INTO COMMERCE.

It is important that the federal government, private industry,
and the scientific community at large integrate current methodology
to put risk assessments on a more sound scientific basis. We have
definitive proof that exposure to certain chemicals induces cancer
in the human population [2], but no proof that chemicals induce
heritable mutations that can lead to serious genetic disorders.
However, there are strong suggestive indications that they do. First,
there already exists a high incidence of diseases in humans which are
caused by heritable mutations [3]. Second, within the last decade
specific locus data have shown that chemicals can cause heritable
mutations in mammals [4], and, more recently, ethyl nitrosourea has
been shown to cause serious dominant skeletal disease in mice [5].
Third, the main germinal target molecule is DNA in both humans and
mice. In light of these observations, it is reasonable to assume
that exposure to chemicals shown to be mutagenic in mouse test systems would also have the potential to be mutagenic in humans and
could lead to serious debilitating heritable genetic disease.

The EPA currently operates under seven statutes, which not only
provide the basis for regulatory action, but also provide the Agency
with the authority to perform health risk assessments. The wording
in these statutes provides for the use of mutagenicity data on environmental pollutants to assess heritable genetic disease. For example, the Toxic Substances Control Act [6] includes mutagenesis
along with teratogenesis and carcinogenesis as toxicological endpoints of concern and for which testing standards are to be developed.

As a first step towards fulfilling these mandates, the EPA proposed general guidelines for mutagenicity assessment in which test
systems were described that would be acceptable for determining the
"intrinsic" mutagenicity of a chemical, its ability to reach the
germ cells, and possible quantitative approaches to estimate human
heritable disease incidences. These guidelines, published in the

Table 2. Weight of Evidence Approach

I. QUALITATIVE ASSESSMENT

    A. DOES THE CHEMICAL HAVE INTRINSIC MUTAGENICITY?

    B. DOES EVIDENCE EXIST TO SHOW THAT THE CHEMICAL REACHES GONADAL TISSUE?

II. QUANTITATIVE ASSESSMENT

    A. TO WHAT EXTENT ARE HERITABLE MUTATIONS INDUCED BY ENVIRONMENTAL EXPOSURES TO THE CHEMICAL?

    B. WHAT FRACTIONS OF THESE INDUCED MUTATIONS RESULT IN GENETIC DISEASE?

Federal Register [7] in 1980, constituted the first attempt by any regulatory agency to consider mutagenicity as a distinct toxicological endpoint rather than merely using mutagenicity as an indicator of carcinogenicity.

The weight-of-evidence approach outlined in the guidelines (Table 2) is composed of two parts. The first part is a qualitative assessment of whether or not a chemical substance causes mutations and whether or not it reaches germinal tissue in an active form. If the data allow, a quantitative assessment of the risk may be performed, which can be based on whole mouse heritable mutagenicity data.

Public comments were solicited on the proposed mutagenicity guidelines and the major concern was that the guidelines were premature. This was based, first, on the lack of human data demonstrating that exposure to chemical agents increases the incidence of heritable genetic diseases and, second, on the lack of sufficient experimental mammalian studies to make quantitative mutagenicity assessments. These criticisms are valid. With respect to the first, it seems unlikely that in the near future we will be able to obtain sufficient human data to determine genetic risk resulting from exposure to environmental chemicals. However, as discussed previously there is indirect evidence indicating that exposure to chemical mutagens has the potential to increase the incidence of genetic disease in the population. Because of the potential of environmental chemicals to cause heritable genetic diseases, current regulatory statutes provide for use of mutagenicity data to protect human health. The remainder of this report will provide some thoughts on how the use of whole mammal mutagenicity tests may be enhanced to better estimate the possible risk of increasing heritable genetic diseases in the human population.

MOUSE BIOASSAYS

     Whole mammal mutagenicity tests can demonstrate that a chemical
has mutagenic activity, reaches germinal tissues, and causes mutations that are transmitted to the next generation, i.e., are heritable. However, one has to realize that, because of interspecies differences, any mammalian system can only approximate the various steps
between exposure (inhalation, ingestion, or dermal absorption) and
the fixation of mutations in human germinal cells. If one assumes
that these steps in animal models, such as the mouse specific locus
tests and mouse dominant skeletal and cataract tests, are representative of events occurring in humans, then extrapolations can be
made from animal results to predict the occurrence of similar events
in humans. This can be accomplished by using the "relative mutation
risk approach," which involves multiplying the ratio of induced mutations per locus caused by chemical exposure to the spontaneous rate
of the test system by the spontaneous genetic disease incidences in
the current human population. When animals tests score for actual
genetic disease caused by chemical exposure, such as in the test for
serious skeletal defects or cataracts, then a more direct method can
be used to predict heritable disease in man. Details of both the
relative risk and direct methods have been extensively described in
the National Academy of Sciencies' report on the "Effects on Populations of Exposure to Low Levels of Ionizing Radiation" [8]. Depending on the risk estimation method employed, different assumptions regarding spontaneous and induced mutation frequencies and disease incidences in both mice and humans are made. If results of the
two methods are similar, then more confidence could be placed in
their use. By employing the relative mutation risk approach using
the mouse visible specific locus test, it was estimated that exposure to 1 rem of ionizing radiation would increase the incidence of
serious dominant heritable disorders per million first-generation
live births between 8 and 40, whereas, with the direct method using
the mouse dominant skeletal test, a range of 5 to 65 was estimated.
It appears that both approaches are in agreement for ionizing radiation. Unfortunately, comparisons between these two approaches are
not as rigorous for chemical mutagenesis since only 25 chemicals
have been tested in the mouse specific locus test [4], and only one
chemical, ethyl nitrosourea, has been tested in the dominant skeletal
assay thus allowing application of the direct method. There is a
need to increase the data base on environmental chemical mutagens
in whole mouse bioassays, and also a need to compare both risk estimation methods for different classes of mutagenic chemical substances.

     The whole mouse test systems provide the best data for use in
estimation of genetic risk. However, there are some drawbacks to
the systems as they are currently employed. It should be noted that
the mouse specific locus test was not designed for risk estimation
purposes. Humans are most likely to be exposed to environmental

toxicants at "relatively" low levels and for long periods of time. In the whole mammalian bioassays used to measure mutagenicity, very high and often unrealistic exposures are given. The shape of the exposure/response curve at actual environmental levels cannot be determined from these studies. This means that, if the animal data are to be used for risk assessment purposes, extrapolations of mutagenicity risk from high to low exposure must be made. The consequencies of this cannot be determined but serious under- or over-estimates of risk could be made. It would be very costly and impractical to increase the sensitivity of the whole mammal bioassays by greatly increasing the number of animals tested and administering low levels of the test compounds to determine the shape of the exposure/mutagenic effect curve at low exposure levels. Because of this, other types of data, which are more readily obtainable, are needed to provide insight about the shape of the exposure response curve necessary for the estimation of mutagenic risk in humans. One way this might be accomplished is to expose animals to radioactively-labeled mutagens both at high levels where mutations are being scored and at the environmental levels to which humans are exposed to obtain a relationship between external exposure and the amount of the substance reaching the gonads. Since gene mutations appear to be directly related to the amount of mutagen bound to DNA [9], this curve could be used as a surrogate for the exposure/ mutagenic effect curve at low exposure levels where mutations cannot be scored. Thus, some of the uncertainties and criticisms associated with extrapolation from high doses, where the effect of mutations are observed, to low doses, where effects cannot be observed, can be eliminated. Therefore, from the standpoint of risk assessment, any future mutagenicity testing of environmental chemicals should utilize approaches enabling an estimation of the shape of the lower end of the exposure/response curve.

USE OF GENETIC TOXICITY TESTS WITH WHOLE MAMMAL BIOASSAYS

There are several biological indicators of genetic toxicity available that can be measured in both humans and test animals (Table 3). They can provide evidence that a mutagen enters the body and has the capacity to cause genetic damage in either somatic or germinal tissues. Similar genetic responses obtained in both mice and humans would indicate that the chemical's pharmacokinetic properties in both organisms result in the same effect. Such supportive interspecies data provide more confidence that use of an animal model is a reasonable approach to estimate human mutagenic risk. As combinations of these tests are utilized and supportive interspecies data become more developed, it is hoped that some of the biological indicators will be sensitive and predictive enough to be used as environmental monitors for humans exposed to potential mutagens. Such tests could also play an important role for quantifying human mutagenic risk in the future. For example, the whole mammal bioassay could be performed in such a manner that indicators of genetic dam-

Table 3. Some Genetic Toxicity Tests that Can
Be Used in Both Humans and Mice

TESTS THAT MEASURE SOMATIC EFFECTS	TESTS THAT MEASURE GERMINAL EFFECTS
HGPRT MUTATIONS IN LEUCOCYTES	ALTERED SPERM COUNTS/ MORPHOLOGY/MOTILITY
RBC VARIANTS	LDH$_X$ "MUTATIONS"
SISTER CHROMATID EXCHANGE	
CHROMOSOMAL ABERRATIONS	

age are measured in the exposed parents and heritable mutations are
scored in the offspring. The measured induced mutation frequency
or increased genetic disease incidence observed in the experimental
animals could be expressed in terms of incidences of genetic damage
[e.g., RBC variants, HGPRT mutations, LDH$_X$ "mutations," sister
chromatid exchanges (Table 3)], rather than exposure. When such
data are available, an increase in one (or more) of the biological
indicators observed in humans can potentially be used to estimate
human genetic risk based on the measured relationship between the
same biological indicator in mice and mouse mutation frequency.
Because the occurrence of the indicator endpoints represents an integrated measurement of all events between external exposure and damage to the cells scored, many assumptions about absorption and
metabolism between the species are avoided in this approach. This
concept is illustrated in Fig. 1.

SUMMARY

In the last two decades humans have become critically aware of
the threat of cancer as a consequence of exposure to certain environmental agents. More recently, people are becoming concerned
about the potential adverse effects of environmental chemicals capable of causing serious genetic diseases. This awareness and concern comes primarily from findings in "short-term" mutagenicity
screening tests, which indicate that many environmental chemicals
have the potential to cause germinal mutations in humans. Regulatory agencies with legislative mandates to protect human health
are often faced with making decisions based on toxicological data

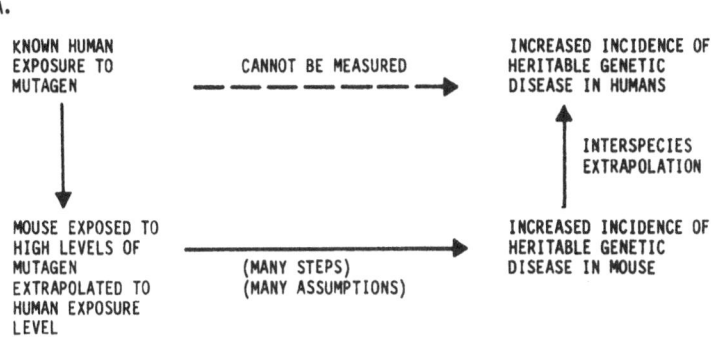

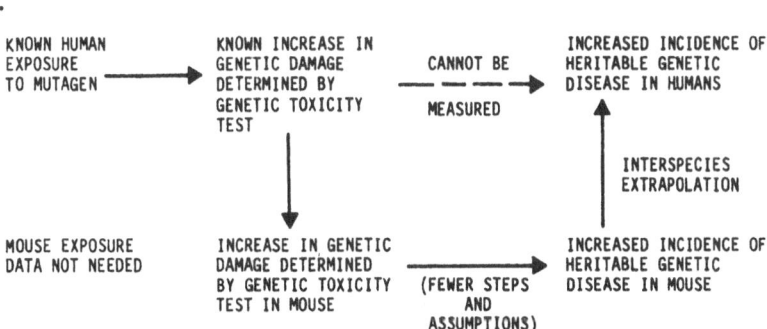

Fig. 1.   A potential use for genetic toxicity tests in quantitative
          mutagenic risk assessment. A) Short-term genetic tests
          are not used; B) short-term genetic tests are used.

that are inadequate for estimating the severity of the genetic
risk, even in qualitative terms. In this paper we suggest that two
additional types of measurements be made parallel with whole mouse
mutagenicity tests when possible. First, the shape of the germinal
DNA dose/external exposure curve should be determined for all chemicals being tested. These data would aid in extrapolating the likely
genetic risk from the high experimental to the low environmental
levels of exposure. Second, biological tests that indicate genetic
damage should be performed in both experimental animals and humans
to eliminate some of the assumptions used in interspecies extrapolations. These approaches have the potential to improve our mutagenicity risk assessment capabilities, but well-designed and carefully
coordinated experimentation is needed. This will require expert in-
put from the federal government, universities, private industry, and
the scientific community at large. Hopefully, recent scientific advances described in other papers in this symposium will stimulate

the necessary research support needed to improve current methodologies for assessing the mutagenic risk of chemicals.

REFERENCES

1.   U.S. Environmental Protection Agency. Chemicals Found in
     Waste Dumps (Love Canal): Results of Toxicity Testing (inter-
     nal document) (1980).
2.   World Health Organization, IARC Monographs on the Evaluation
     of the Carcinogenic Risk of Chemicals to Humans: Chemical and
     Industrial Processes Associated with Cancer in Humans, IARC
     Monographs Supplement 1 (September, 1979).
3.   V. A. McKusick, Mendelian Inheritance in Man: Catalog of
     Autosomal, Dominant, Autosomal Recessive, and x-Linked Pheno-
     type, 5th ed., The Johns Hopkins University Press, Baltimore
     (1978).
4.   L. B. Russell, P. B. Selby, E. Von Halle, W. Sheridan, and L.
     Valcovic, The mouse specific locus test with agents other than
     radiation: Interpretation of data and recommendations for
     future work, Mutat. Res., 86:329-354 (1981).
5.   Paul B. Selby, Risk Assessment in Humans, in: "Chemical Muta-
     genesis: Human Population Monitoring and Genetic Risk Assess-
     ment," K. S. Bora, G. R. Douglas, and E. R. Nestmann, eds.,
     pp. 275-288, Elsevier Biomedical Press, New York (1982).
6.   Public Law 94-469, 94th Congress, An Act, to regulate commerce
     and protect human health and the environment by requiring test-
     ing and necessary use restrictions on certain chemical sub-
     stances, and for other purposes, 90 STAT 2003-90 STAT 2051
     (October 11, 1976).
7.   U.S. Environmental Protection Agency, Mutagenicity Risk Assess-
     ment; Proposed Guidelines, Federal Register, 45(221), 74984-
     74988 (1980).
8.   Biological Effects of Ionizing Radiation III Committee, The
     Effects on Populations of Exposure to Low Levels of Ionizing
     Radiations, pp. 91-180, National Academy Press, Washington,
     D.C. (1981).
9.   W. R. Lee, Dosimetry of Chemical Mutagens in Eukaryote Germ
     Cells, in: "Chemical Mutagens: Principles and Methods for
     Their Detection," A. Hollaender and F. J. de Serres, eds.,
     Vol. 5, pp. 177-202, Plenum Press, New York (1978).

CONTRIBUTORS

Richard Albertini, Center for Health Sciences, University of Wis-
    consin-Madison, University Hospital and Clinics, 600 Highland
    Avenue, Madison, Wisconsin 53792, U.S.A.

April J. Bandy, Biology Division, Oak Ridge National Laboratory,
    P.O. Box Y, Oak Ridge, Tennessee 37830, U.S.A.

William Bigbee, Biomedical Division, Lawrence Livermore Laboratory,
    P.O. Box 5507, Livermore, California 94550, U.S.A.

Jack B. Bishop, Division of Mutagenic Research, National Center for
    Toxicological Research, Jefferson, Arkansas 72079, U.S.A.

Arthur D. Bloom, Director Genetics Clinic, Department of Pediatrics,
    College of Physicians & Surgeons of Columbia University, 630
    West 168th Street, New York, New York 10032, U.S.A.

Elbert W. Branscomb, Biomedical Division, Lawrence Livermore Lab-
    oratory, P.O. Box 5507, Livermore, California 94550, U.S.A.

Katherine T. Cain, Biology Division, Oak Ridge National Laboratory,
    P.O. Box Y, Oak Ridge, Tennessee 37830, U.S.A.

Arland Carsten*, Medical Department, Brookhaven National Laboratory,
    Upton, New York 11973, U.S.A.

Daniel A. Casciano, Division of Mutagenic Research, National Center
    for Toxicological Research, Jefferson, Arkansas 72079, U.S.A.

Ernest H. Y. Chu*, Department of Human Genetics, The University of
    Michigan, 1137 E. Catherine Street, Ann Arbor, Michigan 48109,
    U.S.A.

Robert R. Delongchamp, Division of Mutagenic Research, National
    Center for Toxicological Research, Jefferson, Arkansas 72079,
    U.S.A.

Frederick J. de Serres*, Office of the Director, National Institute
    of Environmental Health Sciences, P.O. Box 12233, Research
    Triangle Park, North Carolina 27709, U.S.A.

Udo H. Ehling, Institut für Genetik, Gesellschaft für Strahlen-Und
    Umweltforschung MBH, D-8042 Neuherberg, Post Oberschleissheim,
    München, Germany.

Ritchie J. Feuers, Division of Mutagenic Research, National Center
    for Toxicological Research, Jefferson, Arkansas 72079, U.S.A.

John R. Fowle III, U.S. Environmental Protection Agency, RD-689,
    401 M. Street, SW, Washington, D.C. 20460, U.S.A.

Walderico Generoso, Biology Division, Oak Ridge National Laboratory,
    P.O. Box Y, Oak Ridge, Tennessee 37830, U.S.A.

Kazuo Goriki, Department of Human Genetics, The University of Michi-
    gan, Ann Arbor, Michigan 48109, U.S.A.

Douglas Grahn*, Division of Biological and Medical Research, Argonne
    National Laboratory, 9700 South Cass Avenue, Argonne, Illinois
    60439, U.S.A.

Samir Hanash, Department of Human Genetics, The University of Michi-
    gan, Ann Arbor, Michigan 48109, U.S.A.

Ernest R. Jackson, U.S. Environmental Protection Agency, RD-689,
    401 M. Street, SW, Washington, D.C. 20460, U.S.A.

Ronald H. Jensen, Biomedical Division, Lawrence Livermore Laboratory,
    P.O. Box 5507, Livermore, California 94550, U.S.A.

Franklin M. Johnson, Laboratory of Genetics, National Institute of
    Environmental Health Sciences, P.O. Box 12233, Research Triangle
    Park, North Carolina 27709, U.S.A.

Todor Krastiff, Department of Human Genetics, The University of
    Michigan, Ann Arbor, Michigan 48109, U.S.A.

Raymond Krzesicki, Department of Human Genetics, The University of
    Michigan, Ann Arbor, Michigan 48109, U.S.A.

Susan E. Lewis, Senior Geneticist, Chemistry and Life Sciences
    Group, Research Triangle Institute, Research Triangle Park,
    North Carolina 27709, U.S.A.

Michael Long, Department of Human Genetics, The University of Michi-
    gan, Ann Arbor, Michigan 48109, U.S.A.

Mary F. Lyon, Radiobiology Unit, Medical Research Council, Harwell Didcot, Oxon OX11 ORD, United Kingdom.

Harvey Mohrenweiser, Department of Human Genetics, The University of Michigan, 1137 E. Catherine Street, Ann Arbor, Michigan 48109, U.S.A.

James V. Neel, Department of Human Genetics, The University of Michigan, 1137 E. Catherine Street, Ann Arbor, Michigan 48109, U.S.A.

Peter E. Nute, Department of Medicine, Division of Medical Genetics, University of Washington, Seattle, Washington 98195, U.S.A.

Thomas Roderick, Jackson Laboratory, Bar Harbor, Maine 04609, U.S.A.

Barnett Rosenburn, Department of Human Genetics, The University of Michigan, Ann Arbor, Michigan 48109, U.S.A.

Edward Rothman, Department of Human Genetics, The University of Michigan, Ann Arbor, Michigan 48109, U.S.A.

Liane B. Russell, Biology Division, Oak Ridge National Laboratory, P.O. Box Y, Oak Ridge, Tennessee 37830, U.S.A.

William L. Russell, Biology Division, Oak Ridge National Laboratory, P.O. Box Y, Oak Ridge, Tennessee 37830, U.S.A.

Chiyoko Satoh, Department of Human Genetics, The University of Michigan, Ann Arbor, Michigan 48109, U.S.A.

William J. Schull*, Center for Demography and Population Genetics, P.O. Box 20334, Houston, Texas 77025, U.S.A.

Anthony G. Searle, Radiobiology Unit, Medical Research Council, Harwell, Didcot, Oxon OX11 ORD, United Kingdom.

Paul B. Selby, Biology Division, Oak Ridge National Laboratory, P.O. Box Y, Oak Ridge, Tennessee 37830, U.S.A.

William Sheridan, Office of the Associate Director for Genetics, National Institute of Environmental Health Sciences, P.O. Box 12233, Research Triangle Park, North Carolina 27709, U.S.A.

Andrew Sivak*, Vice President, Arthur D. Little, Inc., 30 Memorial Drive, Massachusetts 02142, U.S.A.

Michael Skolnick, Department of Human Genetics, The University of Michigan, Ann Arbor, Michigan 48109, U.S.A.

George Stamatoyannopoulos, Department of Medicine, Division of Medical Genetics, University of Washington, Seattle, Washington 98195, U.S.A.

Stanley Sternberg, Department of Human Genetics, The University of Michigan, Ann Arbor, Michigan 48109, U.S.A.

Peter Voytek, U.S. Environmental Protection Agency, RD-689, 401 M. Street, SW, Washington, D.C. 20460, U.S.A.

Salome G. Waelsch, Albert Einstein College of Medicine of Yeshiva University, 1300 Morris Park Avenue, Bronx, New York 10461, U.S.A.

H. Glenn Wolfe*, Division of Biological Sciences, The University of Kansas, Lawrence, Kansas 66044, U.S.A.

Karl-Hans Wurzinger, Department of Human Genetics, The University of Michigan, Ann Arbor, Michigan 48109, U.S.A.

---

*Session Chairmen